The Origins of Totalitarianism

The Current Crisis in American Politics

The Current Crisis in American Politics

WALTER DEAN BURNHAM

New York   Oxford
OXFORD UNIVERSITY PRESS
1982

Library of Congress Cataloging in Publication Data
Burnham, Walter Dean.
The current crisis in American politics.
Bibliography: p.   Includes index.
1. Political parties—United States—Addresses, essays, lectures.
2. Elections—United States—Addresses, essays, lectures.
3. Voting—United States—Addresses, essays, lectures.
4. Political participation—United States—
Addresses, essays, lectures. I. Title
JK2265.B87   324'.0973   82–6511
ISBN 0–19–503219–5   AACR2

Printing (last digit): 9 8 7 6 5 4 3 2 1

Printed in the United States of America

For Tish

PREFACE

This volume is entitled *The Current Crisis in American Politics*. Yet—as immediately becomes evident with the first essay, which was written in 1964 and published in 1965—many of the data that are generated and analyzed are not current at all. In fact, they go back in some places more than a century into the past. But they have not thereby lost a very considerable relevance, I would argue, to the conundrums with which American politics abounds in the 1980s.

There is justification for such an argument, and it ultimately derives from certain special realities of American politics over time—especially those associated with its "nondevelopmental" character. These realities have been perceived and discussed extensively in the writings of cultural and political historians such as Louis Hartz and Lee Benson.* The work of such authors also makes crystal clear why the past so often seems immediately to confront the present in the United States, and hence why it is so important to include materials from the past in dealing with today's problems and issues. Comparative analysis is also often of crucial importance in understanding a problem which at first blush might be thought to be pure "native American." Hartz in particular makes a powerful case that only thus can one become fully aware of just how special some of these realities of American politics are—how profoundly different that politics and its history have been from those of other countries with broadly similar advanced industrial-capitalist economies and societies.

It is perhaps worth noting even in this brief preface that the first of these essays—among the most influential of the lot, it would seem—was written

---

* Louis Hartz, *The Liberal Tradition in America* (New York: Harcourt, Brace, 1955). For his statement of the "fragment society/ culture" hypothesis, see idem, *The Founding of New Societies* (New York: Harcourt, Brace, 1964), Ch. 1. Lee Benson, *The Concept of Jacksonian Democracy* (Princeton: Princeton University Press, 1961), especially Ch. 13, "Outline for a Theory of American Voting Behavior."

at a time when the bulk of the research mainstream in American politics had very different agendas and problematics. An imminent, system-wide crisis of political articulation, representation, and public support was in no way generally visible in 1964. This was the age, after all, of Gabriel Almond and Sidney Verba's *The Civic Culture*, the Michigan group's *The American Voter*, and Robert E. Lane's "The Politics of Consensus in an Age of Affluence."** Nor, for that matter, was any such crisis visible to me during the writing of the "Changing Shape" essay in 1964. To be sure, President Kennedy had just been assassinated, an infinitely tragic and shocking event. But presidential assassinations had happened before, and more than once, in our political history without general crises accompanying or following them.

What was visible to me in 1964 (and even more so later, after the crisis did begin to surface) was that the political universe presented by then current academic and journalistic accounts could not plausibly be fitted into the past. Moreover, that past appeared to me then, as it does now, to be deeply relevant to the present. It somehow had to be included, then, if the present was to be described adequately. Several questions were thus ripe for the asking. Has it always been so? Must it always be so? And what substantive difference to American politics, or to the study of American politics, does it make?

Whatever else may have happened in the many years since then, my view that the past must be linked systematically with the present has not changed. When Oxford University Press approached me with the idea of producing a collection of my essays, it seemed appropriate for many reasons to select those which, incorporating such an approach, particularly centered in one way or another on "the current crisis." More especially, they mostly center on one dimension of this crisis, the degeneration of the contemporary American electoral market and its institutions. A fuller discussion of these issues is found in the introduction which follows. It suffices to say here that crisis sequences in any political system are to a peculiar degree chains of events which demand analysis and theorizing about the statics and dynamics of the system itself. The success of any such effort is largely revealed by the extent to which interested readers find that they gain insights and perspectives about today's political world that they did not have before.

A particular note of thanks and appreciation is due to Mr. Sheldon Meyer, senior vice-president of Oxford University Press, and to his capable staff. My relationships with him and the Press are of very long standing. They go back to the collaboration between him, Professor William N. Chambers, and myself in producing *The American Party Systems* in its two editions (1967 and 1975). When Mr. Meyer informed me that he thought there would be a market for a work along the lines of the present volume, and that there were

---

** Gabriel Almond and Sidney Verba, *The Civic Culture* (Boston: Little Brown, 1963); Angus Campbell, Philip E. Converse, Warren E. Miller, and Donald E. Stokes, *The American Voter* (New York: Wiley, 1960); Robert E. Lane, "The Politics of Consensus in an Age of Affluence," *American Political Science Review* 59 (1965): 874–95.

substantive justifications for doing it besides, it was easy for me to believe him, and to agree to undertake the project.

The intellectual debts which one acquires over the years are enormously large. Many of them are acknowledged in the Introduction which follows— in particular, those I owe to the scholarship of V. O. Key, Jr., E. E. Schattsch- neider, Louis Hartz, Lee Benson, and the Michigan team: the late Angus Campbell, Philip E. Converse, Warren E. Miller, and Donald E. Stokes. In more recent years I have profited greatly from my relationships with Profes- sors Douglas A. Hibbs, the late Jeffrey L. Pressman, and my colleague Thomas Ferguson. Nor should I omit the exceptional importance of intel- lectual give-and-take with my graduate students over the years. It is scarcely possible to name them all here, and it seems invidious to name some without including many others as well. But they should know that their intellectual dialogue has been important in forming the changing shape of my ideas about American politics.

Grateful acknowledgment is made to the following organizations for per- mission to reprint certain essays in this volume: to the American Political Science Association for "The Changing Shape of the American Political Uni- verse" and "Theory and Voting Research," which appeared in the 1965 and 1974 volumes of *The American Political Science Review* respectively, as well as permission to publish an American Political Science Association 1981 con- vention paper, "Shifting Patterns of Congressional Voting Participation in the United States," to which the Association holds the copyright; to the American Bar Association for permission to reprint "The Disappearance of the American Voter" from the proceedings of the Association's 1978 sym- posium on voting participation; to the Academy of Political Science for my essay, "Insulation and Responsiveness in Congressional Elections," which first appeared in the 1975 volume of its publication, *Political Science Quar- terly;* to MIT Press for permission to reprint my essay, "The 1976 Election: Has the Crisis Been Adjourned?" which first appeared in a book published by them in 1978 and edited by myself and Martha W. Weinberg, *American Pol- itics and Public Policy,* and to the editors of *Dissent* for permission to reprint my essay, "American Politics in the 1980s," which first appeared in the Spring 1980 issue of that journal.

Finally, I should acknowledge a debt, both intellectual and intensely per- sonal, to my wife Patricia. She has been, in St. Paul's phrase, a "true yoke- fellow" in life's journey. Her willingness to carry many burdens of family life has entailed no small sacrifice as she has sought to pursue her own profes- sional career. Her energy, cheerfulness, and support have been, most literally, a sine qua non for me.

It should go without saying that no one except the author should be held accountable for any errors of fact or any insufficiencies of analysis.

Cambridge, Massachusetts
Summer 1982

WALTER DEAN BURNHAM

CONTENTS

The Current Crisis in American Politics

Introduction
The Current Crisis

When Oxford University Press expressed interest in publishing a collection of some of my essays, I was flattered; but it also represented a serious challenge. A mundane but very important set of decisions had to be made at the outset of this project: which of my essays to include and which not. Beyond that preliminary stage, the question arose whether the essays selected for inclusion here had any common theme or themes, or on the other hand were related to each other only ad hoc and by virtue of a common authorship. Such a question very naturally arises. These essays were written over a seventeen-year period, from 1964 through 1981, and as is evident from their titles and contents, many of them were written in response to the stimulus of specific events and developments in American electoral politics during that period. This fact is related to another procedural decision that was made early on. The essays in this volume are published as they were originally written: they have not been reworked in order to gloss over analyses, perspectives, or predictions made at the time that today the author might find irrelevant, superseded, or plainly wrong. They, and he, must take their lumps accordingly, in light of what we know empirically and analytically in the early 1980s.

But this decision serves an important purpose that, I think, speaks directly to the underlying question of the unity of the work as a whole. For a unity does exist after all, and in spite of certain changes in the author's perspectives and research agendas over this period of nearly two decades. To explore this argument fully, something of an intellectual autobiography seems required in this introduction, as a means of evaluating the position of these essays in a long flow of scholarly activity centered on American electoral politics. This collection reflects the confrontation of one man with his times on the one hand, and with the dominant intellectual trends within American political science on the other.

3

I grew up during the Great Depression and the Second World War, in the Pittsburgh area. My parents—"middling" middle class, no more and no less—were the embodiment of the Protestant work ethic; I early, if imperfectly, learned that life was difficult and ringed about by scarcity, but could be mastered by determination and hard work. It was evident very early to them and to others that I had a marked intellectual bent that set me sharply apart from the business-class residential milieu in which I grew up. It is eternally to their credit that they managed the resulting tensions and dissonances with intelligence and understanding.

My first conscious memories outside the family-neighborhood milieu were of the Great Depression itself, the Nazi invasion of Norway in April 1940 (which engulfed my mother's cousins and shortened their lives), and the 1940 presidential election. As Bruce Stave has pointed out, unemployment in the Pittsburgh area in 1934 stood at about one-third of the labor force, and it remained extremely high until World War II engaged the services of the "arsenal of democracy."[1] While we never joined the ranks of this "reserve army," one could not live without being aware of its existence and forming some reaction to it. At the time, this reaction importantly included opposition to Roosevelt and the New Deal, though my parents were considerably more moderate in their Republican sympathies than were most of their friends and neighbors. The town in which I grew up gave Wendell Willkie 85 percent of its vote in the 1940 election. As with the parents and the milieu, so with the son. Political socialization and the shaping of party identification worked very much as spelled out later by the authors of *The American Voter*[2] —except that, in retrospect, it seems rather clear that party commitments in our household were tightly woven into a general view of the world and a perception of political and economic goals congruent with party identification.

One day in the fall of 1940 I took the bus to downtown Pittsburgh, arrayed with several Willkie buttons. (In those days, no one seemed the slightest bit anxious about a 10-year-old making a journey by himself into the core of the "central city.") No sooner had I gotten off at my destination than I was surrounded by several men, far more shabbily dressed than my father. They gave me to understand very clearly that my Willkie buttons were not welcome. I removed them, put them in my pocket, and the scene dissolved without further incident. But this encounter made a vivid and life-long impression. When, later, I read *The People's Choice*[3] and learned from Robert Alford's *Party and Society*[4] that the index of class polarization reached an all-time high in the 1940 election, the arguments of both works fitted readily into this remembered experience. Perhaps this experience shaped the skepticism I was later to feel about the universal generality of certain propositions advanced by academic survey research in the 1950s and 1960s; perhaps not. In any case, this tale reveals my keen interest in electoral politics from early on, as well as the intense stimulus of concrete experience in shaping one man's consciousness over the very long term.

My friends and colleagues have often noted in me a bent for historical and comparative analysis that is rather unusual among practitioners of American political science—so much so that not a few of them have asked why I didn't choose a career as a historian rather than as a political scientist. There is little doubt that such influence as my work may have had over the years has been as great among American historians as among political scientists, and it is hard to give an answer to the question that does not in the end turn upon the influence of specific intellectual mentors at decisive moments in my graduate education.

The Nazi invasion of Norway, and the enormous reaction to it in my household, was the first event that, as it were, propelled me out of the neighborhood and the city and into awareness of the much larger world outside. World War II completed this side of my education over the next five years in ways too complex to unravel even decades later. By the time I was fifteen, Hitler's Great German Empire had gone down in fire, ashes, and total destruction for which the fall of the Roman Empire alone might provide some parallel, and the nuclear age had begun with the bombing of Hiroshima and Nagasaki. It is hardly to be wondered at that such overwhelming events should have prompted me to consult the meta-historical reflections of Toynbee, Spengler, and others in a search for clues as to what had happened, and was continuing to happen, to the world. I find little reason to doubt, in retrospect, that my preoccupation with long historic sweeps, dramatic events, and the decoding of underlying patterns of meaning in a confusing melee of events had its germination in this early period of my life. That part of my work which has dealt with critical realignments and their systemic importance in the history of American politics very probably grew out of this seed.[5] By the same token, much of the professional work reflected in these pages is shaped by my long-held view that the problems of American politics are often best grasped and analyzed in a comparative context.

My higher education was obtained at, successively, Johns Hopkins University, the United States Army, and Harvard University. At Johns Hopkins I pursued what was essentially a double major in political science and history. From Carl B. Swisher I learned American constitutional law and jurisprudence and also something about what made a great scholar and his scholarship; from Malcolm Moos, a wealth of detail about American politics as well as something of his infectious enthusiasm about it; and Thomas I. Cook gave me my first exposure to comparative politics and political thought.

Toward the end of my career at Johns Hopkins, one of my other peculiar talents came to the fore: that of amassing large quantities of data that interested me and, as it turned out, appealed to a larger audience as well. During the early 1950s the computerization of political science was still pretty much in its infancy, and, indeed, the so-called behavioral revolution had also just gotten under way. To an extent that would be considered incredible today, the basic quantitative facts about American politics were scattered around the countryside. There was no question at that time of constructing a machine-readable archive of basic historical electoral data. That was to come much later, in a project in which I had a hand during the 1963–64 academic year: the construction of the election archives at the Inter-University Consortium for Political and Social Research at the University of Michigan.[6]

But in the 1940s and early '50s the only effort to construct a county-level compendium of American presidential election returns had been Edgar Eugene Robinson's two-volume work, *The Presidential Vote, 1896–1932,* and *They Voted for Roosevelt,*[7] which continued the series through 1944. It occurred to me around 1950 that the series could well be extended backwards through the nineteenth century to the earliest date for which county-level presidential returns were pretty universally available, i.e., 1836. I undertook this labor, which was virtually completed when I entered the army in 1953. Two years later the work was published by Johns Hopkins University Press, and two decades thereafter was reprinted by the *New York Times*–Arno Press in its American historical documents series.[8] Very much is owed to Professor Moos's consistent encouragement and support for this project, which included lining up external financial backing, then available only in extremely limited form but still crucial to the work's completion.

For a variety of reasons my three-year service in the United States Army, beginning in the last phases of the Korean War, was also a most important "higher education" influence in my development. In many respects I had hitherto led much too sheltered a life. This of course was now totally a thing of the past. Not only was I introduced to, and challenged by, a much wider range of human experience than I had known, but this experience changed many of my perspectives on the nature of social reality in the United States. The result was the beginning of a deep skepticism about many of the manifest values that were commonplaces among my generation and social class. Things were obviously not what that cosmology had taught me to expect them to be. This distancing was reinforced by a profound, year-long exposure to Russian language and culture given me by the Army Language School. The "mission" of this training, from the army's point of view, was to produce people who could knowledgeably participate in the communications side of its Cold War activities. But the Language School had other effects too. It was crucial in broadening my own horizons from the rather narrow, America-centered perspectives I had had when I entered the service. The culture to which we were exposed is profoundly different from our own, and the language—also different in fundamental ways from English—provided the key to some of the puzzles involved. While so very different from our own, this Russian culture has many rich and many sympathetic features, which my fellow trainees and I came to appreciate keenly as the year passed.

Some of my friends were eventually to make careers that, in one way or another, employed these skills and perspectives. I did not, though I gave the matter serious thought on entering graduate school in political science at Harvard University at the close of my army service. The Soviet political system was not a congenial subject of study for me, to put it mildly, and in the 1950s much Soviet-area analysis struck me as seriously deficient in important ranges of data, and as tending to lead its practitioners onto political and scholarly paths I did not care to travel. At the same time, it is hard to overemphasize the importance of this experience and others that built on it later in my graduate career at Harvard. Again I was confronted with world history at its most cosmic, this time with the Bolshevik Revolution and its consequences at the center. My own aptitudes and talents continued to propel me in the direction of the study of American politics, but in a changed and more intellectually troubled context. This context came to include not only historical analysis of American political evolution, but a continuous internal incentive to place such study in a comparative context in order to deal with its problems and riddles, and in order to make it comprehensible.

Leaving the army in 1956, I entered the graduate program in political science at Harvard University. The most profound and far-reaching influences on my subsequent intellectual development were exerted by V. O. Key, Jr., Louis Hartz, and Barrington Moore, Jr.—the first energizing my interest in American electoral politics; the second opening my eyes to the dominance of the "liberal tradition" in American political culture and consciousness; and the last, in what was quite simply the greatest course I ever attended, giving me my first sustained exposure to the ideas of key modern social theorists and critics. All of them—and others too—cumulatively deepened my commitment to work with a historical, comparative, and theoretical focus.

For a time I thought that the specialization to which I would devote my subsequent career lay in the field of American public law (after all, I had been preoccupied with parties and elections for a very long time already!). For a number of reasons this did not work out well in the end. Then the Social Science Research Council approached me and inquired if I would be interested in spending the 1963–64 academic year at the University of Michigan, working on feasibility studies for retrieving the mass of electoral data for major offices across the political history of the United States. We moved rapidly from feasibility studies to being "present at the creation." The project rapidly turned into the groundwork for the historical data archive of the Inter-University Consortium for Political and Social Research at Ann Arbor. In the process it became very clear to me that in all likelihood my best work in the years immediately ahead would be done in the analyses of electoral phenomena in the United States. Only somewhat later did it begin to occur to me that such analysis could be the point of entry for a new understanding of the ways the American political system as a whole functioned as pressures for change grew with changes in economy and society.

II

Mine was a rather peculiar entry into the study of American voting. This field had been, preeminently, the center of the "behavioral revolution" in American political science that had gotten fully under way around 1950. The citadel of training in this area had long been the Survey Research Center (later, the Center for Political Studies) at the University of Michigan, led by that remarkable team of scholars Angus Campbell, Philip E. Converse, Warren E. Miller, and Donald E. Stokes.[9] Their work forms an essential part of the classic corpus of contemporary voting analysis. In many respects praise is therefore superfluous, and criticism should be read with the constituent importance of the Michigan group's work always in mind.

It is, I think, accurate to say of this work that both its intellectual strengths and its limitations arose from the conceptual models derived from social psychology and small-group research that informed these authors. Their work added immeasurably to the identification of the attitudinal characteristics of the mass electorate. To some extent, however, this amplification was purchased at a price: a view of electoral politics that was singularly suited to the specialized case of the United States in the "Augustan age" of Dwight D. Eisenhower. In its initial stages, quite naturally, this work was not oriented toward comparative perspectives—that was to come later, and to full flower. It is scarcely surprising either that, with no national survey of fully acceptable research quality existing prior to the first Survey Research Center study of 1952, any longer-term time dimension was essentially absent. This also was to come later, notably in the "replication" study of 1976 performed by Norman H. Nie, Sidney Verba, and John R. Petrocik, *The Changing American Voter*,[10] and in many other studies. Ultimately, however, the largely micro-oriented research strategies of modern survey research in its first years made comparative sense, as Erwin Scheuch pointed out in 1966, only in a country that was politically organized so as to lack the kinds of fundamental macro-level problematics that were commonplace experiences in European electoral politics.[11] As such problematics inescapably and continuously bring major issues of value to the fore of analysis, the capacity of American political scientists to ignore them (or, the same thing, treat them as givens rather than as problematics) was congenial to a behaviorist orientation that sought to develop only value-free generalizations.

Even in this early period, criticisms and reservations were expressed by political scientists deeply concerned by the spread of campaign technologies that were programmed to engineer consent while debasing the currency of political discourse. In particular, the late V. O. Key, Jr., with his profound commitment to democracy and his sensitivity to the multitude of devices by which it can be undermined, raised the alarm toward the end of his life about the growing triumph of market-research perspectives in political campaign and scholarship alike.[12] Key's own work was sui generis: an extraordinary mixture of wit, the telling anecdote, rigorous analysis (with heavy reliance on simple but powerful aggregate data bases), and preoccupation with macro-level issues affecting the health and the future of democracy itself in the United States. My own intellectual debt to him—and to a few others, such as E. E. Schattschneider—is as obvious as it is incalculable. He was, of course, the first to set forth clearly the phenomenon of the critical election as an analytic problem, and to suggest something of its potential importance as an entering point for the study of system-level political dynamics. His classic *Southern Politics* remains not merely an example of what the anthropologist Clifford Geertz has called "thick description" at its best but a major state-ment of the consequences for policy when party competition is destroyed and electorates are shrunk to a fraction of their potential size.[13] By the same token, his analysis of *American State Politics* evaluates the incredible array of constitutional and legal devices by which Americans, ever distrustful of political power and ever hostile to the development of a true state, have frag-mented political resources and hamstrung the democratic representation of interests that only political parties appear capable of providing.[14]

Needless to say, epigones must find their own more humble place among the massive structures that giants built before them. What Key did was to suggest, in an integrated way, a whole series of research agendas that had a very different focus, orientation, and purpose from those that dominated much of the field of voting behavior and electoral politics at that time. My own work has been designed to carry some of them further along into a later generation. As it happened, American politics moved into an explosive crisis very shortly after Key died in 1963. This crisis, in its successive phases, has dominated that politics almost from that day to this, and seems certain to continue to unfold during the 1980s as well. Inevitably the developing crisis of regime in the United States has served to bring major, if not fundamental, conflicts over political values to the fore in its wake, and, as the authors of *The Changing American Voter* demonstrate, to reveal a much wider range of potentialities for political attention and consciousness among a mass elec-torate than might have been imagined around 1960. As the field of study begins to reveal sudden and profound changes in some of its most important parameters, analytic and descriptive challenges increase. Maybe these large changes have increased the salience and utility of my own work among those with particular interest in American politics, maybe not. In any case, they have had the effect of shaping the nature and focus of this work over time as the crisis matures and deepens.

III

By about 1964, then, it had become clear to me that certain important ques-tions needed to be addressed. Probably the most enduring and important of all these questions turned on the causes, characteristics, and implications (theoretical and empirical) of the periodically recurring critical-realignment sequence in American political history.[15]

Critical realignments are extraordinary upheavals in the flow of American electoral and policy history that occur under conditions of abnormal and general crisis. Realignment episodes involve a major increase in ideological polarizations among parties and political elites, more or less abrupt but thereafter durable shifts in the nature and social location of party coalitions in the electorate, and major changes in the shape and direction of public policy. Elsewhere I have referred to such sequences as "America's surrogate for revolution," and in fact they have had many revolutionary characteristics. Fundamentally different, then, from the electoral norm in the United States, critical realignments have historically recurred at remarkably regular intervals across American political history, at least until the Second World War. Demonstration that this sequence has empirical reality—and not merely in patterns of voting—implies that there are also historically specific "party systems" occupying the space between one realignment and the next. This in turn leads to the hypothesis that these party systems and the realignments themselves can be subjected to systematic comparison with each other.

Space will not permit here an extensive review of this set of phenomena. I have written one book on the subject, and propose to produce in the near future a more extensive and integrated study with this at its core.[16] Others have done so as well, notably James D. Sundquist.[17] And it does seem to be the case that at least some historicans and political scientists have found the party-systems model of use in organizing their work.[18] It is perhaps enough to say here that this area of inquiry is fairly fully reflected in the essay "Party Systems and the Political Process," Chapter Three in this volume. And though all of the essays can be read independently of each other, the others are also linked together in one way or another by critical-realignment and party-system questions.

But no less important than these issues are those that surround the evolution of the American electorate (and American democracy as a whole) over the past century. In this respect, my earlier work compiling the basic data of nineteenth-century American elections provided the perfect "deep background" for skepticism about the mainstream arguments that appeared to extrapolate certain timeless general properties of "mass electorates" from contemporary survey-research concepts and instruments. For it was evident that the shape of the nineteenth-century data was not readily compatible with many assertions in the mainstream literature. For example, levels of formal education are often asserted to be a powerful associative explanatory variable—pointing toward high levels of political cognition and increasing propensity to vote as education level rises. But then how do we account for the fact that, while the general level of education sharply declines as we recede into the past, the turnout rates go up, reaching a maximum a century ago (outside the South, anyway) that compares very favorably with the fully mobilized electoral politics of advanced industrial-capitalist countries in Western Europe? If "surge and decline" help to explain Eisenhower's victory in 1952 and the subsequent Democratic victories in the 1954 congressional elections, why is it that no such phenomenon could be detected in the aggregate data before 1900?[19] A full analysis of these longitudinal files reveals, in addition, not only that electoral pariticpation began a spectacular and protracted decline after 1900, but that parties began visibly to disintegrate at the same time.

What were the explanations for all these changes? What is the balance of probabilities when it comes to accounting for the behavior patterns of nineteenth-century electorates—so far as we can retrieve evidence of such behavior—in terms of the basic frames of reference found in survey-research models of the 1950s and 1960s? A preliminary, and in many ways imperfect, attempt was made to address some of these issues in the first chapter in this collection, "The Changing Shape of the American Political Universe." This article occasioned some comment and not a little criticism, especially from Professor Philip E. Converse and his student Professor Jerrold G. Rusk.[20] Eventually these criticisms prompted a rejoinder, "Theory and Voting Research" (Chapter Two in this collection), which is best understood as a rather sustained balance-of-probabilities argument based on circumstantial (if pretty extensive) evidence. Some may regard such controversies as mere ego trips by the participants. This would, I believe, be an unnecessarily personalist view of the matter. A reasonably accurate view of the past is a matter of exceptional importance, not only in bounding comparisons between it and the present, but also in ordering a clear statement of the temporal and sociological limits of any would-be generalization based on the present alone. To my mind, at any rate, the balance of probabilities supports the view that nineteenth-century Americans behaved politically in very different ways from those of today, and that a primary reason for this could be found in sociological and economic conditions that were optimal for participatory electoral politics on a scale that has no counterpart today. As always, the reader is invited to make his or her own judgments and balances of probabilities.

As is now very well known, the United States has by far the lowest electoral participation rates to be found in any Western country. They are not only the lowest, they are incomparably the most class-skewed. While the average turnout among Swedish or German manual workers approximates 90 percent, recent presidential-year turnouts among their American counterparts hover around the 50 percent mark. Among the upper-middle classes, on the other hand, the participation rates would tend to cluster around 90 percent and 80 percent, respectively. In off-year elections (when, as it happens, most American governors and other state officers are now elected, along with the U.S. House of Representatives and one-third of the Senate), these turnouts are more abysmal still, averaging outside the South about two-fifths of the potential electorate in 1974 and in 1978. This was not always the case: a century ago, outside the South and perhaps a few pockets elsewhere, turnouts could easily reach the 85–90 percent range. The implication is that class skew in participation among such electorates was essentially nonexistent a hundred years ago (which is *not* the same thing as asserting that political oppositions were organized in class terms!).

The comparative point of interest here is that in the United States, uniquely, the "movie has run backwards" over the past century. That is, the modal pattern of historical evolution in European electoral markets during this period has been toward the creation of elaborately organized parties of mass representation and the progressive elimination of legal barriers to universal suffrage, and a long-term movement toward an institutionalized "saturation" level following World War II, with turnouts now in the 75–92 percent range, depending on the country and the election. (It is worthy of note that no evidence can be found for periodically- recurring critical-realignment sequences in such electoral markets either, but that is another—if related—story.) In the United States, by contrast—again, excluding the special case of blacks in the states of the old Confederacy—such legal barriers against major population groups had been eliminated very early in our history except for women, and nationwide woman suffrage was adopted here at almost the same historical moment that it was adopted there. Voting participation—episodically high, frequently very low before the democratizing revolution under Jackson—rose to the awesome levels mentioned above during the 1840–1900 period. Thereafter it sank precipitately across the "system of 1896" (1896–1930), to rise again toward modern peaks in the New Deal era and after (1940, 1952, 1960), and then to decline over the past twenty years nearly to the all-time lows plumbed in the 1920s.

How and why has all this come about? One set of suggested answers is provided in Chapter Four, "The Appearance and Disappearance of the American Voter," and Chapter Five, "Shifting Patterns of Congressional Voting Participation in the United States." There seems to be no question at all that a crucial dimension to the problem lies in the structure—or degeneration—of political parties as vehicles for the representation of broad mass interests in an electoral market. It seems that not a few scholars view the participation question in essentially nonproblematic terms—that is, as presenting no *general theoretical* difficulty.[21] But, issues of aesthetics and national image quite apart, the continued growth of the "party of nonvoters" in the United States in current elections can only be evaluated as a critical and major limit on democracy itself. It marks the degeneration of an electoral market deriving from the forces that have led to the degeneration of its primary institutions, the parties. More precisely, as "Shifting Patterns" suggests, the degeneration seemed to be much more strikingly associated with the Democratic party than with the Republican as the 1980 election approached. It is once again worth noting that Ronald Reagan won his landslide, and his opportunity for the most sweeping policy revolution in a half-century, with 28.0 percent of the potential electorate voting for him—and that this compares with the 40.2 percent that Valéry Giscard d'Estaing received from the French electorate in May 1981 while going down to defeat at the hands of his Socialist rival, François Mitterrand.

In retrospect, American politics in the 1970s were marked by a persistent power vacuum that developed in the spring of 1973, when the Watergate affair burst into full view, and lasted until the inauguration of Ronald Reagan in 1981. This period, with its three rejected presidents (Nixon, Ford, and Carter), was one of growing crisis affecting the foundations of America's economic, social, political, and international institutions. The crisis was reflected at the top by a genuine interregnum. The 1970s were marked by the efflorescence of the "interest-group liberal" (or political-capitalist) state and its programs as parties degenerated, effective presidential leadership ceased to exist, and the political economy moved from the growth on which political capitalism was dependent into stagnation and incipient decline. The contradictions involved in all this multiplied (it appears, looking back, at an exponential rate). The essays in Part Three of this volume, "Insulation and Responsiveness in Congressional Elections" and "The 1976 Election," captured some (though only some) aspects of these emerging contradictions.

A noteworthy feature of modern decomposition of the party-in-the-electorate has been the emergence of incumbency as a major autonomous variable in congressional elections. Canny incumbents, of course, do everything they can to promote their reelection. This extends from traditional techniques of "working the district" to heavy stress on their roles as ombudsmen representing the interests of constituents against the federal bureaucracy and as defenders of the district as a whole against the claims of all other districts and their incumbents. But important institutional pathologies may well develop in the wake of the creation of discrete, office-specific coalitions in the American electoral market. Such developments serve to intensify and deepen the fragmentations of political power that it was the intent of the framers of the American Constitution to build into its structure. Some attempt is made in these essays to deal with the implications of this emergent "governability problem." It can at least be said that the essay on the 1976 election also prefigured the chaotic ineffectiveness of the Carter administration, even if it did not predict it or its electoral consequence in 1980. We may add that the policy revolution (or, more precisely, counterrevolution) of 1981 was based importantly upon powerful presidential leadership and upon partisanship in Congress, not on constituency service or ombudsmen roles among legislators. It remains to be seen whether this was a temporary "aberration" from the general model of partisan decline and institutional compartmentalization that both essays in Part Three stress as essential parts of the 1973–81 interregnum era.

By 1979–80 the crisis of American politics had such an across-the-board character that it seemed necessary to extend the analysis of that politics into a much more theoretically based, less empirically focused, and rather broader frame of reference than had marked most of my earlier essays. A first pass in this direction was made in "American Politics in the 1980s" (Chapter Eight in this collection). The second, "Into the 1980s with Ronald Reagan" (Chapter Nine), was written specifically for this volume, well after the beginning of the 1981 policy counterrevolution in Washington. It therefore deals less with the details of the 1980 election than does an essay that falls chronologically between these two, written immediately after the event and published in 1981 in Thomas Ferguson and Joel Rogers's *The Hidden Election*.[22] "Into the 1980s with Ronald Reagan," therefore, reflects my current assessment of the epochal changes in American politics and political economy through which we are now passing. It will surely not be the last—if for no other reason than that these changes are both fundamental to and unprecedented in the modern history of politics and policymaking in the United States, and that their impact is as a result literally incalculable at the time of this writing.

The conviction grows that the 1980 election as such was not overwhelmingly studded with characteristics of traditional critical realignments—but, equally, that the winners of that election have used the political resources it gave them to produce a realignment after the fact. In the shorter term, this point may appear to recede into the background in the cross-cutting welter of events, which—depending on the state of the economy and other things— may or may not include major Democratic victories in the 1982 or even 1984 elections. But basic to the Reagan counterrevolution is not merely the roll-back of the domestic political-capitalist state. Its core is the repeal of both the theory and practice of social harmony that has dominated public policy under administrations of both political parties for the past half-century. The right is once again attempting to prove that Marx lives: a one-sided class struggle on behalf of the rich has been proclaimed in place of this social harmony. But struggles in politics do not typically remain one-sided indefinitely, even in the United States. At some point in the 1980s this will become fully obvious. And if so, the degenerate organization of the electoral market will, sooner or later, be replaced by something else. Whatever happens, and whatever course the continuing regime crisis takes in the years ahead, it is hard to imagine that a return to the pre-1980 situation is possible for the foreseeable future. And this would imply a critical realignment, all right— though very possibly one of a fundamentally different character from any of its predecessors. If everything in history, ultimately, is transition from something to something else, the present moment seems more than usually so. It also seems a time when, if ever, Antonio Gramsci's mordant comment applies: "The crisis consists precisely in the fact that the old is dying and the new cannot be born. In this interregnum a great variety of morbid symptoms appears."[23]

IV

What, then, are the common themes of this collection of essays? The first theme, clearly, is that it is essential both for scholarship and for action that the very special political history of the United States be recovered and—so far as possible—comprehended. The American philosopher George Santayana was entirely correct in his observation that those who do not understand their history are condemned to repeat it. This has peculiar force in a country that, like the United States, has been dominated by a single "master value system," that of Lockean or proprietarian liberalism. For in such a setting, as Louis Hartz and many others have perceived, the major issues of politics recede into a kind of intellectual twilight. Political and economic assumptions, having nothing seriously to contest or resist them at home, fall beneath the threshold of conscious articulation or defense into a kind of Platonic cave of inarticulate major premises or dogmatic "self-evident" prejudices.

This setting can and does result in a certain intellectual impoverishment of American research in the social sciences in general, and in research into electorates and voting behavior in particular. History as such is no magic talisman, whether or not it comes armed with "locomotives," "laws," or the rest of the post-Hegelian historicist apparatus. It is in general true that each generation seeks in history what interests it, what history may tell that generation about the past handling of major problems that preoccupy it. Moreover, there seems to me no escaping the fact that there must be art as well as science in dealing with historical materials if they are to be utilized effectively. Much of this art can be taught to others directly, but some, surely, must as it were be communicated by example. There is, alas, no mechanical device—say, derived from the application of scientific methods drawn from the arsenal of extreme logical positivism—by which these historical materials can be employed to produce "definitive" conclusions. A certain agnosticism, therefore, had best accompany those who would work with them.

But this is very far from saying that such materials can or should be disregarded by those who work in the social sciences, particularly on the assumption that only those data sources should be used that can be subjected to "rigorous" treatment. This results in what Abraham Kaplan has correctly identified as the "law of the instrument"—letting the intersection between one's value commitments to a certain kind of "science" and one's tools determine what is or is not to be researched. Methodologists have long discovered a whole array of "fallacies" that researchers commit in the course of their work: ecological fallacies, individualist ones, and others.[24] To this list must surely be added two more that have been a significant element in American social science research: the *presentist* fallacy and the *isolationist* fallacy. From the presentist fallacy is derived the striking tendency to develop false—or at least premature—generalizations about the properties of human research subjects without adequate regard for the impact of the contexts that shape their social existence. From the isolationist fallacy, where comparative perspectives are not integrated into the work where needed—often evident, it seems to me by the way, in work that purports to compare the American electorate and its properties with others—comes the failure to appreciate adequately the possibility of *alternative states of being* to that which exists or is defined by the research in question.

It would seem, then, that presentism and isolationism—especially when joined together—create the profound danger of a "science" that confuses the familiar with the necessary. It was precisely this point, I think, that most exercised V. O. Key, Jr., in the years just before his death. For what history teaches us, if it teaches anything, is that "human nature" is remarkably plastic, immensely influenced by the social contexts in which it is articulated, and —so far as politics is concerned—powerfully shaped by the alternatives given to it by entrepreneurial individuals and institutions. For just this reason, Key argued, public opinion is not a fixed, timeless, contextless entity: it is an echo chamber whose outputs are heavily influenced, if not determined, by its inputs. Moreover, as Key perceived, this plasticity is so great that it matters immensely to the future of democracy what messages are communicated to active political entrepreneurs by the findings of research. To put the matter at its crudest, if political entrepreneurs operate on the assumption that, in H. L. Mencken's words, "Bosh is for boobs," then they will organize the electoral market to produce the largest possible quantity of bosh, and in the process help to shape an electorate with many features of a "booboisie." A recent study by Benjamin I. Page suggests some of the practical consequences.[25] It would be more than a pity if we accepted such a corruption of democracy—particularly if, after all, it proved to be the case that there was no such thing as *the* nature of belief-systems-in-general in mass publics-in-general.

Detailed analysis of long-term patterns of behavioral aggregates—no less than comparative analysis of issues such as the applicability of the Downs-Hotelling rational-actor model to the empirical data on electoral markets and their consumers—inevitably raises the gravest skepticism about such generalizations.[26] Moreover, because the United States is the kind of political society it is, the uncovering of systematic, macro-level regularities in its historic past has offered unusual analytic opportunities for identifying some of its important "basement" properties as a system of collective action. My own work, then, has been devoted to the exploration of the apparently existing order in the context of alternative states of being, on one side, and, on the other, the elucidation of specific macro-level properties that are a uniquely American blend of stasis and dynamic action.

This leads to the next major theme of my work, including the essays in this volume. This, growing in relative weight over the years in my own thinking, is the dominance of the "liberal tradition in America," and the political consequences—especially in this century, and most especially in recent years—of that dominance. The late E. E. Schattschneider was no doubt right to argue some years ago that there was nothing in the history of the post-1896 Republican party to cause us to doubt that Americans could build the kinds of political parties they wanted. On the other hand, Karl Marx was equally—and more fundamentally—correct in his view that although men make their own history, they do not make it as they please. Rather, they make this history in the context of what they have inherited (in culture and in consciousness) from the past generations of their own society.

Political parties come into existence and are maintained as important vehicles for collective action when they are needed to build the political resources required to perforn a major collective political task. In no small measure, the quite extraordinary cohesion displayed by the Republicans in the policy battles of 1981 reflected the fact that this party was engaged in a task requiring such collective, purposive action: the reversal of some fifty years of policy history and the dismantling of the political-capitalist state. By the same token but on a much larger (and probably more enduring) scale, the "golden age" of the American party system, and of democratic participation as well, occurred in the last half of the nineteenth century—the direct consequence, it seems to me, of the greatest, most epic collective political struggle in American history so far, the Civil War.

However things may turn out from 1982 onwards, the twentieth-century history of American electoral politics has been essentially—and I think increasingly—a history of excluded political alternatives. These exclusions have been perhaps less a result of purposive (if not conspiratorial) elite behavior aimed at curbing mass democracy, and voiding it of much of its substantive content, than was perhaps initially suggested in "The Changing Shape of the American Political Universe." They have almost certainly been more a matter of common consent among politically organizable parts of the American citizenry than that article may have implied. But the central comparative question remains that of defining what political oppositions have been effectively organizable in a supercapitalist, superdeveloped political economy and society dominated (as no other such economy or society is) by a single liberal-capitalist political culture.

It may be said that the political history of America in the twentieth century falls into three parts. In the first of these, the "system of 1896," organizable electoral politics pivoted on a struggle that came to be defined as pitting regionalist and parochial "peripheries" and "colonies" against the interests of a central "metropole." For much or most of the more socioeconomically "developed" parts of the country, the Republican party became in such a context the only game in town, with the Democrats a largely "unusable" opposition. In the underdeveloped Southern colonial periphery, bound by racism and the pathos of the "lost cause," the antidemocratic thrust of the "politics of excluded alternatives" reached its absolute maximum—and indeed survived the fall of the "system of 1896" by many years. The "colonial" West remained participatory but episodically insurgent against domination by Eastern capital and its hegemonic Republican political instrument. Quite naturally, this region in many respects pioneered the destruction of the preexisting major party channels, leading the way to "curing the ills of democracy by more democracy."

The second phase, the New Deal revival of party and mass electoral participation, was inaugurated by the shattering trauma of the Great Depression and consolidated by Franklin Roosevelt's masterly political entrepreneurship. This phase lasted from 1932 until about 1960. To the extent that the Democratic party became infused with social-democratic elements in response to the crisis of laissez faire capitalism, this period demonstrated genuine and very major conflicts among real political alternatives appropriate to the unequal distribution of wealth, power, and life opportunities in a "modernized" advanced industrial-capitalist political economy. It is hardly surprising that the secular trend of this century toward the disappearance of partisan channels in the electoral market and of the lower classes from the active electorate was substantially reversed in this era everywhere but in the South.

But the Democratic party was not then and never became a social-democratic party in any meaningful comparative sense of the term. For one thing, the international effects of the Bolshevik Revolution, the results of World War II, and the national-liberation struggles in the so-called Third World thereafter were to create a kind of worldwide sectionalism. Vertical conflicts over allocation within the domestic political economy were increasingly transmuted (helped by the occasional outburst of repression) into horizontal conflicts between the United States as the "pure" liberal-capitalist society on one hand and so-called "Marxist" states and Third World claimants in the North-South world dialogue on the other. After World War II this struggle took on a classic Manichaean form, pitting the "free world" against "world Communism" centered in the USSR. The restricting effects on the development of political alternatives at home were, and remain, enormous; they rarely get the attention they deserve in literature dealing with American domestic politics. They were particularly concentrated on the Democratic coalition inherited from New Deal days, and eventuated in the Vietnam War under Lyndon Johnson, which vastly accelerated the process of intraparty disruption. To these international-imperial considerations were added two other erosive changes: the development of relatively high levels of mass affluence (at least among whites), and the saturation of communications with divers entertainments.

The third era of change began in the early 1960s, and leads us to the present. It has been discussed in some detail in "Into the 1980s with Ronald Reagan." We will only pause here to note that the politics of excluded alternatives (at least so far as any left-democratic perspective is concerned) again mushroomed into full view. This was associated with the creation of an immense political vacuum on the ruins of the Democratic party, enormous increases in negative attitudes toward political institutions and leaders among the mass electorate, and—by no means coincidentally—the decay of participation and the concentration of this decay among the working classes of the country. Party was increasingly replaced by interest-group ascendancy in the formulation of public policy, and, with the passage of time, the results came to be condemned explicitly by conservatives and, more or less tacitly, by liberals as well. The stage was set for the filling of this vacuum by Ronald Reagan and the policies of the economic and cultural Right. The politics of excluded alternatives, based upon the uncontested hegemony of the "liberal tradition in America," had come full circle. And, merely to mention a third theme of the later essays in this volume, the intersection of this uncontested hegemony with the political system and its shaping contexts has created a general crisis of regime. This crisis was ratified by Ronald Reagan's elevation to power in 1980. He is committed to resolving it through revitalization of capitalism at home and the American empire abroad. In many respects, I think, these essays provide some clues to the deeper background of this epochal counterrevolution; how many, and how useful, readers must of course decide for themselves.

Individuals make their own contributions to scholarship, but it is manifestly true that scholarship is a social activity; so it is trite but also necessary to repeat the observation that we all stand on the shoulders of the giants who have preceded (or accompanied) us. This introduction has attempted to acknowledge some part of the author's intellectual debts to others involved in this common enterprise. Among these, his critics such as Philip E. Converse, Jerrold Rusk, and James F. Ward should receive explicit thanks and acknowledgment.[27] I anticipate that, so long as strength and intellectual capacity endure, my work in the future will move forward along the lines indicated in this essay. But it has not been unaffected by critiques of its inadequacies, and it is sure to be shaped importantly—as it has in the past—by the intellectual challenges presented by continuing change and crisis in the American political system. There will surely be work enough for all in this collective, and supremely important, enterprise in the years ahead.

W.D.B.

NOTES

1. Bruce Stave, *The New Deal and the Last Hurrah* (Pittsburgh: University of Pittsburgh Press, 1967), p. 46. The official figure for 1934 was 31.9 percent.
2. Angus Campbell, Philip E. Converse, Warren E. Miller, and Donald E. Stokes, *The American Voter* (New York: Wiley, 1960).
3. Paul F. Lazarsfeld et al., *The People's Choice* (New York: Duell, Sloan & Pearce, 1944).
4. Robert R. Alford, *Party and Society* (Chicago: Rand McNally, 1963).
5. On reflection, I am struck by the remarkable gap between the genuinely cosmic perspectives of a Spengler, a Sorokin, or a Toynbee and the very narrow focus of my own work. This

may indeed make a rather ludicrous impression on some, but if so, *tant pis*. For what it is worth, quite apart from the concrete character of such gifts as I may possess, I have always thought it best to attempt where possible to rely upon concrete (preferably, but by no means necessarily, quantifiable) data and to reduce the amount of heteronomy ("That's just your opinion!") in research accordingly.

6. One small fragment of this work is contained in an essay on American data sources in Jerome M. Clubb, William Flanigan, and Nancy Zingale, eds., *Guide to American Electoral History* (Beverly Hills: Sage, 1981).

7. Edgar Eugene Robinson, *The Presidential Vote, 1896–1932* (Stanford: Stanford University Press, 1934); idem, *They Voted for Roosevelt* (Stanford: Stanford University Press, 1947).

8. Walter Dean Burnham, *Presidential Ballots, 1836–1892* (Baltimore: Johns Hopkins University Press, 1955); reprinted as a volume in the Arno Press collection, *America in Two Centuries* (New York: Arno Press, 1976).

9. In addition to their *The American Voter* (see above, note 2), the collective studies of this group include Angus Campbell, Gerald Gurin, and Warren E. Miller, *The Voter Decides* (Evanston: Row, Peterson, n.d. [1954?]), which maps out in detail findings of the 1952 survey on which *The American Voter* was partly based; and Campbell, Converse, Miller, and Stokes, *Elections and the Political Order* (New York: Wiley, 1966). Their individual contributions to the literature, which have been voluminous and enormously influential, need not be cited in detail here.

10. Norman H. Nie, Sidney Verba, and John R. Petrocik, *The Changing American Voter* (Cambridge: Harvard University Press, 1976; 2d ed., 1979).

11. Erwin Scheuch, "Cross-National Comparisons Using Aggregate Data: Some Substantive and Methodological Problems," in Richard L. Merritt and Stein Rokkan, eds., *Comparing Nations* (New Haven: Yale University Press, 1966), pp. 131–67, at pp. 131–34.

12. Most explicitly, perhaps, in V. O. Key, Jr., "Public Opinion and the Decay of Democracy," *Virginia Quarterly Review* 37 (1961): 481–94; but see also his *Public Opinion and American Democracy* (New York: Knopf, 1961), especially pp. 536–58. And his posthumously published study, *The Responsible Electorate* (Cambridge: Harvard University Press, 1966), is notably preoccupied with this problem from beginning to end.

13. V. O. Key, Jr., *Southern Politics in State and Nation* (New York: Knopf, 1949).

14. V. O. Key., Jr., *American State Politics: An Introduction* (New York: Knopf, 1956).

15. A recent collection of essays on this subject, with an excellent bibliography included, is Bruce A. Campbell and Richard J. Trilling, eds., *Realignment in American Politics: Toward a Theory* (Austin: University of Texas Press, 1980).

16. See Walter Dean Burnham, *Critical Elections and the Mainsprings of American Politics* (New York: Norton, 1970). Forthcoming will be my amplified analysis, *The Dynamics of American Politics* (New York: Basic Books).

17. James D. Sundquist, *Dynamics of the Party System* (Washington: Brookings, 1973).

18. See, for example, Paul Kleppner et al., *The Evolution of American Electoral Systems* (Westport, Conn.: Greenwood Press, 1981), especially Professor Kleppner's essay, "Critical Realignments and Electoral Systems," pp. 3–32.

19. Perhaps it is more accurate to say that, as suggested in Chapter One, "The Changing Shape of the American Political Universe," the aggregate processes then were *inverse* to those that are linked to Campbell's 1952–54 "surge and decline" sequence. That is, in the nineteenth century, exceptional "deviating election" landslides tended to be associated with temporary *decreases* in turnout, with these decreases almost entirely concentrated among the voters of one of the two parties (as in Georgia, 1852, and Pennsylvania, 1872). This would make sense, after all, since in a context of extremely full mobilization and evidence of very powerful attachments by the large majority of voters to one or the other party, the necessary electoral "space" for landslides of the modern pattern would not exist—any more than such space has existed, say, in the flow of Italian electoral politics since the early 1950s.

20. See the citations in Chapter Two, "Theory and Voting Research."
21. See, for example, Raymond Wolfinger and Steven J. Rosenstone, *Who Votes?* (New Haven: Yale University Press, 1980).
22. Walter Dean Burnham, "The 1980 Earthquake: Realignment, Reaction or What?" in Thomas Ferguson and Joel Rogers, eds., *The Hidden Election* (New York: Pantheon, 1981), pp. 98–140.
23. Antonio Gramsci, *Selections from the Prison Notebooks of Antonio Gramsci,* ed. Q. Hoare and G. N. Smith (New York: International, 1971), p. 276.
24. A splended review and analysis of these fallacies is contained in Hayward R. Alker, Jr., "A Typology of Ecological Fallacies," in Mattei Dogan and Stein Rokkan, eds., *Quantitative Ecological Analysis in the Social Sciences* (Cambridge: MIT Press, 1969), pp. 69–86.
25. Benjamin I. Page, *Choices and Echoes in Presidential Elections* (Chicago: University of Chicago Press, 1978).
26. Compare, for example, Donald E. Stokes, "Spatial Models of Party Competition," in Campbell, Converse, Miller, and Stokes, *Elections and the Political Order,* pp. 161–79, with the very different conclusions that Samuel H. Barnes reaches for Italy in his study, *Representation in Italy* (Chicago: University of Chicago Press, 1977), especially pp. 97–116.
27. Cf. James F. Ward, "Toward a Sixth Party System? Partisanship and Political Development," *Western Political Quarterly* 36 (1973): 385–413.

I

Some Themes in American Electoral History

1

The Changing Shape of the American Political Universe
1965

In the infancy of a science the use even of fairly crude methods of analysis and description can produce surprisingly large increments of knowledge if new perspectives are brought to bear upon available data. Such perspectives not infrequently require both a combination of methodologies and a critical appraisal of the limitations of each. The emergence of American voting behavior studies over the last two decades constitutes a good case in point. Studies based on aggregate election statistics have given us invaluable insights into the nature of secular trends in the distribution of the party vote, and have also provided us with useful theory concerning such major phenomena as critical elections.[1] Survey research has made significant contributions to the understanding of motivational forces at work upon the individual voter. As it matures, it is now reaching out to grapple with problems which involve the political system as a whole.[2]

Not at all surprisingly, a good deal of well-publicized conflict has arisen between aggregationists and survey researchers. The former attack the latter for their failure to recognize the limitations of an ahistorical and episodic method, and for their failure to focus their attention upon matters of genuine concern to students of politics.[3] The latter insist, on the other hand, that survey research alone can study the primary psychological and motivational building blocks out of which the political system itself is ultimately constructed. Not only are both parties to the controversy partly right, but each now seems to be becoming quite sensitive to the contributions which the other can make. As survey scholars increasingly discover that even such supposedly well-established characteristics of the American voter as his noto-

From *The American Political Science Review* 59, no. 1 (March 1965): 7–28. Copyright © The American Political Science Review

25

riously low awareness of issues can be replaced almost instantaneously under the right circumstances by an extremely pronounced sensitivity to an issue, the importance of the time dimension and factors of social context so viewed become manifest.[4] Students of aggregate voting behavior, on the other hand, are turning to the data and methods of survey research to explore the structure and characteristics of contemporary public opinion.[5] A convergence is clearly under way. One further sign of it is the construction of the first national election data archive, now underway at the Survey Research Center of the University of Michigan.[6] The completion of this archive and the conversion of its basic data into a form suitable for machine processing should provide the material basis for a massive breakthrough in the behavioral analysis of American political history over the last century and a half.

If controversies over method accompany the development of disciplines, so too does the strong tendency of the research mainstream to bypass significant areas of potential inquiry, thus leaving many "lost worlds" in its wake. One such realm so far left very largely unexplored in the literature of American politics centers around changes and continuities in the gross size and shape of this country's active voting universe over the past century. Key, to be sure, made contributions of the greatest significance to our understanding of the changing patterns of party linkage between voters and government. Moreover, he called attention to the need for quantitative analysis of political data other than the partisan division of the vote for leading offices.[7] E. E. Schattschneider's discussion of the struggle over the scope of political conflict and his functional analysis of the American party system remain a stimulus to further research—not least in the direction of examining the aggregate characteristics of the American electorate over time.[8] Other recent studies, for example of the turnout of voters in Canada and Indiana, have added to our knowledge of contemporary patterns of mass political involvement.[9] The fact remains, however, that no systematic analysis over lengthy time periods has yet been made of the massive changes of relative size and characteristics in the American voting universe, despite their obvious relevance to an understanding of the evolving political system as a whole.

This article does not purport to be that systematic study. It is, rather, a tentative reconnaissance into the untapped wealth of a whole range of political data, undertaken in the hope of showing concretely some of the potentialities of their study. The primary objective here is the preliminary exploration of the scope of changes since the mid-nineteenth century in turnout and other criteria of voting participation, and the possible substantive implications of such changes.

There is also a second objective. The day is not far distant when a major effort will be undertaken to relate the findings of survey research to contemporary aggregate data and then to examine the aggregate data of past generations in the light of these derived relationships. Before such inquiry is undertaken, it will be a matter of some importance to ascertain whether and to what extent the basic findings of survey research about the present American electorate are actually relevant to earlier periods of our political history. Firm conclusions here as elsewhere must await much more comprehensive and detailed study. Even so, enough can be learned from the contours of the grosser data to warrant posting a few warning signs.

I

Several criteria of voting participation have been employed in this analysis: (1) estimated turnout; (2) drop-off; (3) roll-off; (4) split-ticket voting; (5) mean partisan swing. Turnout, the most indispensable of these criteria, is also unfortunately the "softest." A number of errors of estimate can arise from the necessary use of census data. For example, interpolations of estimates for intercensal years can produce significant error when abnormally large increases or decreases in population are bunched together within a few years. Estimates of the alien component in the total adult male population must also necessarily remain quite speculative for the censuses from 1880 through 1900, and are impossible to secure from published census data prior to 1870. No doubt this helps explain why students of voting behavior research have avoided this area. But we need not reject these admittedly imprecise data altogether, because of their imperfections, when secular changes in turnout levels and variabilities from election to election are of far too great a magnitude to be reasonably discounted on the basis of estimate error.[10]

Moreover, the other criteria employed in this study not only share a very similar directional flow over time, but are directly derived from the voting statistics themselves. Free from the estimate-error problem, they are ordinarily quite consistent with the turnout data.[11] What is called "drop-off" here is the familiar pattern of decline in the total vote between presidential and succeeding off-year elections. The drop-off figures usually presented below are reciprocals of the percentage of the presidential-year total vote which is cast in the immediately following off-year election. If the total vote for the two successive elections is the same, drop-off is zero; if the total vote in the off-year election exceeds that cast in the immediately preceding presidential election, drop-off is negative. Secular increases in the amplitude of drop-off could be associated with such factors as a declining relative visiblity or salience of off-year elections, or with an increasing component of active voters who are only marginally involved with the voting process as such.

"Roll-off" measures the tendency of the electorate to vote for "prestige" offices but not for lower offices on the same ballot and at the same election. If only 90 percent of those voting for the top office on the ticket also vote for the lesser statewide office receiving fewest votes at the same election, for example, the roll-off figure stands at 10-percent. Secular increases in this criterion of voting participation could be associated with such variables as a growing public indifference to elections for administrative offices which might well be made appointive, or with a growing proportion of peripheral voters in the active electorate, or with changes in the form of ballots. Split-ticket voting has been measured rather crudely here as the difference between the highest and lowest percentages of the two-party vote cast for either party among the array of statewide offices in any given election. Zero on this scale would correspond to absolute uniformity in the partisan division of the vote for all offices at the same election. The amplitude of partisan swing is computed in this study without reference to the specific partisan direction of the swing, and is derived from the mean percentage of the two-party vote cast for either party among all statewide races in the same election. Both of these latter criteria are more directly related to changes in the strength of partisan linkage between voters and government than are the others employed in this study.

Two major assumptions underlie the use of these criteria. (1) If a secular decline in turnout occurs, and especially if it is associated with increases in drop-off and roll-off, we may infer that the active voting universe (a) is shrinking in size relative to the potential voting universe; and (b) is also decomposing as a relative increase in its component of peripherally involved voters occurs. Opposite implications, of course, would be drawn from increases in turnout accompanied by decreases in these rough indices of voter peripherality. (2) If split-ticket voting and the amplitude of partisan swings are also increasing over time, we may infer that a decline in party-oriented voting is taking place among a growing minority of voters. Reductions in these criteria would suggest a resurgence of party-oriented voting.

A recent study by Angus Campbell tends to support the view that the above criteria are actually related to the component of marginal voters and voters with relatively weak partisan attachments in today's active electorate.[12] Campbell argues that surge and decline in voting particpation and in partisan distribution of the vote result from two major factors: the entrance into the active electorate of peripherally involved voters who tend to vote disproportionately for such beneficiaries of partisan surges as President Eisenhower, and then abstain from the polls in subsequent low-stimulus elections; and the temporary movement of core voters with relatively low levels of party iden-tification away from their nominal party allegiance, followed by their return to that allegiance in subsequent low-stimulus elections. Campbell's study reveals that split-ticket voting in the 1956 election tended to be heavily con-centrated among two groups of voters: those who voted Republican for pres-ident in 1956 and did not vote in 1958, and those who voted Republican in 1956 but Democratic in 1958—in other words, among those with peripheral involvement in the political process itself and those with borderline partisan commitments. Moreover, roll-off—the failure to vote a complete ticket in 1956—was heavily concentrated among the nonvoters of 1958. It is also suggestive that the level of drop-off in Campbell's panel from 1956 to 1958, 23 percent, very closely approximates the level of drop-off as measured by the aggregate voting data.[13]

II

Even the crudest form of statistical analysis makes it abundantly clear that the changes which have occurred in the relative size and shape of the active electorate in this country have not only been quantitatively enormous but have followed a directional course which seems to be unique in the contemporary universe of democratic polities. In the United States these transformations over the past century have involved devolution, a dissociation from politics as such among a growing segment of the eligible electorate, and an apparent deterioration of the bonds of party linkage between electorate and government. More precisely, these trends were overwhelmingly prominent between about 1900 and 1930, were only very moderately reversed following the political realignment of 1928–36, and now seem to be increasing once again along several dimensions of analysis. Such a pattern of development is pronouncedly retrograde compared with those which have obtained almost everywhere else in the Western world during the past century.

Probably the best-known aspect of the changing American political universe has been the long-term trend in national voter turnout: a steep decline from 1900 to about 1930, followed by a moderate resurgence since that time.[14] As the figures in Table 1 indicate, nationwide turnout down through 1900 was quite high by contemporary standards—comparing favorably in presidential years with recent levels of participation in Western Europe—and was also marked by very low levels of drop-off. A good deal of the precipitate decline in turnout after 1896 can, of course, be attributed to the disfranchisement of Negroes in the South and the consolidation of its one-party regime. But as Table 2 and Figure 1 both reveal, non-Southern states not only shared this decline but also have current turnout rates which remain substantially below nineteenth-century levels.[15]

The persistence of mediocre rates of American voting turnout into the present political era is scarcely news. It forms so obvious and continuing a problem of our democracy that a special presidential commission has

Table 1. Decline and partial resurgence: Mean levels of national turnout and drop-off, 1848–1962 (Percentages)

In presidential years	Mean estimated turnout	In off-years	Mean estimated turnout	Mean drop-off
1848–72	75.1	1850–74	65.2	7.0
1876–96	78.5	1878–98	62.8	15.2
1900–16	64.8	1902–18	47.9	22.4
1920–28	51.7	1922–30	35.2	28.7
1932–44	59.1	1934–46	41.0	27.8
1948–60	60.3	1950–62	44.1	24.9

Note: Off-year turnout data based on total vote for congressional candidates in off years.

Table 2. Sectionalism and participation: Mean
turnout in presidential elections, 1868–1960
(Percentages)

Period	Eleven southern states	Period	Non-southern states
1868–80	69.4	1868–80	82.6
1884–96	61.1	1884–96	85.4
1900 (transition)	43.4	1900	84.1
1904–16	29.8	1904–16	73.6
1920–48	24.7	1920–32	60.6
1952–60	38.8	1936–60	68.0

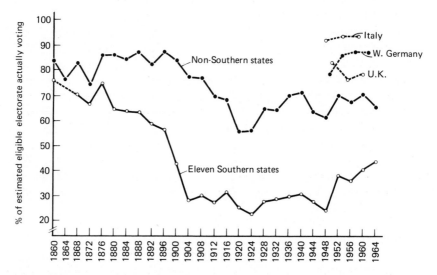

Figure 1. Patterns of turnout: United States, 1860–1964, by region, and selected Western European nations, 1948–1961.

recently given it intensive study.[16] Two additional aspects of the problem, however, emerge from a perusal of the forgoing data. In the first place, it is quite apparent that the political realignment of the 1930s, while it restored two-party competition to many states outside the South, did not stimulate turnout to return in most areas to nineteenth-century levels. Even if the mere existence of competitiveness precludes such low levels of turnout as are found in the South today, or as once prevailed in the Northern industrial states, it falls far short of compelling a substantially full turnout under present-day conditions. Second, drop-off on the national level has shown markedly little tendency to recede in the face of increases in presidential-year turn-out over the last thirty years. The component of peripheral voters in the active electorate has apparently undergone a permanent expansion from about one-sixth in the late nineteenth-century to more than one-quarter in recent decades. If, as seems more than likely, the political regime established after 1896 was largely responsible for the marked relative decline in the active voting universe and the marked increase in peripherality among those who still occasionally voted, it is all the more remarkable that the dramatic political realignment of the 1930s has had such little effect in reversing these trends.

At least two major features of our contemporary polity, to be sure, are obviously related to the presently apparent ceiling on turnout. First, the American electoral system creates a major "double hurdle" for prospective voters which does not exist in Western Europe: the requirements associated with residence and registration, usually entailing periodic re-registration at frequent intervals, and the fact that elections are held on a normal working day in this employee society rather than on Sundays or holidays.[17] Second, it is very probably true that nineteenth-century elections were major sources of entertainment in an age unblessed by modern mass communications, so that it is more difficult for politicians to gain and keep public attention today than it was then.[18] Yet if American voters labor under the most cumbersome sets of procedural requirements in the Western world, this in itself is a datum which tends to support Schattschneider's thesis that the struggle for democracy is still being waged in the United States and that there are profound resistances within the political system itself to the adoption of needed procedural reforms.[19] Moreover, there are certain areas—such as all of Ohio outside the metropolitan counties and cities of at least 15,000 population—where no registration procedures have ever been established, but where no significant deviation from the patterns outlined here appears to exist. Finally, while it may well be true that the partial displacement by TV and other means of entertainment has inhibited expansion of the active voting universe during the past generation, it is equally true that the structure of the American voting universe—i.e., the adult population—as it exists today was substantially formed in the period 1900–20, *prior* to the development of such major media as the movies, radio, and television.

III

As we move below the gross national level, the voting patterns discussed above stand out with far greater clarity and detail. Their divergences suggest something of the individual differences which distinguish each state subsystem from its fellows, as their uniformities indicate the universality of the broader secular trends. Five states have been selected for analysis here. During the latter part of the nineteenth-century two of these, Michigan and

Pennsylvania, were originally competitive states which tended to favor the Republican party. They developed solidly one-party regimes after the realignment of 1896. These regimes were overthrown in their turn and vigorous party competition was restored in the wake of the New Deal realignment. In two other states, Ohio and New York, the 1896 alignment had no such dire consequences for two-party competition on the state level. These states have also shown a somewhat different pattern of development since the 1930s than Michigan and Pennsylvania. Our fifth state is Oklahoma, where a modified one-party system is structured heavily along sectional lines and operates in a socioeconomic context unfavorable to the classic New Deal articulation of politics along ethnic-class lines of cleavage.

Michigan politics was marked from 1894 through 1930 by the virtual eclipse of a state Democratic Party which had formerly contested elections on nearly equal terms with the Republicans. The inverse relationships developing between this emergent one-partyism on the one hand, and both the relative size of the active voting universe and the strength of party linkage on the other, stand out in especially bold relief.

A decisive shift away from the stable and substantially fully mobilized voting patterns of the nineteenth-century occurred in Michigan after the realignment of 1896, with a lag of about a decade between that election and the onset of disruption in those patterns. The first major breakthrough of characteristics associated with twentieth-century American electorates occurred in the presidential year 1904, when the mean percentage Democratic for all statewide offices reached an unprecedented low of 35.6 and the rate of split-ticket voting jumped from almost zero to 17.1 percent. A steady progression of decline in turnout and party competition, accompanied by heavy increases in the other criteria of peripherality, continued down through 1930.

The scope of this transformation was virtually revolutionary. During the Civil War era scarcely 15 percent of Michigan's potential electorate appears

Table 3. Michigan, 1854–1962: Decay and resurgence? (Percentages)

Period	Mean turnout		Mean drop-off	Mean roll-off	Mean split-ticket voting	Mean partisan swing	Mean % D of two-party vote
	Presidential years	Off years					
1854–72	84.8	78.1	7.8	0.9	0.8	3.2	43.9
1878–92	84.9	74.9	10.7	0.8	1.6	2.2	48.0
1894–1908	84.8	68.2	22.3	1.5	5.9	4.7	39.6
1910–18	71.4	53.0	27.2	3.0	9.8	4.1	40.4*
1920–30	55.0	31.5	42.9	6.0	10.0	7.3	29.8
1932–46	63.6	47.3	25.9	6.7	6.0	7.4	47.9
1948–62	66.9	53.6	19.1	4.1	5.8	4.9	51.0

*Democratic percentage of three-party vote in 1912 and 1914.

to have been altogether outside the voting universe. About 7 percent could be classified as peripheral voters by Campbell's definition, and the remainder—more than three-quarters of the total—were core voters. Moreover, as the extremely low nineteenth-century level of split-ticket voting indicates, these active voters overwhelmingly cast party-line ballots. By the 1920s less than one-third of the potential electorate were still core voters, while nearly one-quarter were peripheral and nearly one-half remained outside the political system altogether. Drop-off and roll-off increased sixfold during this period, while the amplitude of partisan swing approximately doubled and the split-ticket voting rate increased by a factor of approximately eight to twelve.

For the most part these trends underwent a sharp reversal as party competition in Michigan was abruptly restored during the 1930s and organized in its contemporary mode in 1948. As the mean Democratic percentage of the two-party vote increased and turnout—especially in off-year elections—showed a marked relative upswing, such characteristics of marginality as drop-off, roll-off, split-ticket voting, and partisan swing declined in magnitude. Yet as the means for the 1948–62 period demonstrate, a large gap remains to be closed before anything like the status quo ante can be restored. Our criteria—except, of course, for the mean percentage Democratic of the two-party vote—have returned only to the levels of the transitional period 1900–1918. As is well known, exceptionally disciplined and issue-oriented party organizations have emerged in Michigan since 1948, and elections have been intensely competitive throughout this period.[20] In view of this, the failure of turnout in recent years to return to something approaching nineteenth-century levels is all the more impressive, as is the continuing persistence of fairly high levels of drop-off, roll-off, and split-ticket voting.[21]

The Michigan data have still more suggestive implications. Campbell's discussion of surge and decline in the modern context points to a cyclical process in which peripheral voters, drawn into the active voting universe only under unusual short-term stimuli, withdraw from it again when the stimuli are removed. It follows that declines in turnout are accompanied by a marked relative increase in the component of core voters in the electorate and by a closer approximation in off years to a "normal" partisan division of the vote.[22] This presumably includes a reduction in the level of split-ticket voting as well. But the precise opposite occurred as a secular process—not only in Michigan but, it would seem, universally—during the 1900–30 era. Declines in turnout were accompanied by substantial, continuous increases in the indices of party and voter peripherality among those elements of the adult population which remained in the political universe at all. The lower the turnout during this period, the fewer of the voters still remaining who bothered to vote for the entire slate of officers in any given election. The lower the turnout in presidential years, the greater was the drop-off gap between the total vote cast in presidential and succeeding off-year elections. The lower the turnout, the greater were the incidence of split-ticket voting and the amplitude of partisan swing. Under the enormous impact of the forces which produced these declines in turnout and party competitiveness after 1896, the component of highly involved and party-oriented core voters in the active electorate fell off at a rate which more than kept pace with the progressive shrinking of that electorate's relative size. These developments necessarily imply a limitation upon the usefulness of the surge-decline model as it relates to secular movements prior to about 1934. They suggest, moreover, that the effects of the forces at work after 1896 to depress voter participation and to dislocate party linkage between voters and government were even more crushingly severe than a superficial perusal of the data would indicate.

Pennsylvania provides us with variations on the same theme. As in Michigan, the political realignment centering on 1896 eventually converted an industrializing state with a relatively slight but usually decisive Republican bias into a solidly one-party GOP bastion. To a much greater extent than in Michigan, this disintegration of the state Democratic party was accompanied by periodic outbursts of third-party ventures and plural party nominations of major candidates, down to the First World War. Thereafter, as in Michigan, the real contest between competing candidates and political tendencies passed into the Republican primary, where it usually remained until the advent of the New Deal. In both states relatively extreme declines in the rate of turnout were associated with the disappearance of effective two-party competition, and in both states these declines were closely paralleled by sharp increases in the indices of peripherality.

As Table 4 demonstrates, the parallel behavior of the Michigan and Pennsylvania electorates has also extended into the present; the now familiar pattern of increasing turnout and party competition accompanied by marked declines in our other indices has been quite visible in the Keystone State since the advent of the New Deal. On the whole, indeed, a better approximation to the status quo ante has been reached in Pennsylvania than in Michigan or

Table 4. Voting patterns in Pennsylvania, 1876–1962: Decline and resurgence? (Percentages)

	Mean turnout		Mean drop-off	Mean roll-off	Mean split-ticket voting	Mean partisan swing	Mean % D of two-party vote
	Presidential years	Off years					
1876–92	78.5	69.3	9.4	0.6	0.6	1.4	47.7
1894–1908	75.7	64.7	12.2	5.2	1.3	6.3	38.5
1910–18	64.0	51.4	20.0	4.3	4.7	5.8	43.6*
1920–30	50.4	39.5	28.0	5.2	8.9	7.1	32.8
1932–48	61.5	51.9	14.9	2.2	1.4	6.1	49.0
1950–62	67.5	56.3	12.2	1.8	3.1	3.3	49.3

*Combined major anti-Republican vote (Democrat, Keystone, Lincoln, Washington).

perhaps in most other states. But despite the intense competitiveness of its present party system, this restoration remains far from complete.

A more detailed examination of turnout and variability in turnout below the statewide level raises some questions about the direct role of immigration and woman suffrage in depressing voter participation. It also uncovers a significant transposition of relative voter involvement in rural areas and urban centers since about 1930.

It is frequently argued that declines in participation after the turn of the century were largely the product of massive immigration from Europe and of the advent of woman suffrage, both of which added very large and initially poorly socialized elements to the potential electorate.[23] There is no question that these were influential factors. The data in Table 5 indicate, for example, that down until the Great Depression turnout was consistently higher and much less subject to variation in rural counties with relatively insignificant foreign-stock populations than in either the industrial-mining or metropolitan counties.

Yet two other aspects of these data should also be noted. First, the pattern of turnout decline from the 1876–96 period to the 1900–1916 period was quite uniform among all categories of counties, though the rank order of their turnouts remained largely unchanged. It can be inferred from this that, while immigration probably played a major role in the evolution of Pennsylvania's political system as a whole, it had no visible direct effect upon the secular decline in rural voting participation. Broader systemic factors, including but transcending the factor of immigration, seem clearly to have been at work. Second, a very substantial fraction of the total decline in turnout from the 1870s to the 1920s—in some rural native-stock counties more than half— occurred *before* women were given the vote. Moreover, post-1950 turnout levels in Pennsylvania, and apparently in most other non-Southern states, have been at least as high as in the decade immediately preceding the general enfranchisement of women. If even today a higher percentage of American than European women fail to come to the polls, the same can also be said of such population groups as the poorly educated, farmers, the lower-income classes, Negroes, and other deprived elements in the potential electorate.[24] In such a context woman suffrage, as important a variable as it certainly has been in our recent political history, seems to raise more analytical problems than it solves.

Particularly suggestive for our hypothesis of basic changes in the nature of American voting behavior over time is the quite recent transposition of aggregate turnout and variations in turnout as between our rural sample and the two metropolitan centers. In sharp contrast to the situation prevailing before 1900, turnout in these rural counties has tended during the past generation not only to be slightly lower than in the large cities but also subject to far wider oscillations from election to election. In Bedford County, for example, turnout stood at 82.5 percent in 1936, but sagged to an all-time low

Table 5. Differentials in aggregate turnout and variations of turnout in selected Pennsylvania counties: Presidential elections, 1876–1960

County and type	N	% Foreign stock, 1920	1876–96		1900–1916		1920–32		1936–60	
			Mean turnout (%)	Coef. var.*	Mean turnout (%)	Coef. var.*	Mean turnout (%)	Coef. var.*	Mean turnout (%)	Coef. var.*
Urban										
Allegheny	1	56.6	71.8	6.75	56.7	2.45	43.8	10.11	68.9	5.82
Philadelphia	1	54.3	85.2	4.61	72.9	6.42	50.5	12.57	68.8	4.40
Industrial-Mining	4	49.0	88.1	4.48	72.8	4.41	54.2	11.63	64.7	10.88
Rural	8	13.5	88.5	3.12	76.4	3.63	56.0	8.09	65.2	13.20

*The coefficient of variability is a standard statistical measure; see V. O. Key, Jr., *A Primer of Statistics for Political Scientists* (New York, 1954), pp. 44–52. Since secular trends, where present, had to be taken into account, this coefficient appears abnormally low in the period 1900–1916. During this period many counties registered a straight-line decline in turnout from one election to the next.

Table 6. Urban-rural differences in stability of political involvement in Pennsylvania

County and type	N	1936–60 turnout 1876–96 turnout (%)	1936–60 variability 1876–96 variability (%)
Urban			
Allegheny	1	95.9	86.2
Philadelphia	1	80.8	95.4
Industrial-mining	4	73.4	249.6
Rural	8	73.7	447.4

of 41.2 percent in 1948. The comparable figures in Philadelphia were 74.3 and 64.8 percent, and in Allegheny County 72.5 percent (in 1940) and 60.6 percent.

A major finding revealed by survey research is that the "farm vote" is currently one of the most unstable and poorly articulated elements in the American electorate.[25] It is said that since rural voters lack the solid network of group identifications and easy access to mass-communication media enjoyed by their city cousins, they tend to be both unusually apathetic and exceptionally volatile in their partisan commitments. As rural voting turnout was abnormally low in 1948, its rate of increase from 1948 to 1952 was exceptionally large and—fully consistent with Campbell's surge-decline model—was associated with a one-sided surge toward Eisenhower. A restatement of the data in Table 5 lends strong support to this evaluation of the relative position of the rural vote as a description of the *current* American voting universe.

But the data strongly imply that virtually the opposite of present conditions prevailed during the nineteenth century. Such variables as education level, communications and non-family-group interaction were probably much more poorly developed in rural areas before 1900 than they are today. Not only did this leave no visible mark on agrarian turnout; it seems extremely likely that the nineteenth-century farmer was at least as well integrated into the political system of that day as any other element in the American electorate. The awesome rates of turnout which can be found in states like Indiana, Iowa and Kentucky prior to 1900 indicate that this extremely high level of rural political involvement was not limited to Pennsylvania.[26] As a recent study of Indiana politics demonstrates, the primarily rural "traditional vote" in that state was marked prior to 1900 by an overwhelming partisan stability as well.[27]

Perhaps, following the arguments of C. Wright Mills and others, we can regard this extraordinary change in rural voting behavior as a function of the conversion of a crackerbarrel society into a subordinate element in a larger mass society.[28] In any event, this rural movement toward relatively low and widely fluctuating levels of turnout may well be indicative of an emergent political alienation in such areas. It is suggestive that these movements have been accompanied generally in Pennsylvania as in states like West Virginia by a strongly positive Republican trend in these agrarian bailiwicks during the last thirty years.[29] The impression arises that the political realignment of the 1930s, which only imperfectly mobilized and integrated urban populations into the political system, had not even these limited positive effects in more isolated communities.

The behavior of the Ohio electorate down to about 1930 closely paralleled the patterns displayed in its neighbor states, Michigan and Pennsylvania. Since then a marked divergence has been manifest.

Two-party competition here was far less seriously affected by the sectional political alignment of 1896–1932 than in most other northern industrial states. Of the eighteen gubernatorial elections held in Ohio from 1895 to 1930, for example, Democrats won ten. But here as elsewhere are to be found the same patterns of decline in turnout and sharp increases in indices of voter peripherality after 1900. Indeed, while turnout bottomed out during the 1920s at a point considerably higher than in Michigan or Pennsylvania, it had also been considerably higher than in either of them during the nineteenth century. Here too such variables as woman suffrage seem to have played a smaller role as causal agents—at least so far as they affected the growing tendencies toward peripherality among active voters—than is commonly supposed. Drop-off from presidential to off-year elections began to assume its modern shape in Ohio between 1898 and 1910. As Figure 2 shows, roll-off—an especially prominent feature in contemporary Ohio voting behavior—emerged in modern form in the election of 1914.

Ohio, unlike either Michigan or Pennsylvania, has demonstrated only an extremely limited resurgence since the realignment of the 1930s. Presiden-

Table 7. Patterns of voter participation in Ohio, 1857–1962: Decline without resurgence?

Period	Mean turnout Presidential years (%)	Off years (%)	Mean drop-off	Mean roll-off	Mean split-ticket voting
1857–79	89.0	78.4	9.7	0.6	0.5
1880–1903	92.2	80.5	11.2	0.8	0.6
1904–18	80.4	71.2	9.2	2.5	3.3
1920–30	62.4	45.8	24.1	7.9	9.9
1932–46	69.9	49.1	27.2	7.6	6.5
1948–62	66.5	53.3	19.0	8.2	11.1

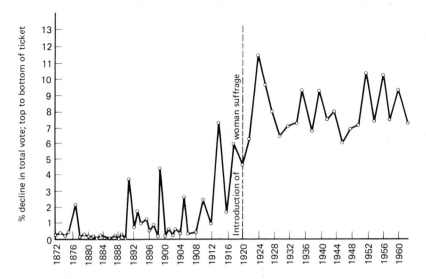

Figure 2. Increases in roll-off: The case of Ohio, 1872–1962.

tial-year voting turnout in the period 1948–60 actually declined from the mean level of 1932–44, and was not appreciably higher than it had been in the trough of the 1920s. If mean drop-off has declined somewhat in recent years, it still stands at a level twice as high as in any period before 1920. Moreover, roll-off and the rate of split-ticket voting have actually increased to unprecedented highs since 1948. By 1962 the latter ratio touched an all-time high of 21.3 percent (except for the three-party election of 1924), suggesting that Ohio politics may be becoming an "every man for himself" affair. This pattern of behavior stands in the sharpest possible contrast to nineteenth-century norms. In that period turnout had reached substantially full proportions, drop-off was minimal and well over 99 percent of the voters cast both complete ballots and straight party tickets—an achievement that may have been partly an artifact of the party ballots then in use.[30] The political reintegration which the New Deal realignment brought in its wake elsewhere has scarcely become visible in Ohio.

Two recent discussions of Ohio politics may shed some light upon these characteristics. Thomas A Flinn, examining changes over the past century in the partisan alignments of Ohio counties, concludes that until the first decade of the twentieth century the state had a set of political alignments based largely on sectionalism within Ohio—a product of the diverse regional backgrounds of its settlers and their descendants. This older political system broke down under the impact of industrialization and a national class-ethnic partisan realignment, but no new political order of similar coherence or par-

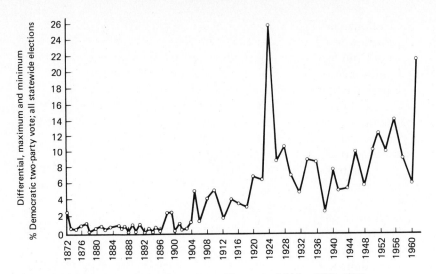

Figure 3. Increases in split-ticket voting: The case of Ohio, 1872–1962.

tisan stability has yet emerged to take its place.[31] Flinn's findings and the conclusions which Lee Benson has drawn from his study of New York voting behavior in the 1840s are remarkably similar.[32] In this earlier voting universe the durability of partisan commitment and the extremely high levels of turnout appear to have had their roots in a cohesive and persistent set of positive and negative group referents. These, as Flinn notes, provided "no clear-cut class basis for statewide party following from the time of Jackson to that of Wilson."[33]

John H. Fenton, discussing the 1962 gubernatorial campaign, carries the argument one step further.[34] Basic to Ohio's social structure, he argues, is an unusually wide diffusion of its working-class population among a large number of middle-sized cities and even smaller towns. The weakness of the labor unions and the chaotic disorganization of the state Democratic Party seem to rest upon this diffusion. Ohio also lacks agencies which report on the activities of politicians from a working-class point of view, such as have been set up by the United Automobile Workers in Detroit or the United Mine Workers in Pennsylvania or West Virginia. The result of this is that to a much greater extent than in other industrial states, potential recruits for a cohesive and reasonably well-organized Democratic party in Ohio live in an isolated, atomized social milieu. Consequently they tend to vote in a heavily personalist, issueless way, as the middle and upper classes do not. Such a state of affairs may provide clues not only for the relative failure of voter turnout to increase during the past generation, but for the persistent and growing indications of voter peripherality in Ohio's active electorate as well.

The development of the voting universe in New York is more analogous to the situation in Ohio than in either Michigan or Pennsylvania. In New York, as in Ohio, two-party competition was not as dislocated by the 1896–1930 alignment as a hasty survey of the presidential election percentages during that period might suggest. Democrats remained firmly in control of New York City, and this control helped them to capture the governorship eight out of eighteen times from 1896 through 1930. There were other parallels with Ohio as well, for here too this persistence of party competition did not prevent the normal post-1896 voting syndrome from appearing in New York. Nor has there been any pronounced resurgence in turnout levels or convincing declines in the other variables since the 1930s. Drop-off, roll-off, split-ticket voting, and partisan swing are not only quite high in New York by nineteenth-century standards, but have been twice as great as in neighboring Pennsylvania during the past decade. This relative failure of political reintegration is revealed not only by the data presented in Table 8 but—in much more dramatic fashion—by the rise and persistence of labor-oriented third parties which are centered in New York City and have enjoyed a balance-of-power position between the two major party establishments. The existence of the American Labor and Liberal parties, as well as the continuing vitality of anti-Tammany "reform" factions, are vocal testimony to the failure of the old-line New York Democratic party to adapt itself successfully to the political style and goals of a substantial portion of the urban electorate.

Curiously enough, examination of the data thus far presented raises some doubt that the direct primary has contributed quite as much to the erosion of party linkages as has been often supposed.[35] There seems to be little doubt that it has indeed been a major eroding element in some of the states where it has taken root—especially in states with partially or fully one-party sys-

Table 8. New York voting patterns, 1834–1962: Decline without resurgence? (Percentages)

	Mean turnout (presidential years)	Mean drop-off	Mean roll-off	Mean split-ticket voting	Mean partisan swing	Mean % D of two-party vote
1834–58	84.8	3.3	1.6	1.2	1.7	50.9*
1860–79	89.3	7.9	0.4	0.6	2.6	50.1
1880–98	87.9	10.4	1.2	1.6	5.0	50.5
1900–1908	82.5	8.3	1.1	2.2	3.7	47.2
1910–18	71.9	10.9	5.1	3.3	3.8	46.2
1920–30	60.4	17.3	5.5	9.5	8.3	49.6
1932–46	71.3	22.5	4.9	3.4	3.2	53.2**
1948–62	67.8	20.6	3.6	6.5	5.8	47.3**

*Elections from 1854 to 1858 excluded because of major third-party vote.
**The American Labor Party, 1936–46, and the Liberal Party, 1944–62, are included in Democratic vote when their candidates and Democratic candidates were the same.

tems where the primary has sapped the minority party's monopoly of opposition. But comparison of New York with our other states suggests the need of further concentrated work on this problem. After a brief flirtation with the direct primary between 1912 and 1921, New York resumed its place as one of the very few states relying on party conventions to select nominees for statewide offices, as it does to this day. Despite this fact, the post-1896 pattern of shrinkage in turnout and increases in our other indices of political dissociation was virtually the same in New York as elsewhere. To take a more recent example, New York's split-ticket voting ratio was 16.1 percent in 1962, compared with 21.3 percent in Ohio, 7.1 in Michigan, and 6.8 in Pennsylvania. The overall pattern of the data suggests that since 1932 the latter two states may have developed a more cohesive party politics and a more integrated voting universe with the direct primary than New York has without it.

If the data thus far indicate some link between the relative magnitude of voter nonparticipation and marginality with the cohesiveness of the local party system, even greater secular trends of the same sort should occur where one of the parties has continued to enjoy a perennially dominant position in state politics. Oklahoma, a Border state with a modified one-party regime, tends to support such an assumption.[36] The relatively recent admission of this state to the union naturally precludes analysis of its pre-1896 voting behavior. Even so, it is quite clear that the further back one goes toward the date of admission, the closer one comes to an approximation to a nineteenth-century voting universe. In Oklahoma, curiously enough, the secular decline in turnout and increases in the other indices continued into the New Deal era itself, measured by the off-year elections when—as in a growing number of states[32]—a full slate of statewide officers is elected. Since 1946 very little solid evidence of a substantial resurgence in turnout or of major declines in dropoff, roll-off, or split-ticket voting has appeared, but there is some evidence that the minority Republican party is atrophying.

Table 9. Oklahoma: 1907–62: Voter peripherality and party decay? (Percentages)

	Mean turnout (off years)	Mean drop-off	Mean roll-off	Mean split-ticket voting	State and congressional elections uncontested by Republicans	
					%	Mean N*
1907–18	52.9	12.1	6.1	3.6	2.1	32
1922–30	40.1	13.0	13.9	9.7	2.1	31
1934–46	37.1	32.2	16.4	8.1	14.8	32
1950–62	44.5	26.3	14.0	10.5	41.3	29

Note: Roll-off and split-ticket voting are computed for contested elections only.
*Mean number of state and congressional races in each off-year election.

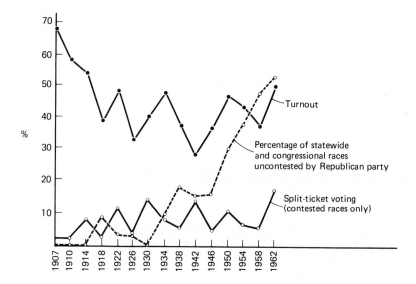

Figure 4. Patterns of political evolution: The case of Oklahoma, 1907–62.

The magnitude of drop-off and roll-off has become relatively enormous in Oklahoma since the 1920s, with a very slight reduction in both during the 1950–62 period. While turnout has correspondingly increased somewhat since its trough in the 1934–46 period, at no time since 1914 have as many as one-half of the state's potential voters come to the polls in these locally decisive off-year elections. Still more impressive is the almost vertical increase in the proportion of uncontested elections since the end of World War II. The 1958 and 1962 elections, moreover, indicate that the trend toward decomposition in the Republican party organization and its linkage with its mass base is continuing. In 1958 the party virtually collapsed, its gubernatorial candidate winning only 21.3 percent of the two-party vote. Four years later the Republican candidate won 55.5 percent of the two-party vote. The resultant partisan swings of 34.2 percent for this office and 22.0 for all contested statewide offices was the largest in the state's history and one of the largest on record anywhere. But while 1962 marked the first Republican gubernatorial victory in the state's history, it was also the first election in which the Republican party yielded more than half of the statewide and congressional offices to its opposition without any contest at all. Even among contested offices, the Oklahoma electorate followed a national trend in 1962 by splitting its tickets at the unprecedented rate of 17.3 percent.

As Key has suggested, the direct primary has almost certainly had cumulatively destructive effects on the cohesion of both parties in such modified one-party states as Oklahoma.[38] The rapidly spreading device of "insulating" state politics from national trends by holding the major state elections in off years has also probably played a significant role. Yet it seems more than likely that these are variables which ultimately depend for their effectiveness upon the nature of the local political culture and the socioeconomic forces which underlie it. Pennsylvania, for example, also has a direct primary. Since 1875 it has also insulated state from national politics by holding its major state elections in off years. Yet since the realignment of the 1930s, both parties have contested every statewide office in Pennsylvania as a matter of course. Indeed, only very infrequently have elections for seats in the state legislature gone by default to one of the parties, even in bailiwicks which it utterly dominates.[39]

These five statewide variations on our general theme suggest, as do the tentative explorations below the statewide level in Pennsylvania, that an extremely important factor in the recent evolution of the voting universe has been the extent to which the imperatives of the class-ethnic New Deal realignment have been relevant to the local social structure and political culture. In the absence of an effectively integrating set of state political organizations, issues, and candidates around which a relatively intense polarization of voters can develop, politics is likely to have so little salience that very substantial portions of the potential electorate either exclude themselves altogether from the political system or enter it in an erratic and occasional way. As organized and articulated in political terms, the contest between "business" and "government" which has tended to be the linchpin of our national politics since the 1930s has obviously made no impression upon many in the lowest income strata of the urban population. It has also failed to demonstrate sustained organizing power in areas of rural poverty or among local political cultures which remain largely preindustrial in outlook and social structure.

IV

The conclusions which arise directly out of this survey of aggregate data and indices of participation seem clear enough. On both the national and state levels they point to the existence and eventual collapse of an earlier political universe in the United States—a universe in many ways so sharply different from the one we all take for granted today that many of our contemporary frames of analytical reference seem irrelevant or misleading in studying it. The late nineteenth-century voting universe was marked by a more complete and intensely party-oriented voting participation among the American electorate than ever before or since. Approximately two-thirds of the potential national electorate were then "core" voters, one-tenth fell into the peripheral category, and about one-quarter remained outside. In the four Northern states examined in this survey the component of core elements in the potential electorate was even larger: about three-quarters core voters, one-tenth peripherals, and 15 percent nonvoters.

In other ways too this nineteenth-century system differed markedly from its successors. Class antagonisms as such appear to have had extremely low salience by comparison with today's voting behavior. Perhaps differentials in the level of formal education among various groups in the population contributed to differentials in nineteenth-century turnout as they clearly do now. But the unquestionably far lower *general* level of formal education in America during the last century did not preclude a much more intense and uniform mass political participation than any which has prevailed in recent decades. Though the evidence is still scanty, it strongly implies that the influence of rurality upon the intensity and uniformity of voting participation appears to have been precisely the opposite of what survey-research findings hold it to be today. This was essentially a preindustrial democratic system, resting heavily upon a rural and small-town base. Apparently it was quite adequate, both in partisan organization and dissemination of political information, to the task of mobilizing voters on a scale which compares favorably with recent European levels of participation.

There is little doubt that the model of surge and decline discussed above casts significant light upon the behavior of today's American electorate as it responds to the stimuli of successive elections. But the model depends for its validity upon the demonstrated existence of very large numbers both of peripheral voters and of core voters whose attachment to party is relatively feeble. Since these were not pronounced characteristics of the nineteenth-century voting universe, it might be expected that abnormal increases in the percentage of the vote won by either party would be associated with very different kinds of movements in the electorate, and that such increases would be relatively unusual by present-day standards.

Even a cursory inspection of the partisan dimensions of voting behavior in the nineteenth century tends to confirm this expectation. Not only did the amplitude of partisan swing generally tend to be much smaller then than now,[40] but nationwide landslides of the twentieth-century type were almost nonexistent.[41] Moreover, when one party did win an unusually heavy majority, this increase was usually associated with a pronounced and one-sided *decline* in turnout. Comparison of the 1848 and 1852 elections in Georgia and of the October gubernatorial and November presidential elections of 1872 in Pennsylvania, for example, makes it clear that the "landslides" won by one of the presidential contenders in 1852 and 1872 were the direct consequence of mass abstentions by voters who normally supported the other party.[42] Under nineteenth-century conditions, marked as they were by substantially full mobilization of the eligible electorate, the only play in the system which could provide extraordinary majorities had to come from a reversal of the modern pattern of surge and decline—a depression in turnout which was overwhelmingly confined to adherents of one of the parties.[43]

This earlier political order, as we have seen, was eroded away very rapidly after 1900. Turnout fell precipitately from nineteenth-century levels even before the advent of woman suffrage, and even in areas where immigrant elements in the electorates were almost nonexistent. As turnout declined, a larger and larger component of the still active electorate moved from a core to a peripheral position, and the hold of the parties over their mass base appreciably deteriorated. This revolutionary contraction in the size and diffusion in the shape of the voting universe was almost certainly the fruit of the heavily sectional party realignment which was inaugurated in 1896. This "system of 1896," as Schattschneider calls it,[44] led to the destruction of party competition throughout much of the United States, and thus paved the way for the rise of the direct primary. It also gave immense impetus to the strains of antipartisan and antimajoritarian theory and practice which have always been significant elements in the American political tradition. By the decade of the 1920s this new regime and business control over public policy in this country were consolidated. During that decade hardly more than one-third of the eligible adults were still core voters. Another one-sixth were peripheral voters and fully one-half remained outside the active voting universe altogether. It is difficult to avoid the impression that while all the forms of political democracy were more or less scrupulously preserved, the functional result of the "system of 1896" was the conversion of a fairly democratic regime into a rather broadly based oligarchy.

The present shape and size of the American voting universe are, of course, largely the product of the 1928–36 political realignment. Survey-research findings most closely approximate political reality as they relate to this next broad phase of American political evolution. But the characteristics of the present voting universe suggest rather forcefully that the New Deal realignment has been both incomplete and transitional. At present about 44 percent of the national electorate are core voters, another 16 percent or so are peripheral, and about 40 percent are still outside the political system altogether. By nineteenth-century standards, indices of voter peripherality stand at very high levels. Party organizations remain at best only indifferently successful at mobilizing a stable, predictable mass base of support.

The data which have been presented here, though they constitute only a small fraction of the materials which must eventually be examined, tend by and large to support Schattschneider's functional thesis of American party politics.[45] We still need to know a great deal more than we do about the specific linkages between party and voter in the nineteenth century. Systematic research also remains to be done on the causes and effects of the great post-1896 transition in American political behavior. Even so, it seems useful to propose a hypothesis of transition in extension of Schattschneider's argument.

The nineteenth-century American political system, for its day, was incomparably the most thoroughly democratized of any in the world. The development of vigorous party competition extended from individual localities to the nation itself. It involved the invention of the first organizational machinery—the caucus, the convention, and the widely disseminated party press—which was designed to deal with large numbers of citizens rather than with semiaristocratic parliamentary cliques. Sooner than the British, and at a time when Prussia protected its elites through its three-class electoral system, when each new change of regime in France brought with it a change in the size of the electorate and the nature of *le pays légal,* and when the basis of representation in Sweden was still the estate, Americans had elaborated not only the machinery and media of mass politics but a franchise which remarkably closely approached universal suffrage. Like the larger political culture of which it was an integral part, this system rested upon both broad consensual acceptance of middle-class social norms as ground rules and majoritarian settlement (in "critical" elections from time to time), once and for all, of deeply divisive substantive issues on which neither consensus nor further postponement of a showdown was possible. Within the limits so imposed it was apparently capable of coherent and decisive action. It especially permitted the explicit formulation of sectional issues and—though admittedly at the price of civil war—arrived at a clear-cut decision as to which of two incompatible sectional modes of social and economic organization was henceforth to prevail.

But after several decades of intensive industrialization a new dilemma of power, in many respects as grave as that which had eventuated in civil war, moved toward the stage of overt crisis. Prior to the closing years of the century the middle-class character of the political culture and the party system, coupled with the afterglow of the Civil War trauma, had permitted the penetration and control of the cadres of both major parties by the heavily concentrated power of our industrializing elites. But this control was inherently unstable, for if and when the social dislocations produced by the industrial revolution should in turn produce a grassroots counterrevolution, the party whose clienteles were more vulnerable to the appeals of the counterrevolutionaries might be captured by them.

The takeoff phase of industrialization has been a brutal and exploitative process everywhere, whether managed by capitalists or commissars.[46] A vital functional political need during this phase is to provide adequate insulation of the industrializing elites from mass pressures, and to prevent their displacement by a coalition of those who are damaged by the processes of capital accumulation. This problem was effectively resolved in the Soviet Union under Lenin and Stalin by vesting a totalitarian monopoly of political power in the hands of Communist industrializing elites. In recent years developing nations have tended to rely upon less coercive devices such as nontotalitarian single-party systems or personalist dictatorship to meet that need, among others. The nineteenth-century European elites were provided a good deal of insulation by the persistence of feudal patterns of social deference and especially by the restriction of the right to vote to the middle and upper classes.

But in the United States the institutions of mass democratic politics and universal suffrage uniquely came into being *before* the onset of full-scale industrialization. The struggle for democracy in Europe was explicitly linked from the outset with the struggle for universal suffrage. The eventual success of this movement permitted the development in relatively sequential fashion of the forms of party organization which Duverger has described in detail.[47] In the United States—ostensibly at least—the struggle for democracy had already been won, and remarkably painlessly, by the mid-nineteenth century. In consequence, the American industrializing elites were, and felt themselves to be, uniquely vulnerable to an anti-industrialist assault which could be carried out peacefully and in the absence of effective legal or customary sanctions by a citizenry possessing at least two generations' experience with political democracy.

This crisis of vulnerability reached its peak in the 1890s. Two major elements in the population bore the brunt of the exceptionally severe deprivations felt during this depression decade: the smaller cash-crop farmers of the Southern and Western "colonial" regions and the ethnically fragmented urban working class. The cash-crop farmers, typically overextended and undercapitalized, had undergone a thirty-year decline in the prices for their commodities in the face of intense international competition. With the onset of depression in 1893, what had been acute discomfort for them became disaster. The workers, already cruelly exploited in many instances during this "takeoff" phase of large-scale industrialization, were also devastated by the worst depression the country had thus far known. Characteristically, the farmers resorted to political organization while the workers sporadically resorted to often bloody strikes. The industrializers and their intellectual and legal spokesmen were acutely conscious that these two profoundly alienated groups might coalesce. Their alarm was apparently given quite tangible form when the agrarian insurgents captured control of the Democratic party in 1896.

But the results of that great referendum revealed that the conservatives' fears and the anti-industrialists' hopes of putting together a winning coalition on a Jacksonian base were alike groundless. Not only did urban labor *not* flock to William Jennings Bryan, it repudiated the Democratic party on an unprecedented scale throughout the industrialized Northeast. The intensity and permanence of this urban realignment was paralleled by the Democrats' failure to make significant inroads into Republican strength in the more diversified and depression-resistant farm areas east of the Missouri River, and by their nearly total collapse in rural New England. The Democratic-Populist effort to create a coalition of the dispossessed created instead the most enduringly sectional political alignment in American history—an alignment which eventually separated the Southern and Western agrarians and transformed the most industrially advanced region of the country into a bulwark of industrialist Republicanism.

This realignment brought victory beyond expectation to those who had sought to find some way of insulating American elites from mass pressures without formally disrupting the preexisting democratic-pluralist political structure, without violence and without conspiracy. Of the factors involved in this victory three stand out as of particular importance.

(1) The depression of 1893 began and deepened during a Democratic administration. Of course there is no way of ascertaining directly what part of the decisive minority which shifted its allegiance to the Republican party reacted viscerally to the then incumbent party and failed to perceive that Cleveland and Bryan were diametrically opposed on the central policy issues of the day. But contemporary survey findings would tend to suggest that such a component in a realigning electorate might not be small. In this context it is especially worth noting that the process of profound break with traditional voting patterns began in the fall of 1893, not in 1896. In a number of major states like Ohio and Pennsylvania the voting pattern of 1896 bears far more resemblance to those of 1893–95 than the latter did to pre-1893 voting patterns. Assuming that such visceral responses to the Democrats as the "party of depression" did play a major role in the realignment, it would follow that the strong economic upswing after 1897 would tend to strengthen this identification and its cognate, the identification of the Republicans as the "party of prosperity."

(2) The Democratic platform and campaign were heavily weighted toward the interests and needs of an essentially rural and semicolonial clientele. Considerably narrowed in its programmatic base from the farmer-labor Populist platform of 1892, the Democratic party focused most of its campaign upon monetary inflation as a means of redressing the economic balance. Bryan's viewpoint was essentially that of the smallholder who wished to give the term "businessman" a broader definition than the Easterners meant by it, and of an agrarian whose remarks about the relative importance of farms and cities bespoke his profound misunderstanding of the revolution of his time. Silver mine owners and depressed cash-crop farmers could greet the prospect of inflation with enthusiasm, but it meant much less to adequately capitalized and diversified farmers in the Northeast, and less than nothing to the depression-ridden wage earners in that region's shops, mines, and factories. Bryan's appeal at base was essentially Jacksonian—a call for a return to the simpler and more virtuous economic and political arrangements which he identified with that bygone era. Such nostalgia could evoke a positive response among the native-stock rural elements whose political style and economic expectations had been shaped in the far-away past. But it could hardly seem a realistic political choice for the ethnically pluralist urban populations, large numbers of whom found such nostalgia meaningless since it related to nothing in their past or current experience. Programmatically, at least, these urbanites were presented with a two-way choice only one part of which seemed at all functionally related to the realities of an emergent industrial society. With the Democrats actually cast in the role of reactionaries despite the apparent radicalism of their platform and leader, and with no socialist alternative even thinkable in the context of the American political culture of

the 1890s, the Republican party alone retained some relevance to the urban setting. In this context, its massive triumph there was a foregone conclusion.

(3) An extremely important aspect of any political realignment is the unusually intense mobilization of negative-reference-group sentiments during the course of the campaign. The year 1896 was typical in this respect. Profound antagonisms in culture and political style between the cosmopolitan, immigrant, wet, largely non-Protestant components of American urban populations and the parochial, dry, Anglo-Saxon Protestant inhabitants of rural areas can be traced back at least to the 1840s. Bryan was virtually the archetype of the latter culture, and it would have been surprising had he not been the target of intense ethnocultural hostility from those who identified with the former. He could hardly have appeared as other than an alien to those who heard him in New York in 1896, or to those who booed him off the stage at the Democratic convention—also in New York—in 1924. Moreover, his remarks about the Northeast as "the enemy's country"—anticipating Senator Goldwater's views about that region in 1964—could only intensify a broadly sectional hostility to his candidacy and deepen the impression that he was attacking not only the Northeast's industrializing elites but the Northeast itself. Both in 1896 and 1964 this region gave every visible evidence of replying in kind.

As Schattschneider has perceptively observed, the "system of 1896" was admirably suited to its primary function. One of its major working parts was a judiciary which proceeded first to manufacture the needed constitutional restraints on democratic political action—a development presaged by such decisions as the Minnesota railroad case of 1890[48] and the income tax cases of 1894–95[49]—and then to apply these restraints against certain sensitive categories of national and state economic legislation.[50] Another of the new system's basic components was the control which the sectional alignment itself gave to the Republican Party, and through it the corporate business community, over the scope and direction of national public policy. Democracy was not only placed in judicial leading strings, it was effectively placed out of commission—at least so far as two-party competition was concerned—in more than half of the states. Yet it was one of the greatest, if unacknowledged, contributions of the "system of 1896" that democratic forms, procedures, and traditions continued to survive.[51] Confronted with a narrowed scope of effective democratic options, an increasingly large proportion of the eligible adult population either left, failed to enter or—as was the case with Southern Negroes after the completion of the 1890–1904 disfranchisement movement in the Old Confederacy—was systematically excluded from the American voting universe. The results of this on the exercise of the franchise have already been examined here in some detail. It was during this 1896–1932 era that the basic characteristics associated with today's mass electorate were formed.

These characteristics, as we have seen, have already far outlived the 1896 alignment itself. There seems to be no convincing evidence that they are being progressively liquidated at the present time. If the reemergence of a competitive party politics and its at least partial orientation toward the broader needs of an urban, industrialized society were welcome fruits of the New Deal revolution, that revolution has apparently exhausted most of its potential for stimulating turnout or party-oriented voting in America. The present state of affairs, to be sure, is not without its defenders. The civic-minded have tended to argue that the visible drift away from party-oriented voting among a growing minority of voters is a sign of increasing maturity in the electorate.[52] Others have argued that mediocre rates of turnout in the United States, paralleled by the normally low salience of issues in our polit-ical campaigns, are indicative of a "politics of happiness."[53] It is further con-tended that any sudden injection of large numbers of poorly socialized adults into the active voting universe could constitute a danger to the Republic.[54]

But there is another side to this coin. The ultimate democratic purpose of issue formulation in a campaign is to give the people at large the power to choose their and their agents' options. Moreover, so far as is known, the blunt alternative to party government is the concentration of political power, locally or nationally, in the hands of those who already possess concentrated economic power.[55] If no adequate substitute for party as a means for mobi-lizing nonelite influence on the governing process has yet been discovered, the obvious growth of "image" and "personality" voting in recent decades should be a matter of some concern to those who would like to see a more complete restoration of the democratic process in the United States.

Moreover, recent studies—such as Murray Levin's examinations of the attitudes of the Boston and Massachusetts electorate—reveal that such phe-nomena as widespread ticket splitting may be associated quite readily with pervasive and remarkably intense feelings of political alienation.[56] Convinced that both party organizations are hopelessly corrupt and out of reach of pop-ular control, a minority which is large enough to hold the balance of power between Republicans and Democrats tends rather consistently to vote for the lesser, or lesser known, of two evils. It takes a mordant variety of humor to find a kind of emergent voter maturity in this alienation. For Levin's data are difficult to square with the facile optimism underlying the civics approach to independent voting. So, for that matter, are the conclusions of survey research about the behavior of many so-called independent voters.[57]

Findings such as these seem little more comforting to the proponents of the "politics of happiness" thesis. Granted the proposition that most people who have been immersed from birth in a given political system are apt to be unaware of alternatives whose explicit formulation that system inhibits, it is of course difficult to ascertain whether their issueless and apathetic political style is an outward sign of "real" happiness. We can surmise, however, that the kind of political alienation which Levin describes is incompatible with political happiness, whether real or fancied. A great many American voters, it would seem, are quite intelligent enough to perceive the deep contradiction which exists between the ideals of rhetorical democracy as preached in school and on the stump, and the actual day-to-day reality as that reality intrudes on his own milieu. Alienation arises from perception of that contradiction, and from the consequent feelings of individual political futility arising when the voter confronts an organization of politics which seems unable to produce minimally gratifying results. The concentration of socially deprived characteristics among the more than forty million adult Americans who today are altogether outside the voting universe suggests active alienation—or its passive equivalent, political apathy—on a scale quite unknown anywhere else in the Western world. Unless it is assumed as a kind of universal law that problems of existence which can be organized in political terms must fade out below a certain socioeconomic level, this state of affairs is not inevitable. And if it is not inevitable, one may infer that the political system itself is responsible for its continued existence.

Yet such an assumption of fade-out is clearly untenable in view of what is known about patterns of voting participation in other democratic systems. Nor need it be assumed that substantial and rapid increases in American voting participation would necessarily, or even probably, involve the emergence of totalitarian mass movements. The possibility of such movements is a constant danger, to be sure, in any polity containing so high a proportion of apolitical elements in its potential electorate. But it would be unwise to respond to this possibility by merely expressing the comfortable hope that the apoliticals will remain apolitical, and by doing nothing to engage them in the system in a timely and orderly way. It is much more to the point to seek a way, if one can be found, to integrate the apolitical half of the American electorate into the political system before crisis arises.[58] The United States, after all, enjoyed intense mass political involvement without totalitarian movements during the last part of the nineteenth century, as do other Western democracies today.

No integration of the apoliticals can be carried out without a price to be paid. Underlying the failure of political organizations more advanced than the nineteenth-century middle-class cadre party to develop in this country has been the deeper failure of any except middle-class social and political values to achieve full legitimacy in the American political culture. It may not now be possible for our polity to make so great a leap as to admit non-middle-class values to political legitimacy and thus provide the preconditions for a more coherent and responsible mode of party organization. But such a leap may have to be made if full mobilization of the apolitical elements is to be achieved without the simultaneous emergence of manipulative radicalism of the left or the right. The heart of our contemporary political dilemma appears to lie in the conflict between this emergent need and the ideological individualism which continues so deeply to pervade our political culture. Yet the present situation perpetuates a standing danger that the half of the American electorate which is now more or less entirely outside the universe of active politics may someday be mobilized in substantial degree by totalitarian or quasi-totalitarian appeals. As the late President Kennedy seemed to intimate in his executive order establishing the Commission on Registration and Voting Participation, it also raises some questions about the legitimacy of the regime itself.[59]

NOTES

1. The leading work of this sort thus far has been done by the late V. O. Key, Jr. See, for example, his "A Theory of Critical Elections," *Journal of Politics* 17 (1955): 3–18; and his *American State Politics* (New York, 1956). See also such quantitatively oriented monographs as Perry Howard, *Political Tendencies in Louisiana, 1812–1952* (Baton Rouge, 1957).
2. The most notable survey-research effort to date to develop politically relevant theory regarding American voting behavior is Angus Campbell, Philip E. Converse, Warren E. Miller, and Donald E. Stokes, *The American Voter* (New York, 1960), especially ch. 20.
3. V. O. Key, Jr., "The Politically Relevant in Surveys," *Public Opinion Quarterly* 24 (1960): 54–61: V. O. Key, Jr., and Frank Munger, "Social Determinism and Electoral Decision: The Case of Indiana," in Eugene Burdick and Arthur J. Brodbeck, eds., *American Voting Behavior* (Glencoe, Ill., 1959), pp. 281–99.
4. Warren E. Miller and Donald E. Stokes, "Constituency Influence in Congress," *American Political Science Review* 57 (1963): 45–56. The authors observe that in the 1958 Hays-Alford congressional race in Arkansas, the normally potential nature of constituency sanctions against representatives was transferred under the overriding pressure of the race issue into an actuality which resulted in Hays' defeat by a write-in vote for his opponent. The normally low issue and candidate consciousness among the electorate was abruptly replaced by a most untypically intense awareness of the candidates and their relative postions on this issue. For an excellent cross-polity study of voting behavior based on comparative survey analysis, see Robert R. Alford, *Party and Society* (Chicago, 1963).
5. V. O. Key, Jr., *Public Opinion and American Democracy* (New York, 1961), based largely on survey-research data at the University of Michigan.
6. This effort, to which the author was able to contribute thanks to a Social Science Research Council grant for 1963–64, has been supported by the Council and by the National Science Foundation. This article is in no sense an integral part of that larger project. But it is proper to acknowledge gratefully here that the SSRC, by making it possible for me to spend a year at the Survey Research Center, has helped to provide conditions favorable to writing it. Thanks are also due to Angus Campbell, Philip E. Converse, Donald E. Stokes, and Warren E. Miller for their comments and criticisms. They bear no responsibility for the defects of the final product.
7. Key, *American State Politics*, pp. 71–73, 197–216.
8. E. E. Schattschneider, *Party Government* (New York, 1942) and *The Semi-Sovereign People* (New York, 1960), pp. 78–96.
9. Howard A. Scarrow, "Patterns of Voter Turnout in Canada," *Midwest Journal of Political Science* 5 (1961): 351–64; James A. Robinson and William Standing, "Some Correlates of Voter Participation: The Case of Indiana, *Journal of Politics* 22 (1960): 96–111. Both articles—one involving a political system outside of but adjacent to the United States—indicate patterns of contemporary participation which seem at variance with the conclusions of survey studies regarding the behavior of the American electorate. In Canada rural turnout is higher than urban, and no clear-cut pattern of drop-off between federal and provincial elections exists. Voter participation in Indiana apparently does not increase with the competi-

tiveness of the electoral situation, and does increase with the rurality of the election jurisdiction. With the possible exception of the relationship between competitiveness and
turnout, all of these are characteristics associated with nineteenth-century voting behavior
in the United States.

10. In computing turnout data, note that until approximately 1920 the criteria for eligibility to
vote differed far more widely from state to state than they do now. In a number of states
west of the original thirteen—for example, in Michigan until 1894 and in Wisconsin until
1908—aliens who had merely declared their intention to become citizens were permitted to
vote. Woman suffrage was also extended piecemeal for several decades prior to the general
enfranchisement of 1920. The turnout estimates derived here have been adjusted, so far as
the census data permit, to take account of such variations.

11. If one computes the off-year total vote of the years 1950–62 as a percentage of the total vote
cast in the preceding presidential election, a virtually identical correspondence is reached
with estimated off-year turnout as a percentage of turnout in the immediately preceding
presidential year.

Year	Total off-year vote	Estimated off-year turnout
1950	82.9	80.4
1954	69.2	67.5
1958	73.9	72.1
1962	74.4	73.6

12. Angus Campbell, "Surge and Decline: A Study of Electoral Change," *Public Opinion Quarterly* 24 (1960): 397–418.

13. Ibid., p. 413. The percentage of drop-off from 1956 to 1958, as computed from aggregate
voting data, was 25.6 percent.

14. See, for example, Robert E. Lane, *Political Life* (Glencoe, Ill., 1959), pp. 18–26.

15. There are, of course, very wide divergences in turnout rates even among non-Southern states.
Some of them, like Idaho, New Hampshire, and Utah, have presidential-year turnouts which
compare very favorably with European levels of participation. A detailed analysis of these
differences remains to be made. It should prove of the utmost importance in casting light
upon the relevance of current forms of political organization and partisan alignments to
differing kinds of electorates and political subsystems in the United States.

16. *Report of the President's Commission on Registration and Voting Participation* (Washington, 1963), esp. pp. 5–9. Hereafter cited as *Report.*

17. Ibid., pp. 11–14, 31–42.

18. See, for example, Stanley Kelley, "Elections and the Mass Media," *Law and Contemporary
Problems* 27 (1962): 307–26.

19. Schattschneider. *Semi-Sovereign People*, pp. 102–3.

20. Joseph La Palombara, *Guide to Michigan Politics* (East Lansing: Michigan State University
Press, 1960), pp. 22–35.

21. This recalls Robinson and Standing's conclusion that voter participation in Indiana does not
necessarily increase with increasing party competition. Of the eight Michigan gubernatorial
elections from 1948 to 1962 only one was decided by a margin of 55 percent or more, while
three were decided by margins of less than 51.5 percent of the two-party vote. Despite this
intensely competitive situation, turnout—while of course much higher than in the 1920s—
remains significantly below normal pre-1920 levels.

22. Campbell, "Surge and Decline," pp. 401–4.

23. Herbert Tingsten, *Political Behavior* (Stockholm, 1937), pp. 10–36. See also Charles E. Mer-
riam and Harold F. Gosnell, *Non-Voting* (Chicago, 1924), pp. 26, 109–22, for a useful dis-

cussion of the effect of woman suffrage on turnout in a metropolitan area immediately following the general enfranchisement of 1920.

24. Survey-research estimates place current turnout among American women at 10 percent below male turnout: Campbell et al., *American Voter,* pp. 484–85. This sex-related difference in participation is apparently universal, but is significantly smaller in European countries which provide election data by sex, despite the far higher European level of participation by both sexes. The postwar differential has been 5.8 percent in Norway (1945–57 mean), 3.3 percent in Sweden (1948–60 mean), and 1.9 percent in Finland (1962 general election). While in 1956 only about 55 percent of American women went to the polls, the mean turnout among women in postwar elections was 76.1 percent in Norway and 79.4 percent in Sweden.

25. Ibid., pp. 402–40.

26. The estimated rates of turnout in presidential elections from 1876 through 1896, mean turnout in the period 1936–60, and estimated turnout in 1964 were as follows in these states:

State	1876	1880	1884	1888	1892	1896	1936–60 (Mean)	1964 (Prelim.)
Indiana	94.6	94.4	92.2	93.3	89.0	95.1	75.0	73.3
Iowa	89.6	91.5	90.0	87.9	88.5	96.2	71.7	72.0
Kentucky	76.1	71.0	68.0	79.1	72.6	88.0	57.6	52.6

27. Key and Munger, "Social Determinism and Electoral Decision," pp. 282–88.

28. C. Wright Mills, *The Power Elite* (New York, 1956), pp. 298–324. See also Arthur J. Vidich and Joseph Bensman, *Small Town in Mass Society* (New York, 1960), pp. 5–15, 202–27, 297–320.

29. John H. Fenton, *Politics in the Border States* (New Orleans; 1957), pp. 117–20.

30. However, Ohio's modern pattern of split-ticket voting, formed several decades ago, seems to have been little (if at all) affected by the 1950 change from party-column to office-block ballot forms. See Figure 3.

31. Thomas A. Flinn, "Continuity and Change in Ohio Politics," *Journal of Politics* 24 (1962): 521–44.

32. Lee Benson, *The Concept of Jacksonian Democracy* (Princeton, 1961), pp. 123–207, 288–328.

33. Flinn, "Continuity and Change," p. 542.

34. John H. Fenton, "Ohio's Unpredictable Voters," *Harper's,* 225 (1962): 61–65.

35. This would seem to suggest a limitation on Key's findings, *American State Politics,* pp. 169–96.

36. This designation is given the state's political system in Oliver Benson, Harry Holloway, George Mauer, Joseph Pray, and Wayne Young, *Oklahoma Votes: 1907–1962* (Norman, Okla., 1964), pp. 44–52. For an extensive discussion of the sectional basis of Oklahoma politics, see Ibid., pp. 32–43, and Key, *American State Politics,* pp. 220–22.

37. In 1936, thirty-four states (71 percent) elected governors for either two- or four-year terms in presidential years, and the three-year term in New Jersey caused major state elections to coincide with every fourth presidential election. By 1964, only twenty-five of fifty states (50 percent) still held some of their gubernatorial elections in presidential years. Two of these, Florida and Michigan, are scheduled to begin off-year gubernatorial elections for four-year terms in 1966.

38. Key, *American State Politics,* pp. 169–96.

39. In the period 1956–62 there have been 840 general election contests for the Pennsylvania House of Representatives. Of these all but six, or 0.7 percent, have been contested by both

major political parties. No Pennslvania state senate seat has been uncontested during this period. Despite the 1962 Republican upsurge in Oklahoma, however, there were no contests between the parties in 11 of 22 senate seats (50.0 percent) and in 73 of 120 house seats (60.9 percent). All the uncontested senate seats and all but two of the uncontested house seats were won by Democrats.

40. Mean national partisan swings in presidential elections since 1872 have been as follows: 1872–92, 2.3 percent; 1896–1916, 5.0 percent; 1920–32, 10.3 percent; 1936–64, 5.4 percent.

41. If a presidential landslide is arbitrarily defined as a contest in which the winning candidate received 55 percent or more of the two-party vote, only the election of 1872 would qualify among the sixteen presidential elections held from 1836 to 1896. Of seventeen presidential elections held from 1900 through 1964, at least eight were landslide elections by this definition, and a ninth—the 1924 election, in which the Republican candidate received 54.3 percent and the Democratic candidate 29.0 percent of a three-party total—could plausibly be included.

42. The total vote in Georgia declined from 92,203 in 1848 to 62,333 in 1852. Estimated turnout declined from about 88 percent to about 55 percent of the eligible electorate, while the Democratic share of the two-party vote increased from 48.5 percent in 1848 to 64.8 percent in 1852. The pattern of participation in the Pennsylvania gubernatorial and presidential elections of 1872 is also revealing:

Raw vote	Governor, Oct. 1872	President, Nov. 1872	Absolute decline
Total	671,147	562,276	−108,871
Democratic	317,760	213,027	−104,733
Republican	353,387	349,249	− 4,138

Estimated turnout in October was 82.0 percent, in November 68.6 percent. The Democratic percentage of the two-party vote was 47.3 percent in October and 37.9 percent in November.

43. The only apparent exception to this generalization in the nineteenth century was the election of 1840. But this was the first election in which substantially full mobilization of the eligible electorate occurred. The rate of increase in the total vote from 1836 to 1840 was 60.0 percent, the largest in American history. Estimated turnout increased from about 58 percent in 1836 to about 80 percent in 1840. This election, with its relatively one-sided mobilization of hitherto apolitical elements in the potential electorate, not unnaturally bears some resemblance to the elections of the 1950s. But the increase in the Whig share of the two-party vote from 49.2 percent in 1836 to only 53.0 percent in 1840 suggests that that surge was considerably smaller than those of the 1950s.

44. Schattschneider, *Semi-Sovereign People*, p. 81.

45. Ibid., esp. pp. 78–113. See also his "United States: The Functional Approach to Party Government," in Sigmund Neumann, ed., *Modern Political Parties* (Chicago, 1956), pp. 194–215.

46. Clark Kerr, John T. Dunlop, Frederick S. Harbison, and Charles A. Myers, *Industrialism and Industrial Man* (Cambridge, Mass., 1960), pp. 47–76, 98–126, 193, 233. Walt W. Rostow, *The Stages of Economic Growth* (Cambridge, 1960), pp. 17–58.

47. Maurice Duverger, *Political Parties*, 2d ed. (New York, 1959), pp. 1–60.

48. *Chicago, Milwaukee & St. Paul Railway Co.* v. *Minnesota*, 134 U.S. 418 (1890).

49. *Pollock* v. *Farmers, Loan & Trust Co.*, 157 U.S. 429 (1895); (rehearing) 158 U.S. 601 (1895).

50. The literature on this process of judicial concept formulation from its roots in the 1870s through its formal penetration into the structure of constitutional law in the 1890s is extremely voluminous. Two especially enlightening accounts are: Benjamin Twiss, *Lawyers*

and the Constitution (Princeton, 1942); and Arnold M. Paul, *Conservative Crisis and the Rule of Law* (Ithaca, N.Y., 1960).

51. Paul, *Conservative Crisis,* pp. 131–58.

52. See, among many other examples, *Congressional Quarterly Weekly Report* 22 (May 1, 1964): 801.

53. Heinz Eulau, "The Politics of Happiness," *Antioch Review* 16 (1956): 259–64; Seymour M. Lipset, *Political Man* (New York, 1960), pp. 179–219.

54. Lipset, *Political Man,* pp. 216–19; Tingsten, *Political Behavior,* pp. 225–26.

55. V. O. Key, Jr., *Southern Politics* (New York, 1949), pp. 526–28; Schattschneider, *Semi-Sovereign People,* pp. 114–28.

56. Murray B. Levin, *The Aliented Voter* (New York, 1960), pp. 58–75, and his *The Compleat Politician* (Indianapolis, 1962), esp. pp. 133–78. While one may hope that Boston and Massachusetts are extreme case studies in the pathology of democratic politics in the United States, it appears improbable that the pattern of conflict between the individual's expectations and reality is entirely unique to the Bay State.

57. Campbell et al., *American Voter,* pp. 143–45.

58. The line of reasoning developed in this article—especially that part of it which deals with the possible development of political alienation in the United States—seems not entirely consistent with the findings of Gabriel A. Almond and Sidney Verba, *The Civic Culture* (Princeton, 1963), pp. 402–69, 472–505. Of course there is no question that relatively high levels of individual satisfaction with political institutions and acceptance of democratic norms may exist in a political system with abnormally low rates of actual voting participation, just as extremely high turnout may—as in Italy—be associated with intense and activist modes of political alienation. At the same time, the gap between American norms and the actual political activity of American individuals does exist, as Almond and Verba point out on pp. 479–87. This may represent the afterglow of a Lockean value consensus in an inappropriate socioeconomic setting, but in a polity quite lacking in the disruptive discontinuities of historical development which have occurred during this century in Germany, Italy, and Mexico. Or it may represent something much more positive.

59. "Whereas less than sixty-five percent of the United States population of voting age cast ballots for Presidential electors in 1960; and

"*Whereas popular participation in Government through elections is essential to a democratic form of Government;* and

"Whereas the causes of nonvoting are not fully understood and more effective corrective action will be possible on the basis of a better understanding of the causes of the failure of many citizens to register and vote . . ." (emphasis supplied). The full text of the executive order is in *Report,* pp. 63–64. Compare with Schattschneider's comment in *The Semi-Sovereign People,* p. 112: "A greatly expanded popular base of political participation is the essential condition for public support of the government. This is the modern problem of democratic government. The price of support is participation. The choice is between participation and propaganda, between democratic and dictatorial ways of *changing consent into support, because consent is no longer enough* (author's emphasis)."

2

Theory and Voting Research
1974

I. Background to a Controversy

In his brilliant and justly famous monograph, *The Structure of Scientific Revolutions,* Thomas S. Kuhn has illuminated the sociology of scientific discovery.[1] Developing an antinomy between "normal science" and "scientific revolutions," he places great stress upon the crucial importance of the discovery of anomalous data—data which do not "fit" and cannot be made to fit—in preparing the ground for the overthrow or supersession of currently dominant scientific paradigms.[2] This dialectical pattern—paradigm, anomaly, resistance, revolution—appears to dominate the history of science.

No one pretends, or should pretend, that most areas of political science, including voting behavior research, have advanced to the point that true paradigms have been constructed, or that professional controversies over the nature of political reality have more than a closely analogous relationship to the kinds of natural-science dynamics which Kuhn has brought to our attention. Ours is a more modest enterprise than this. But the analogies seem close enough to provide a starting point for this explicitly controversial (or controversy-oriented) paper.

In 1965 the *American Political Science Review* published a paper of mine, "The Changing Shape of the American Political Universe."[3] This paper presented a theory of American electoral change which stands in some contrast to the then-current professional emphasis on political system. Moreover, it presented data which were anomalous in light of the then-received theories of American voting behavior—many if not most derived from survey research on the contemporary electorate. Such anomalies included the steep decay in post-1900 voting behavior by nineteenth-century American or con-

From *The American Political Science Review* 68, no. 3 (September 1974): 1002–23. Copyright © The American Political Science Review.

temporary European standards, sharp increases in partisan volatility, and other indicators of growing "marginality" preceding chronologically *pari passu* with the shrinkage of the "voting universe." From this it was argued that the establishment of industrial-capitalist political hegemony after the realignment of 1896 was closely and causally associated with these remarkable movements, movements which were markedly retrograde to electoral development elsewhere in the Western world. The inference was drawn that this erosion of the mass base of American politics was a consequence of the tension between capitalism and democracy, and of the ascendancy of capitalism.

This presentation brought forth other anomalous findings in terms of contemporary survey-research theory. (1) "Landslides" in the nineteenth century tended, as did much else in the period, to be inverses of the contemporary behavioral phenomenon—they were not the product of "surges" as much as of one-sided abstentions by a normally involved partisan electorate. (2) Rural voting behavior in the contemporary period tends to reveal eccentricities in turnout and in partisan alignment. In the last century, by contrast, agrarian turnout tended to be more invariant and normally a good deal higher than in cities; rural partisan shifts were correspondingly very small except during realignment sequences. (3) The oscillatory or "sawtooth" pattern in turnout between presidential and off-year elections tended to be much smaller in relative amplitude before 1900 than is the case today. This is another way of saying that the relative net proportion of marginal or "in and out" voters was much smaller then than it is now, despite the much higher turnout rates in presidential elections. (4) The educational level of the American population has steadily increased over the past century. Yet despite this increase, voting participation went into a steep decline early in this century from which it has only marginally recovered. As is well known, educational level is a strong predictor of voter turnout in the contemporary era. An anomaly exists here which is disturbing to the University of Michigan Survey Research Center's paradigm.

This list could be extended, but it is at least indicative of the analytical problem. Perhaps the most disturbing empirical finding of this piece was that the indicators of nineteenth-century voting behavior, taken as an integrated set of characteristics, point to an electorate much more involved in democratic politics than its present counterpart, and vastly more so than that of the 1920s.

At the very base of these anomalies and the conclusions drawn from them is the author's belief that sociological and historical contexts profoundly shape both political consciousness and political behavior. This suggests a plasticity, as well as a sense of potential cognition of political objects within a mass electorate, which is most uncongenial to the semiparadigms of contemporary survey research.[4] In particular, such a suggestion casts serious doubt upon the generalizability of many—though far from all—survey findings based on contemporary American (and other) voters. Resistance to incorporation of these anomalies and their implications has been correspondingly strong. It has, moreover, been greatly facilitated by a crucial deficiency in the original (1965) presentation: that work failed specifically to incorporate changes in electoral mechanics after 1890, which are of obvious significance in shaping subsequent voting behavior and nonbehavior.

Now when one is confronted with anomalies it is possible to assimilate them into an existing paradigm in several ways. For example, those anomalies which have to do with change toward a present condition can achieve explanation through incorporation of straightforward and "undramatic" structural change unaccounted for in the original presentation of the anomaly. Similarly, those anomalies which have to do with the nature of reality before the onset of system change, in cases where their empirical reality cannot be disputed, may be explained away through hypotheses positing that they are artifacts or "optical illusions." Such efforts will, of course, be resisted by those who discover anomalies. The conflict which necessarily arises is most certainly not a matter either of ill will or of differences in intelligence: it is a matter of differences in reality perception. The conflict between anomaly and existing paradigm is a key vehicle for the development of fruitful scientific controversy, and such controversy in turn is a most valuable, indeed necessary, step in the development of better approximations to reality than any which currently exist.

Philip E. Converse has recently written an essay, "Change in the American Electorate," which has the perceptiveness and elegance of style and insight which we have come to expect from him.[5] The first half of this absorbing study consists of a detailed exposition and critique of my 1965 "Changing Shape" article and some more recent reformulations of its thesis. The differences between us, while most cordial, are profound and substantive enough to merit an examination by me of this counterhypothesis.

Converse's hypotheses fall into two fairly discrete sets: those dealing with intervening legal or rule variables which may explain most of the observable change, and those which raise questions about the politicized character of the nineteenth-century voting universe.

The first set of hypotheses includes the intervening variables of (1) women's suffrage (1914–20); (2) the introduction of the Australian ballot (ca. 1889–96); and (3) the introduction of personal-registration requirements for voting (ca. 1900–1920). Since the first variable, that of women's suffrage, was mentioned and in a sense "controlled for" in the 1965 article, the only point discussed here has to do with its possible contribution to the notorious electoral volatility of the 1920s. The second has partially been discussed elsewhere in light of Jerrold Rusk's extensive review article on the subject in the *American Political Science Review* and will receive limited treatment here.[6] The third—the personal-registration requirement—will receive much closer analysis in this paper.

The second set of Converse's hypotheses, if I have read him correctly, includes: (1) the primitive level of social bookkeeping in the nineteenth century, which casts some doubt upon turnout figures themselves; (2) the widespread existence of corruption at the polls, especially that involving fraudulent surplusage of votes, a corruption which is suggested to have been very widespread in rural as well as urban America before the turn of this century; (3) discussion of the causes of higher rural than urban turnout in contemporary underdeveloped countries (the Philippines and Turkey), based on survey work there, is presented as though somehow relevant to the American nineteenth-century case.[7] Let us evaluate each of these in turn, with such evidence as is currently available.

II. The Way Things Were? An Exploration
of Three Structural-Change Hypotheses

The Influence of Personal-Registration Statutes: An Examination of Turnout Decline in Four States, 1900–1916. There is an extraordinary messiness in the analysis of many variables affecting historical American voting behavior. In the first place, since social bookkeeping in this country is in many respects primitive compared with Europe or even Canada, we are left in ignorance as to the precise number of potential voters, or even of eligibles, in any election.[8] If we are to make any sense of the data at all, it is necessary to construct after-the-fact estimates of the potential electorate at given times, controlling as best we can for large potential inflators, such as noncitizen males in states where only citizens had the right to vote. These estimates must in most cases be based on linear interpolations between decennial federal census years. Thus, in addition to the serious problem of accuracy in enumeration (most serious, apparently, in the censuses of 1870 and 1920), it seems reasonably certain that such estimates of the potential electorate are most likely to be relatively more accurate the nearer in a decade one is to the preceding or succeeding census, and least likely to be so at midpoints. Moreover, as any analyst of historical data can attest, it is not easy to identify the precise legal status of election regulations in the past. Still more problematic, once one has identified at least the approximate date and coverage, say, of personal-registration statutes, is the issue of the local effectiveness of such requirements. These aspects of American bookeeping present such difficulty that one could well be pardoned for abandoning any effort to define change in electoral participation and its components in any way which could generate broad acceptance of the accuracy of the results. Still, best approximations are considerably better than nothing, even if they are relatively crude; and one may hope that whatever errors exist tend randomly to cancel themselves out.

With all of these limitations, the initial structure of the personal-registration legislation adopted at the turn of the century permits a straightforward statistical testing of the hypothesis that it accounted for the lion's share of the post-1900 decline in American voting participation. Such laws, of course, must themselves be integrated into a conceptual pattern of system change in response to stress: they, too, represent change produced by concrete men acting out their ideological biases in concrete legislative form. This is nowhere more evident than in the fact that, in a great many states, the first such laws were applied initially against the populations of cities while nonurban areas were left without such coverage.[9] But this pinpointed registration coverage permits us to decompose changes in the turnout rate by county and thus to establish some relatively satisfactory measures of differential change. Indeed, it permits us to make a first, if crude, approximation to the probable relative contribution of personal-registration requirements to post-1900 turnout declines in the metropolitan areas themselves. Four states have been examined by this method: Pennsylvania, New York, Ohio, and Missouri.

One major point about which we must be clear in measuring longitudinal change is that statistical measures will produce quite discrete findings depending upon the observer's analytical focus. This is not to say, with the conventional popular wisdom, that "there are lies, damned lies, and statistics," but rather that two apparently discordant readings can often be reconciled by, for example, bearing in mind the distinction between longitudinal and cross-sectional orientations. Thus in the case of Pennsylvania we have chosen to begin the discussion with analysis of variance. The analysis of variance is based upon a stratification of the state's sixty-seven counties in terms of the proportion of their populations covered in 1920 by the state's 1906–37 personal-registration requirements (Tables 1 and 2).[10]

While there are some occasional "sports" in the presidential-election array, the values of the F ratio are generally gratifyingly high for the 1908–36 period, that covered by the 1906 statute. Moreover, these values tend to show longitudinal increase from the beginning to the end of the period. They are correspondingly lower—usually well below the level of statistical significance—both for the 1880–1904 and for the 1940–60 periods. On the presidential level, at least, it can be taken as extremely probable that selective personal-registration requirements did produce a statistically significant effect in differentially depressing turnout in the state's urban areas. There is very little doubt that similar analysis-of-variance tests performed for New York, Ohio, and Missouri would show comparable results. So we certainly have an intervening variable of statistical significance for subsequent outcomes. But cross-sectional analysis of variance does not tell us anything about *trends* in any of the categories analyzed. How large a share of the post-1900 turnout decline can be attributed to this variable? How important is it?

Considerations of length preclude an elaborate discussion of Pennsylvania's party history from 1890 through 1932, but we may wish to begin analysis of this question with the Pennsylvania gubernatorial data of Table 2. This is a much shorter analysis-of-variance series, but the table contains a number of obvious anomalies, and masks others. Despite the 1906 statute,

Table 1. Registration and voting turnout: Pennsylvania presidential elections, 1880–1968, by extent of 1920 personal-registration coverage

Year	Means				Variances				$F =$
	M_1	M_2	M_3	M_4	V_1	V_2	V_3	V_4	
1880	90.0	91.8	93.2	91.3	23.67	24.23	30.18	34.27	0.405
1884	81.0	84.7	85.3	86.9	72.77	43.23	13.16	61.68	1.417
1888	76.7	82.0	84.0	85.6	113.86	66.05	9.93	51.48	3.253
1892	73.7	74.7	78.6	79.5	79.43	67.94	18.95	50.28	2.338
1896	82.2	83.1	86.3	85.5	49.84	70.57	5.29	36.58	1.024
1900	75.3	78.4	79.3	81.2	87.97	23.94	10.71	37.58	2.323
1904	72.9	74.2	75.6	76.5	66.96	47.59	56.68	81.12	0.538
1908	68.1	73.5	75.8	77.7	61.25	41.71	130.45	41.14	4.340
1912	61.2	67.0	66.5	70.4	31.96	24.75	76.01	32.45	5.841
1916	60.6	64.3	64.9	69.2	26.78	23.50	23.68	43.69	5.912
1920	40.4	41.3	45.2	48.2	36.67	16.45	25.25	43.65	7.099
1924	42.4	46.5	48.5	52.8	10.23	31.50	10.91	49.68	8.473
1928	58.9	63.0	64.1	65.8	24.56	25.93	79.20	47.12	2.629
1932	48.1	51.4	55.2	61.5	15.07	29.76	16.27	60.13	14.070
1936	68.5	69.5	73.9	79.9	61.13	28.86	10.38	55.15	12.048
1940	65.1	63.3	65.1	68.8	52.58	24.46	11.67	52.80	2.947
1944	57.9	55.9	55.6	58.7	41.45	13.44	13.45	44.93	1.095
1948	53.1	51.1	52.1	52.1	80.54	34.36	26.52	56.83	0.153
1952	64.4	62.0	64.8	64.3	58.60	16.63	26.37	59.22	0.210
1956	63.1	62.8	63.4	66.9	48.32	21.15	14.12	63.70	1.759
1960	69.5	68.4	70.6	71.9	49.61	18.38	20.11	52.99	1.132
1964	66.7	63.1	65.7	67.0	44.48	21.82	7.92	47.69	1.455
1968	64.6	61.2	65.1	64.7	39.29	28.70	8.98	46.54	1.300
$N =$	8	15	7	37	d.f. = 7	14	6	36	

the values of the F ratio do not climb to the significant range until the third election after its enactment, 1914. Indeed, the 1910 value is the lowest of the entire 1890–1918 series. The explanation of this phenomenon ties in closely with Converse's discussion. Let us summarize.

(1) Presidential years were generally years of harmony within the hegemonic Republican majority after 1896, largely because no really important statewide offices were at stake. (2) The profile of off-year electoral politics between 1898 and the First World War was one of extreme party fragmentation and massive but unsuccessful uprisings of Progressives against the machine Regulars who controlled the state's political process. (3) Within Philadelphia particularly, the 1906 registration statute utterly failed to curb fraudulent vote surplusage during the locally important off-year elections. In fact, methods of controlling votes reached new heights of sophistication and comprehensiveness precisely during this period. One result of this was that,

Table 2. Registration and voting turnout: Pennsylvania gubernatorial elections, 1890–1918, by extent of 1920 personal-registration coverage

	Means					Variances				
	M_1	M_2	M_3	M_4		V_1	V_2	V_3	V_4	$F =$
1890	69.4	73.2	77.7	79.1		18.33	18.41	5.60	59.72	4.273
1894	66.6	70.0	72.2	74.4		16.62	32.23	4.90	73.10	1.666
1898	62.4	67.4	69.2	72.3		14.65	20.95	7.29	62.15	3.751
1902	65.1	66.2	68.3	71.3		11.21	21.67	9.93	68.99	1.976
1906	54.1	59.1	60.3	63.7		13.89	21.27	9.95	74.19	2.721
1910	50.6	53.8	56.2	55.8		12.23	13.83	8.64	50.22	1.570
1914	50.7	56.5	59.5	61.6		7.96	15.88	9.18	48.37	7.525
1918	38.0	43.8	43.8	48.7		8.75	13.06	6.02	57.05	6.513
$N =$	8	15	7	37	d.f. =	7	14	6	36	

Note: Throughout this period, d.f. = 3,63. Values of the F-ratio at the several levels of significance with d.f. = 3.63 are:

level	F
.05	2.76
.01	4.13
.001	6.17

The stratification is based on registration coverage as of 1920. M_1 and V_1 are derived from counties 50 percent or more of whose populations were covered by personal-registration requirements. M_2 and V_2 correspond to counties 20.0 to 49.9 percent of whose populations were so covered; M_2 and V_2 correspond to counties 0.1 to 19.9 percent of whose populations were so covered; and M_4 and V_4 correspond to counties which were left unaffected by the 1906 statute (thirty-seven in all).

in the 1902–18 period, this city—which was 100 percent covered by the 1906 statute—had a higher turnout rate in all elections but one than did a *majority* of the thirty-seven counties where no personal registration was required at all (Table 3).[11]

Another major anomaly can be sensed in the array of Table 2, and it is important enough to require further discussion. The strenuous state conflicts of 1902–14 pitted insurgent or Progressive Republicans against Organization Regulars. The Democratic party as such was a hopeless minority throughout the post-1896 period and played an ambiguous role. In 1910, the high-water mark of local Progressive insurgency, it fielded a candidate at the behest of the Republican state machine to siphon off enough votes to permit the Organization's candidate for governor to eke out a plurality.[12] On other occasions, 1906 and 1914, the party fused unsuccessfully with insurgent Republican fragments.

One could expect in this situation that elections would be locally most competitive in areas with partial or no registration, with the machine producing lopsided organization majorities in the big cities to win statewide victory. The partisan data reveal that this was exactly the case (Table 4).

Table 3. Extreme cases: Philadelphia and nonregistration-county turnout 1890–1918

Year	Number of nonregistration counties with turnout:		% Lower than Philadelphia
	Lower than Philadelphia	Higher than Philadelphia	
1890	9	28	24.3
1894	15	22	40.5
1898	5	32	13.5
1902	22	15	59.5
1906	19	18	51.4
1910	33	4	89.2
1914	17	20	45.9
1918	20	17	54.1

Table 4. Partisan distribution of Pennsylvania gubernatorial vote, 1890–1918, by 1920 registration coverage

Area (N)	1890	1894	1898	1902	1906	1910	1914	1918
Philadelphia (1)								
Democrat	44.3	27.6	22.3	29.8	42.4	6.8	25.2	27.5
Republican	55.2	71.1	65.1	68.9	56.1	54.4	72.6	70.7
Progressive			12.2			35.9		
Other	0.4	1.3	0.5	1.3	1.5	2.8	2.2	1.9
50.0–99% (7)								
Democrat	52.6	33.2	39.7	39.0	42.1	12.2	41.6	39.6
Republican	45.5	61.1	42.0	56.1	51.7	44.2	48.6	53.3
Progressive			12.4			32.3		
Other	1.9	5.8	0.9	5.0	6.2	11.2	9.8	7.1
20.0–49.9% (15)								
Democrat	49.8	37.1	39.6	45.1	47.0	13.8	46.2	34.4
Republican	47.9	57.3	45.1	48.9	48.6	38.4	47.4	59.8
Progressive			15.1			41.9		
Other	2.3	5.6	0.3	6.0	4.4	5.9	6.4	5.7
0.1–19.9% (7)								
Democrat	53.5	39.3	41.7	46.3	45.9	12.7	44.9	37.9
Republican	44.4	55.5	42.1	46.2	48.2	33.7	47.2	56.7
Progressive			15.5			40.5		
Other	2.1	5.5	0.7	7.6	6.0	13.0	7.9	5.4
No registration (37)								
Democrat	51.3	38.3	41.6	48.1	49.6	20.1	48.3	32.2
Republican	46.6	56.2	45.4	47.3	46.5	32.2	46.0	61.1
Progressive			12.9			41.7		
Other	2.1	5.5	0.6	4.6	3.9	6.0	5.8	6.7

Note: Counties falling in each classification—

    100%: Philadelphia

50.0–99.9%: Allegheny, Berks, Erie, Lackawanna, Lawrence, Lehigh, Venango.

20.0–49.9%: Blair, Butler, Cambria, Clinton, Crawford, Dauphin, Delaware, Lancaster, Lebanon, Luzerne, Lycoming, McKean, Mercer, Northampton, York.

  0.9–19.9%: Chester, Clearfield, Fayette, Northumberland, Schuylkill, Washington, Westermoreland.

    0%: All other counties

We can accept that a strong machine could overcome the intentions of a personal-registration statute. But the literature also emphasizes that closeness of competition in elections tends to facilitate turnout. Here we find the opposite. Despite the closeness of contests in the nonregistration areas—indeed, all four elections from 1902 through 1914 revealed opposition *pluralities* in these counties—their decline in turnout was huge and, except for a small upswing in 1914, continuous from 1890 onward. In fact, the turnout decline accelerated markedly in 1906 and 1910, despite the intensity of local party competition, and opposition success in most of these counties. If we take all eight of the counties, more than half of whose populations were subject to registration requirements, and compare their negative 1890–1918 regression slope with that of the thirty-seven nonregistration counties, we find that the steepness of the former is only 14.5 percent greater than the latter on a standardized measure.

It is also worth noting that this gubernatorial regression line for the nonregistration counties of Pennsylvania is steeper than the 1900–1916 presidential election regression slopes for any category of counties in any other of our three states except for New York City, and that city, in addition to being 100 percent covered by personal registration, was subject to a periodic reregistration requirement as well. The turnout decline in the rural, heavily native-stock Protestant areas of Pennsylvania not covered by the 1906 statute constitutes a major limiting case on any registration-variable hypothesis. So far as we know, it was wholly "behavioral," not an artifact of rules change. Somehow, despite the intensity of insurgent activity, the terms of politics had been redefined from those of the late nineteenth century in such a way as to discourage participation in elections. What other barriers stood in the way for these voters?

From this point we can move to an examination of the longitudinal trends of the 1900–1915 period in our other states. Unfortunately, the criteria of county stratification are not entirely uniform. Accordingly, comparisons may have more limited value than the data might suggest. Still, all of these states permit some differentiation. Until the late 1960s, New York had registration provisions similar to those of Pennsylvania before 1937, and relevant volumes of the state's legislative manual provide a useful breakdown of the registration characteristics of precincts by county. Similar data are available for Ohio back at least to 1932. In the case of Missouri, the turn-of-the-century registration statute initially and for a long time afterward was for all practical purposes confined to the state's two metropolitan areas. Moreover, we are fortunate in being able to deal with secular trends of large size and some uniformity. So far as county-by-county analysis of 1900–1916 turnout rates can be made, we find no exceptions to the *fact* of decline in any of the 217 counties of New York, Ohio, and Pennsylvania during this period. There are numerous exceptions in Missouri, but these may find a subsequent and, I hope, convincing behavioral explanation.

Table 5 presents regressions on time for the turnout rate during this turn-of-century transitional period, stratified by the relative extent of personal-registration coverage in groups of geographical areas (counties). In all categories and in all four states, the regression slopes are negative. But the differences among states in the slopes are considerable. Pennsylvania stands at one extreme, that of maximum decline, and Missouri reveals almost no decline at all, even in its three metropolitan counties. The other two states fall in the middle and approximately together, although somewhat closer to the Pennsylvania end of the spectrum. Indeed, a contemplation of the standardized regression slopes for these states and categories (bX/a) suggests that

Table 5. Turnout data for four states, 1890–1918: Regression metrics

A. Pennsylvania (Gubernatorial), 1890–1918

Registration Coverage, 1920 and N		$Y_c =$	$S_{cy \cdot x} =$	$bX/a =$
0	(37)	$83.75 - 3.98X$	3.516	$- .0475$
0.1–19.9	(7)	$82.14 - 4.16X$	3.569	$- .0507$
20.0–49.9	(15)	$78.46 - 3.82X$	3.023	$- .0487$
50.0–100.0	(8)	$75.63 - 4.12X$	3.659	$- .0544$

B. Pennsylvania (Presidential), 1900–1916

Registration Coverage, 1920 and N		$Y_c =$	$S_{cy \cdot x} =$	$bX/a =$
0	(37)	$84.03 - 3.01X$	1.745	$- .0358$
0.1–19.9	(7)	$83.79 - 3.79X$	2.051	$- .0452$
20.0–49.9	(15)	$82.10 - 3.54X$	1.191	$- .0431$
50.0–100.0	(8)	$79.95 - 4.11X$	1.472	$- .0514$

C. New York (Presidential), 1900–1916

Mean Registration Coverage of Precincts, 1948, and N		$Y_c =$	$S_{cy \cdot x} =$	$bX/a =$
0	(10)	$90.84 - 2.16X$	1.914	$- .0238$
16.7	(14)	$90.89 - 2.23X$	2.256	$- .0245$
55.1	(33)	$91.80 - 3.12X$	2.427	$- .0340$
100.0	(5)*	$85.31 - 4.25X$	1.182	$- .0498$

D. Ohio (Presidential), 1900–1916

Registration category, 1932, and N		$Y_c =$	$S_{y \cdot x} =$	$bX/a =$
None	(57)	$94.44 - 2.26X$	3.874	$- .0239$
Partial	(28)	$93.78 - 3.58X$	5.873	$- .0382$
"Full"**	(3)	$88.25 - 3.19X$	3.401	$- .0361$

E. Missouri (Presidential), 1900–1916

Area and N		$Y_c =$	$S_{cx \cdot y} =$	$bX/a =$
Nonmetropolitan	(112)	$84.67 - 0.29X$	3.358	$- .0034$
Metropolitan***	(3)	$74.68 - 0.72x$	5.280	$- .0096$

*New York City. In this case, to avoid problems associated with the creation of Bronx County in 1914, the citywide total is the base.

**The three metropolitan counties, Cuyahoga, Franklin, and Hamilton. Because most suburbs and outlying areas were not initially covered, these are "relatively" full registration territories.

***St. Louis city and County of Jackson. For all intents and purposes these were the only initial areas of personal registration in Missouri.

the *between-state* differentials are at least as significant as the registration-determined *within-state* differentials. In all of these states, to be sure, the posited relationship holds: very generally, the larger the registration coverage in a given set of counties, the steeper the negative regression slope is. But it should also be noted that the relationships are not quite monotonic either in Pennsylvania or in Ohio.

The idiosyncracies of the early personal-registration statutes make it possible to derive a first approximation of the relative contribution of the personal-registration statute to this nearly universal turnout decline. Leaving Converse's theory of apparently general rural corruption aside for the moment, the assumption is very strong that any secular trend in voting participation to be found in counties without personal-registration requirements must be attributed to behavioral factors. By definition, no known changes in electoral mechanics could be causes of such change in turnout. This being the case, let us assume for each state that the proportion of turnout decline in any area which corresponds to the rate of decline in nonregistration areas also corresponds to a "pure-behavioral" component of change. Thus, whatever residue is left can be provisionally attributed to the personal-registration requirement as an intervening causal variable.

The steps to be followed in analysis are these. First, the 1916 regression-line turnout value in nonregistration territory is calculated as a proportion of the 1900 value. Second, an "expected" 1916 value is computed for all other categories, based on a multiplication of this proportion by the 1900 regression value for each. Third, the differential between the "expected" 1916 value, based on the calculations in the second step, and the actual 1916 regression-line value is derived. Fourth, the absolute difference between the 1900 and 1916 values is computed by subtraction for each category, including the nonregistration category. Finally, the absolute loss on the regression line is divided into the expected-observed differential by a formula such as

$$C_T = \frac{E(T + n) - O(T + n)}{T - T + n} \times 100.$$

The resulting percentage can be described as that part of the regression-line decline in a given area which can be attributed to factors other than the behavioral ones operative in nonregistration areas (Table 6).[13]

Most of these percentages are remarkably small. Leaving aside the Missouri metropolitan counties for the moment, the highest contribution for the 1900–1916 period is found in New York City, where the residue is 53.5 percent. Here we deal with a case of exceptional stringency. In addition to the personal-registration requirement, the state legislature—with malice aforethought—imposed a periodic requirement as well, one which was applied to this city alone until its elimination in the 1950s. For reasons discussed above, the percentage figure in Pennsylvania gubernatorial elections within the eight most urban counties is extremely small. Otherwise, nonbehavioral vari-

Table 6. A crude estimate of the contribution of personal-registration statutes to turnout decay

A. Pennsylvania: Gubernatorial Elections, 1890–1918 (T to $T + 7$)

Cat. of Reg. Coverage 1920	Regression line est.: "Expected" 1918	"Observed" 1918	$E(18) - O(18)$	$T - T + 7$	$E(18) - O(18)$ / $T - T + 7$
0	51.95	51.95	0	27.83	0
0.1–19.9	50.78	48.83	1.95	29.15	6.7
20.0–49.9	48.60	47.87	0.73	26.76	2.7
50.0–100.0	46.57	42.71	3.86	28.80	13.4

B. Pennsylvania: Presidential Elections, 1900–1916

Cat. of Reg. Coverage	Regression line est.: "Expected" 1916	"Observed" 1916	$E(16) - O(16)$	$T - T + 4$	$E(16) - O(16)$ / $T - T + 4$
0	69.0	69.0	0	12.0	0
0.1–19.9	68.2	64.8	3.4	15.2	22.4
20.0–49.9	66.9	64.4	2.5	14.2	27.6
50.0–100.0	64.6	59.4	5.2	16.4	31.7

C. New York: Presidential Elections, 1900–1916 (T to $T + 4$)

Cat. of Reg. Coverage 1948 (Mean for each)	Regression line est.: "Expected" 1916	"Observed" 1916	$E(16) - O(16)$	$T - T + 4$	$E(16) - O(16)$ / $T - T + 4$
0	80.04	80.04	0	8.64	0
16.7	80.02	79.74	0.28	8.92	3.1
55.1	80.04	76.20	3.84	13.48	30.8
100.0	73.16	64.06	9.10	17.00	53.5

D. Ohio: Presidential Elections, 1900–1916

Cat. of 1932 Registration Coverage and N	Regression line est.: "Expected" 1916	"Observed" 1916	$E(16) - O(16)$	$T - T + 4$	$E(16) - O(16)$ / $T - T + 4$
None (57)	83.1	81.3	0	9.1	0
Partial (28)	81.3	75.9	5.4	14.3	37.8
"Full"* (3)	76.7	72.3	4.4	12.8	34.4

E. Missouri: Presidential Elections, 1900–1916

Grouping of Counties and N	Regression line est.: "Expected" 1916	"Observed" 1916	$E(16) - O(16)$	$T - T + 4$	$E(16) - O(16)$ / $T - T + 4$
Metropolitan** (3)	83.2	83.2	0	1.4	0
Other (112)	73.0	71.1	1.9	2.9	63.8

*Cuyahoga, Franklin, Hamilton
**St. Louis City, St. Louis County, Jackson

ables as measured by this indicator tend (except in Missouri) to explain about one-third of the gross turnout decline in urban areas.

The Missouri case is special, and leads us to the next part of this discussion. Here we have a gratifying high figure of 63.8 percent contribution of nonbehavioral components in the metropolitan areas of the state, but the longitudinal movement is extremely slight, falling far below the standard error of estimate for the regression equation. So small is this 1900–1916 turnout shift in Missouri that it may even be nothing more than the product of random errors in data. We are left with the question of positioning all four of our states on a continuum which will help us understand the very wide interstate differentials in our "sample"; Missouri is at one end and Pennsylvania is at the other. What kind of continuum would this be?

My original discussion of the turn-of-the-century shrinkage in the American electorate placed heavy stress upon the immense changes in partisan alignments which followed the critical-realignment sequence of 1893–96.[14] This realignment led, as is now very well known, to several profound consequences for party politics in the United States. One of these was the destruction of competition between the major parties in many states, and a more or less pronounced movement away from close two-party competition in many others. The second, closely related to the first, was the achievement of corporation-capitalist dominance over public policy. This dominance reached its peak in the 1920s. The relationship between elite hegemony, one-party politics, and suppression of the electorate is so conspicuous in the post-1890 South as to require no further discussion here.[15] But perhaps this was a localized, if extreme, relationship. My original argument suggested otherwise, since the focus of attention was on turnout decline in non-Southern states. But perhaps personal-registration requirements vitiated the force of the party-competition argument outside of the South.

The four states we have analyzed here cannot be said to be a representative sample of post-1896 voting behavior outside of the South, though they did contain among them 36.7 percent of the total non-Southern population in 1910. With only four composite data points, the posited relationship can only be said to be suggestive. Still, they remain of sufficient interest and visibility to merit presentation here. If one examines the mean partisan percentages for the five presidential elections before 1896, and the five after for these four states, the following pattern appears (Table 7). The relationship between changes in competitiveness and in the magnitude of the mean turnout slope is both obvious and quite close.[16]

The 1896 realignment produced very different partisan aftermaths in different states. The Democratic party was virtually destroyed in Pennsylvania. Yet it retained great post-1896 vitality in New York City, and as Thomas Flinn and John Fenton have shown, nineteenth-century alignments were not submerged in Ohio until well into the present century.[17] In Missouri, as in the other Border states, the realignment of 1896 had partisan effects exactly the opposite from those affecting states north or south of that region. It produced close two-party competition for the first time since Reconstruction, and this intense party competition endured until the New Deal. The creation of close party competition seems closely related to the virtual absence of major turnout decline in Missouri before 1920. If it be true that personal-registration qualifications in the state's metropolitan centers explained nearly two-thirds of their 1900–1916 decline, the decline there was miniscule compared even to the secular trend in nonregistration territories of New York,

Table 7. Shifts in partisan competition and turnout in four states

	1876–1892			1900–1916			Mean partisan lead	
	% D	% R	% Other	% D	% R	% Other	1876–92	1900–16
Pennsylvania	45.7	51.6	2.7	34.3	60.2	5.5	3.9R	25.9R
New York	49.0	48.3	2.7	42.3	51.8	4.9	0.7D	10.5R
Ohio	47.5	50.0	2.5	43.5	51.2	4.3	2.5R	7.7R
Missouri	52.7	42.6	4.7	48.8	47.8	3.4	10.1D	1.0D

	Net shift in party competitiveness, 1876–92/1900–1916	Mean standardized turnout regression slope, 1900–1916 bX/a
Pennsylvania	−22.0	−.0439
New York	−10.5	−.0330
Ohio	− 5.2	−.0327
Missouri	+ 9.1	−.0065

Ohio, and Pennsylvania. Similarly, one doubts that the relatively very steep decline in Pennsylvania's participation—and in all categories of registration coverage in its counties—is in any way accidentally related to the destruction of major-party competition there.

Let us summarize the discussion so far. Examining the four states which form our data base, we can conclude that the intervention of personal-registration requirements involved the intrusion of a highly significant intervening variable. Of course, this variable must henceforth be explicitly included in any future model of change in the American electorate after 1890. But we can also conclude, if tentatively, that it explains very much less of the decline in turnout during the first two decades of this century than might be supposed. If our inferences from the cross-time political behavior of nonregistration territories are correct, this variable contributed between one-fifth and one-third at the most to the 1900–1916 statewide turnout declines in New York, Ohio, and Pennsylvania. The extremely rapid rate of decay in Pennsylvania counties not covered by registration, particularly when studied together with the local intensity of political opposition to the Republican Organization and the closeness of elections, would reinforce a theory of cumulative political alienation—at least for that state.

Finally, the evidence is suggestive here of the relationship between the gross effects of the 1896 realignment on party competition and the relative magnitude of turnout decline in presidential elections from 1900 through 1916. Here, Missouri seems in particular to be a limiting case, since virtually no decline occurred, while two-party competition became extremely close during the era of the fourth party system. It may also be noted that there was

an enormous gap in the relative level of economic development between
states such as Missouri and Pennsylvania; and, in politics, Missouri retained
(and still retains) a nineteenth-century element which has disappeared in
more "developed" parts of the country.

Thus, on the basis of the forgoing analysis, it must be assumed for now
that the bulk of the decay in participation from the end of the last century
through the enfranchisement of women must be explained in behavioral
terms. The magnitude and—except for Missouri—the universality of this
decay is such that "nondramatic" variables cannot begin to explain it all
away. The same is almost certainly true for our other indicators of electoral
decomposition. Something in the redefined agenda of politics after 1896
drove large masses of voters away from active participation in electoral politics—most conspicuously where the immediate partisan effects of the "system of 1896" were most conspicuous, as in states like Pennsylvania. Alter-
natively, it may well be that there was a severe decay in age-cohort inputs
into the active electorate after 1900; we cannot know for sure at this stage,
if ever. Whatever that something was, it was profound and system-wide—
mediated, of course, by the interposition of new structural variables which
were themselves the result of system stress and system adaptation. I find no
difficulty on this record in adhering to the view that an explicitly urban,
bourgeois elitism—partly corporate-conservative, partly Progressive "reformist"—was intimately and causally associated with these remarkable behavioral transformations.

Effects of Woman Suffrage on Electoral Decomposition in the 1920s. The
preceding analysis points to the conclusion that decomposition in the American electorate had proceeded quite far by the end of the First World War.
Other evidence has been presented elsewhere which points to the same conclusion.[18] If so, and if one can assume that these measures taken together are
indicative of an overwhelming process of "political desocialization," it
would follow that this process would have been well along toward completion before the emergence of a second intervening variable, the enfranchisement of women. It has been customary to accept the view that the addition
of so quantitatively huge a variable explains subsequent anomalies in the
American behavioral universe. One survey-research finding which has apparently universal generality is that women tend to be less politically conscious
than men, are less "socialized" into the world of active political participation, and tend to participate less.[19] Similarly, the further decline in participation during the 1920s finds very much of its explanation by the sheer dou-
bling of the electorate by the addition of this weakly socialized component.
Even today, as Converse points out, the rate of female voting in the United
States runs about 5 to 10 percentage points below male turnout.[20]

While these female political characteristics are well known, it might be
useful to raise some pointed questions about the influence of women's suf-

frage in its early years, particularly in areas of voting behavior other than turnout. In addition to stressing its manifest importance for subsequent turnout rates, Converse evidently uses it to account in part for the peaks in such indicators as split-ticket voting and roll-off which developed in the 1920s. But does it? If the Burnham analyses of the changing American electorate in this period are valid, it could be expected that partisan decomposition and other evidences of "desocialization" were so far advanced by the 1915–20 period that the still more weakly socialized women would add little more to it. This would imply, of course, that most of the observable electoral volatility reached in the 1920s would have been present even if women had not received the vote until the advent of the New Deal.

One notes in the first place one very curious anomaly in statewide turnout rates during this period. In Kentucky—one of the Border states which became intensely competitive during the lifetime of the "system of 1896"— the highest recorded turnout rate in the entire period from 1920 to the present was achieved in 1920, the very first year of women's suffrage in that state.[21] But this curiosity, intriguing as it is, is also unique for statewide presidential-election turnout arrays in the 1920–68 period. There is, however, one place in the United States where detailed and sex-stratified aggregate data are available during this transition period, and to this we now turn.[22]

Fortunately for researchers, Illinois gave partial but not complete suffrage to its women, effective with the 1915 Chicago mayoral election. As a result, detailed breakdowns of voting behavior by sex are available for Chicago from 1915 through 1920, and for the suburban towns of Cook County from 1916 through 1920. This series reveals certain attributes of female voting behavior which are well known. One which was shared in common with the behavior of women in other countries was a marked aversion to voting for leftist parties (Labor, Socialist, Farmer-Labor) by comparison with men. Another, of course, was a stratified differential in turnout. Suburban Cook County was (and is) much more native-stock, middle class, and Republican as a whole than Chicago. Thus the number of women voting in the city was 58.0 percent of the number of men voting in 1915, 59.4 percent in 1916, 58.5 percent in 1919, and 59.7 percent in 1920. By comparison, the female percentage in the suburbs was 69.0 percent in 1916, and 69.2 percent in 1920. These figures correspond to a sex differential in turnout of some 30 percent in the city and 23 percent in the suburbs, compared with the 5–10 percent national differential today. So far, so good; one hardly needs to add that the turnout rates in both jurisdictions underwent a sharp decline when women were enfranchised.

But the rest of the data point to very different conclusions about the electoral significance of this variable in the Chicago metropolitan area. Let us begin by a comparison by sex of the partisan swings from the 1915 to the 1919 Chicago mayoral elections, and from the 1916 to the 1920 presidential elections in both areas (Table 8).

Table 8. Sex and partisan swing: Chicago and its suburbs, 1915–20

Pair of years and sex	Office	Shift in % of total vote			Defection ratios
		D	R	Other	
		A. Chicago			
1915–19 Men	Mayor	− 3.4	−20.5	+23.9	34.8 (R)
1915–19 Women	Mayor	− 0.1	−18.9	+19.0	31.8 (R)
1916–20 Men	President	−22.1	+18.6	+ 3.5	48.4 (D)
1916–20 Women	President	−24.5	+23.0	+ 2.5	51.8 (D)
		B. Suburban Cook			
1916–20 Men	President	−17.3	+15.2	+ 2.1	52.9 (D)
1916–20 Women	President	−19.0	+19.8	− 0.8	58.2 (D)

There is no doubt that the relative velocity of pro-Republican movement from 1916 to 1920 was somewhat greater among women than among men, and greatest of all among suburban women. Nevertheless, stratified by sex, the swings are clearly of the same order of magnitude in each area. Between the two mayoral elections, indeed, the male Republican defection actually *exceeded* that for women. In absolute terms, shifts of 20 percent or so between one election and the next are symptomatic of considerable electoral volatility by any longitudinal standard, particularly where—as here—they are not part of a critical-realignment sequence. If the mean female partisan swing in the 1920 landslide was only 16.5 percent greater than the male swing for both major parties and in both areas, what significance should we attach to women's suffrage as an intervening variable promoting dramatic increases in electoral volatility? The difference between the 1916 and 1920 swing in the total vote and what it would have been had men alone voted can be easily computed. When this is done, the mean sex-related increment for both parties in both areas works out to 5.6 percent (2.9 percent for the Democrats, 8.4 percent for the Republicans). While this variable thus produces a visible differential effect, and in the predicted direction, it is hard to get very excited about it.

The data become more interesting still when one examines split-ticket voting and roll-off, or net aggregate "ballot fatigue" (Table 9). Once again the posited relationships duly materialize. Women in each jurisdiction do reveal somewhat greater propensities to vote split tickets than do men, and rather more marked rates of ballot fatigue. But these differences within each area, particularly in split-ticket voting, are not very large. *Moreover, they pale into insignificance by comparison with the very wide differences in behavior between Chicago and its suburbs.* Again, such differential behavior is anomalous, conspicuously so where roll-off is concerned. For it might well be expected that the suburban areas—relatively affluent, native-stock, and so on—would display less rather than much more ballot fatigue. One quite

Table 9. Sex and Indicators of split-ticket voting and ballot fatigue: Chicago and its suburbs, 1920

Area	Sex	Mean % D	Variance	Standard deviation	Roll-off
Chicago	Men	29.5	28.20	5.31	3.8
Chicago	Women	29.9	33.04	5.75	5.0
Suburban Cook	Men	21.8	82.99	9.11	13.8
Suburban Cook	Women	21.4	102.22	10.11	16.2

Note: Based on percentage vote for seven offices: president, U.S. senator, governor, state's attorney, recorder, clerk, and coroner.

orthodox hypothesis suggests itself: this may well reflect the "workings of the machine in the city, and of post-progressive independence in the suburbs."

Whatever the explanation, one thing seems clear. The sex differential is quite small in each area, while the absolute size of the split-ticket and ballot-fatigue indicators *among men* appears quite substantial by comparative standards, and is conspicuously so for the roll-off indicator. With data which are this seedy along the posited sex dimension of political behavior, we may well suspect that there is less than meets the eye in this famous woman-suffrage variable. It is a pity that so little relatively hard data exist pertaining to this set of issues. What we have forcefully suggests that—if these large values may be taken as evidence of weak political socialization—those women *who voted* in Cook County between 1915 and 1920 were only very marginally weaker in political socialization than men.

Moreover, it is widely agreed that female turnout did not begin to move much closer to male participation rates in the United States until the 1928–36 realignment sequence. Accordingly, one should find further increases in these measures of decomposition during and shortly after this sequence. But precisely the opposite happened: in general, levels of split-ticket voting and ballot fatigue *declined* markedly as turnout sharply *increased*. This 1928–36 influx of new voters was heavily concentrated among working-class and women—particularly ethnic women—voters in large industrial states like Pennsylvania and Illinois. Such voters were hitherto exceptionally weakly socialized, yet their addition to the electorate was associated with *declines,* not increases, in measures of electoral volatility. When these two data points are reviewed together, the argument that behavioral rather than statutory change was the crucially important development in the evolution of the American electorate from 1900 to 1940 seems, if anything, to be strengthened.

The Introduction of the Australian Ballot. The analytic ground entirely shifts when we turn to this posited intervening variable. Since I have recently discussed it at some length in response to an *American Political Science Review* article by Professor Jerrold Rusk, commentary here will be brief.[23] Unlike the other variables discussed here, the Australian ballot, when adopted, tended to be applied universally within each state from the beginning. Moreover, it was adopted virtually nationwide in a very short period. Massachusetts was the first state to adopt it generally, in 1889; by 1896, far more than three-quarters of the states had followed suit. Hence, the analytic strategy of subunit differentiation cannot be pursued in this case.

The basic issues at controversy concerning this variable are conceptual in nature. They boil down to the question: what are we primarily interested in measuring? In this sense, there are close analogies to the problem of analysis posed by the Pennsylvania turnout arrays in Tables 2 and 3. If one is interested in detecting the existence of a statistically significant intervening variable, one can perform a series of cross-sectional analyses of variance and compute the appropriate F-ratios for each of the election years 1 through n. If one is interested in examining longitudinal trends in each grouping used in the preceding analysis, the matrix must be transposed so that regressions on time can be performed for each grouping. As we have seen, the first approach yields the conclusion that turnout decline was of very large magnitude even in counties where no personal-registration requirements were introduced. Both approaches seem to be essential to describe change in this universe with any adequacy.

We are on similar conceptual if not statistical ground when we contemplate the effects of the Australian ballot. Unlike either of the other variables discussed so far, this is in the first place a crucial and universal intervening variable of the sine qua non variety. But in the second, it is a *permissive* variable rather than one which is deterministic or directly influential as such in shaping subsequent behavior.[24] So far as the first point is concerned, the change from the printed party-distributed ticket to the official consolidated ballot in any of its forms was by definition an indispensable precondition for large-scale ballot splitting in the electorate. The old partisan ticket did not make it impossible for voters to split their ballots, but it unquestionably and systematically constrained such behavior in ways which the official ballot was in fact designed by its authors to overcome. This dichotomous change is of such transcendent importance that elaborate statistical demonstrations of it become almost supererogatory.

But if this intervening variable was indispensable to all that followed, it also was clearly less immediately shaping in its effects on partisan voting than either of the other two examined here. This point can, like many others, be most easily perceived by the analysis of specific and probably extreme cases. To take just one of several examples, Massachusetts was the first state to adopt the Australian ballot. Moreover, it chose the office-block format, which has been in use ever since. This format was explicitly designed to confront the voter with each individual candidate on a long ballot, and hence made it as hard as possible for the individual to vote a straight party ticket.

Yet there is virtually no change in the state's split-ticket indicators in the years after the ballot's first general use in the 1889 election. In fact, it took *fifteen annual trials* before split-ticket voting in Massachusetts suddenly began to rise sharply above traditionally low levels.[25] From this fact it must be inferred that traditionally high levels of partisanship survived for half a generation after the ballot was changed, and despite the exceptional procedural difficulties which that change put in the way.

The Australian ballot in this state thus opened the way to much later electoral change, which deeply affected the intensity of partisanship in the electorate. *But that is all.* The events of 1904 and subsequent years must find their explanation in profound change in popular attitudes and behavior, not in any directly measurable partisan effects of the Australian ballot itself. Nor, as I have tried to show elsewhere, is this a unique case in point.[26] There appear to have been three waves of increase in split-ticket behavior throughout the country before the most recent post-1950 upsurge. Each of these three crested at a higher peak than its predecessor, and the present fourth wave has moved to higher levels still. The first of these three occurred in the 1890s, concurrent with the establishment of the Australian ballot in most states. The second fell in the 1904–12 period, and the third during the 1920s. All of them except the first must be explained on grounds other than those of changes in election mechanics; all except the first represent significant behavioral changes within the electorate. If one is primarily concerned with long-term trends within an electorate, he must of course be vitally concerned with such structural changes as the introduction of new ballot forms. *But we are also primarily concerned with the whys and wherefores of subsequent changes in measurable behavior.* These changes—all of them evidently of much greater magnitude than those of the ballot-reform era itself—cannot receive explanation through structural hypotheses.

Moreover, there is a problem even with the first or ballot-reform-era upsurge in split-ticket voting, one which may well turn out to be extremely refractory to quantitative analysis. This concerns the timing of the reform. Unhappily for analytic certitude, the period 1890–96 is not only one of ballot reform but also of nationwide critical realignment. In his recently published study, *The Winning of the Midwest,* Richard Jensen makes the important point that each of the elections in this period—1890, 1892, 1894, and 1896—involved unprecedented partisan volatility which primarily reflected itself in landslides, the first two in favor of the Democrats, the second two in favor of the Republicans.[27] These closely spaced but divergent landslides themselves drastically loosened the bonds of partisan allegiance which had dominated the preceding third party-system era. Moreover, political independence took a second significant form during the period in the "colonial" regions of the country: massive support for the Populist party which ended only with the Bryanite fusion of 1896. It is reasonable to suppose, as Jensen suggests, that these behavioral upheavals themselves produced some irreversible consequences for partisanship in the succeeding era. It would thus be reasonable to posit a cause or causes arising within the American sociopolitical system in this period which generated Populism, massive landslides, critical realignment, and the Australian ballot reform. This of course does not deny the permissive significance of the ballot-reform variable for subsequent electoral change. But it once again calls our attention to two propositions. First, reform was the consequence of behavior undertaken in response to system-wide strain. Second, an interlocking network of unprecedented behavior at the mass base occurred at the same time, and in response to the same system-wide political strain.

The Australian ballot and the Populist revolt grew out of the same primordial sociopolitical reality: the rapidly accelerating dysfunctions of a system of mass electoral politics—including electoral mechanics—which originated in preindustrial, preurban times. In particular, it grew out of the "swamping" of this system by the concentrated power of what our forebears called "combinations" in both politics and economics: the urban and state-wide political machine on the one hand, the corporation and the "trust" on the other, often enough with close links between the two. As the Australian ballot and subsequent reforms of election machinery were promoted by a moralistic urban *bourgeois* revolt against the effects and beneficiaries of urban machines, so the Populist revolt was aimed against the subversion of the national preindustrial political economy by concentrated corporate capital. The two events are, unfortunately, precisely synchronous with each other, and both occur at the approximate terminal date of a dying preindustrial nineteenth-century universe of electoral politics.

Thus one has more than a little reason for supposing that a presently indefinable but probably large component of split-ticket increase even during the first (reform-era) wave can be attributed to behavioral change. While the ballot reform of course facilitated it, this change was at its roots attitudinal rather than structural in origin, as the failure of split-ticket voting in Massachusetts to increase significantly from 1889 through 1903 rather clearly indicates. It would probably be quite fruitful to stratify and analyze the post-ballot-reform split-ticket measures both by region and by the extent of Populist voting in 1892 and 1894, and perhaps as well by the extent and direction of partisan swing in the 1896 election. For in Massachusetts, Populism and "colonial" revolt against Eastern finance capital were vanishingly small. Perhaps this is why we have to wait until the Progressive-era wave of 1904–14 to see any convincing partisan effects of this reform in the Bay State.

III. Implications: Toward a Comparative Sociology
of Past Voting Behavior

By this point the reader may have some sense of enlightenment concerning some specific issues in the controversy but still be quite unpersuaded of its overall substantive importance. If we accept the view that these rules changes should be regarded as wholly mechanical intervenors with no origins of their own in system stress and system adaptation, even so we could surmise that taken together they explain no more than half of the decompositional changes discussed in the 1965 article. But assuming this is true, so what? A reply to this legitimate query must take us to Converse's discussion of the electorate which existed before these changes took place. For not only is this discussion the weakest and most ambiguous link in an otherwise splendid and provocative paper, it also discloses the heart of the controversy.

Converse devotes considerable attention to the possibilities for, and occasional realities of, voter fraud, including repeating votes, in nineteenth-century America.[28] That this was occasionally notorious and even locally widespread cannot be denied. It comes out with compelling vividness in those hidden gold mines for historical electoral research, congressional hearings on contested elections in the South and elsewhere between 1868 and 1901. This too is a controlling variable which, to the extent it was a reality, would serve to depress the magnitude of change after the 1890s by shrinking the initial baseline of measurement. The problem, of course, is the virtual impossibility of ascertaining what, if any, fraudulent component existed at a given time and place, particularly in the rural reaches of nineteenth-century America.

However titillating the exploits of "repeaters," "floaters," and other venal hangers-on appear in the official and unofficial literature of contemporaries, it is simple prudence to avoid a precipitate surrender to Progressive-era political clichés. As the example of Philadelphia and other cities reveals down through the 1920s and later, the many urban machines which continued to thrive after turn-of-the-century reform movements had no difficulty in generating surplus voters on demand despite the statutes. Moreover, outside of the South and Border states, most of the preoccupation with corruption was focused, with good reason, on the cities rather than the countryside. When one considers that turnout rates in rural areas tended to be high over very long periods of time down until about 1900, any speculations on corruption as an artificial inflator of general rather than local significance must be open-ended in time as well as space. The assumption must be made that not only were corrupt surpluses more or less universal on territorial lines, but that these surpluses remained a more or less uniform proportion of the total reported vote for two generations and then, for reasons left most obscure, declined toward insignificant levels after 1900.

The first assumption may have some validity for some nineteenth-century statewide political systems outside the South. In Indiana in particular, a "floating vote" of some 10 percent of the total electorate appears to have existed from the 1870s through the 1890s. Even here, however, it should be pointed out that by no means all of this purchasable vote produced artificial inflations of turnout: a vote which is for sale is not necessarily a case of repeating or fraudulent surplusage. Richard Jensen's recently published study of the issue raises an appropriately cautionary note:

> It is easy to accept the fulminations of embittered losers at face value.... But the historian has to be more careful, especially in view of the long history of a "corruption" theme in American politics.... It is vastly easier to make blanket allegations of illicit, secret activities than to disprove them; almost always there is a grain of truth in the charges.... The myth of massive corruption so cleverly conceived that it cannot be detected is a ghost story.[29]

The conclusion which Jensen derives from a massive sifting of the available evidence is that traceable corruption, being a dangerous enterprise for its practitioners, was at most a marginal phenomenon. Most particularly it was marginal in national elections: the motivation of politicians to engage in such practices was at its highest point in local elections of direct importance to local machines—precisely, it may be added, as the Philadelphia story from 1900 through 1930 amply demonstrates. More significant, in Jensen's view, was the amount of error which could creep into the tabulation and counting of votes. Even so, the overall conclusion which Jensen makes after his survey of the evidence is that corruption was episodic and marginal in the Midwest rather than universal. The implication is strong that "militaristic" mobilizations of voters and an extreme strength and intensity of party identification among the mass electorate were overwhelmingly responsible for the phenomenal participation rates found in the 1880–1900 period.

If corrupt surplusage or deliberate tabulation errors could explain *a part* of Indiana's phenomenal turnout rate between 1876 and 1900, such explanations are both less obvious and—for reasons of local political culture and party strategy alike—much less persuasive in states which politicians considered "safe" for one party, and where moralism was a conspicuous part of the local political culture. States such as Iowa and Wisconsin come to mind. Similarly, it is particularly doubtful—and supremely so in the absence of empirical findings—that rural Pennsylvania or the counties of upstate New York were dominated by pervasive vote frauds and overcounts. If one were to ask for a better and more generally plausible explanation for this mass mobilization, the answer must surely lie in the direction which Jensen suggests: in the re-Christianization of the country through pietistic religious revivals between 1800 and 1850, and the development of increasingly polarized ethnocultural conflict in which political and religious perspectives became closely fused into a coherent framework for individual electoral decisions. One may indeed surmise that pietistic revivalism provided both the absolutism and the moral energy required to bring about the realignments of the 1850s and, hence, the war against secession and slavery.

It seems clear enough that the Civil War trauma was of tremendous polarizing significance in its own right—particularly in the lower Midwest and states like Pennsylvania—and hardly surprising that most elections from 1864 through 1888 were of the "maintaining" variety. It is now still more certain that the struggles between pietists and liturgicals, not only over the Civil War but over local variants of the "social issue" of the time, were intense down until the 1890s. They were central to the building and erosion of political coalitions down to the beginning of the present century. They also were part of a sociological and political context which optimized the probability that any individual member of that "voting universe" would actually cast a ballot; it is this largely vanished context which seems to explain much of the difference between nineteenth- and twentieth-century American electorates and their behavior.

We shall return to this point in a short while. Before doing so, we should examine Converse's curiously opaque references to contemporary survey findings concerning rural voting participation in underdeveloped countries such as Turkey and the Philippines. Since the discussion in this version of Converse's chapter is so brief, our discussion will be correspondingly limited. It should be noted above all that the *general* relevance of these twentieth-century studies in the underdeveloped world to American rural politics in the nineteenth is not merely "limited,"[30] but negative. To be sure, politics in the South was marked by a number of "underdeveloped nation" characteristics. In this region, overwhelming evidence attests to corruption and fraudulence in closely contested elections. Similarly, the voting of hordes of black *campesinos* by their white *jefes* was a conspicuous feature of post-Reconstruction Southern politics. In Alabama, for example, blacks were voted *en bloc* by local white elites to beat back a statewide Populist challenge in 1892 and 1894, and in 1901 they were similarly led to the polls to vote themselves out of political existence.

But these are a few of the many ways in which the South, particularly the Deep South, has been a deviant case in American politics since the Civil War. Even the implication that modern Third World survey evidence might have any applicability to most non-Southern rural voting behavior in the nineteenth century betrays a deep misconception of the historical sociology of the United States. Here the "conventional wisdom" of "impressionistic" American historians and such students of American political culture as Louis Hartz surely leave us on sounder ground.[31] That view argues that except for the very special case of the black population and perhaps some locally significant cases, the nineteenth-century American rural population was composed not of peasants in traditionalist social relations with local aristocracies, but of independent yeomen who were thoroughly bourgeois and "modern" in values and personal goals.

A challenge to the general accuracy of this view, as of any other, is surely possible. But such a challenge would seem to require empirical validation as a first step to acceptance. If Converse does not intend to make such an argument, then inclusion of these references to Turkish and Philippine voting behavior is irrelevant, and irrelevant in an important way. For its inclusion makes literally no sense except as a strained attempt to explain away anomaly. The particular anomaly in question here is evidently the educational variable. The levels of formal education in the American electorate a hundred years ago were vastly lower than they are today. Yet by all aggregate measures that electorate was obviously far more politically mobilized and active than it is today. And since a majority of Americans lived in rural areas until about 1880, this activity was associated both with low formal education and rurality.

So a problem exists. This problem is a real analytical conundrum only in the context of a particular mode of reality perception. While Converse duly denies that modern survey findings are "immutable,"[32] the implicit assumption which informs his entire critique is that of the basic uniformity and generalizability of such findings, at least for the United States. In particular, the broad sociopsychological model of the contemporary American electorate which is laid down in *The American Voter* and in Converse's subsequent works is implicitly adhered to from beginning to end. But this model obviously does not fit the data of change in the American electorate. It is inherently static, without historical depth and, for a variety of reasons relating to the intellectual history of the discipline, as weak in its sociological dimensions as it is strong in its social psychology.[33] Since it does not fit the data of long-term change, and since above all its findings in the here and now can give us rather little guidance in the analysis of past political upheavals such as that surrounding the Civil War, any longitudinal analysis made on its premises will turn up a very long list of very important anomalies.

Above all, the shape of the aggregate data for the last century—and particularly for the crises of the 1860s and 1890s—almost compels the conclusion that the level of political cognition in mass electorates is not more or less uniformly constant and low, but is subject to great fluctuations which are heavily dependent upon historical and sociological contexts. Such a conclusion, of course, leads to relativist skepticism, not about the methods and techniques of contemporary survey research, but about the relevance of the findings derived from them at a given point in time and space to other times, places, and contexts. More than this, however, it requires a complete rethinking of the celebrated pluralism and overarching harmony in the American political system of today, and to necessarily cautionary reflections on the health of that system. It is hardly surprising that such a conclusion would be stoutly resisted, or that major efforts should be mounted to discount or explain away the data which seem to support it.

Let us make the assumption that when politics matters, when it is not a spectator sport but is widely perceived to raise life-and-death issues, people behave as though it matters. In other words, let us assume that a significantly larger fraction of the electorate will be politically active and relatively politically aware in crisis situations than surveys of contemporary American presidential elections would lead us to believe, and that the deeper the crisis the more politicized the electorate.[34] Let us finally assume that the sociological context of electoral politics in America was quite different a century ago from the situation today. When we make such assumptions, the anomalies which have perturbed both Converse and Rusk vanish. In the first place, we find ourselves on the trail of a useful explanation of that immensely important American peculiarity, the periodic critical-realignment sequence. Second, we can accept with entire ease the numerous behavioral properties of nineteenth-century electorates which are inverses of the American electorates behavior in the twentieth century. Third, we can cheerfully get to work on constructing a better model of change than the one which I stated in 1965—a model which incorporates structural and behavioral components, gives each its proper weight, and integrates both into theory.

Let us take a concise example of where this might lead us, the Lincoln-Douglas debates of 1858 and their contextual setting in the American electorate of the time. Historians have repeatedly brought to our attention the intensity of both oral and written political communications in that period, and the great expenditures of time and money which practical politicians put into both. As is well known, practical politicians viewed Horace Greeley's *Weekly Tribune* as indispensable to the Republican mobilization in the 1850s. As is equally obvious, the ideational content of the speeches which Lincoln, Douglas, and others delivered was of a richness and complexity which, say, the Kennedy-Nixon debates of 1960 did not begin to match. These debates and speeches were reprinted in diamond-point type in the *Tribune*—as likely as not along with the dispatches on European politics filed by Greeley's European correspondent Karl Marx.

If we assume that political elites make informed and rational efforts to achieve maximum gain with minimum expenditure of scarce resources, we can deduce a good deal about the people they were attempting to reach and influence. While—like Lincoln himself—they had little formal education, they were mostly literate. Otherwise, the major investment in densely packed written communication which politicians made was an irrational investment. The wide spread of weekly newspapers for which subscriptions had to be paid bespeaks as well some active interest in the external world among this largely rural electorate. The massive physical movements of farmers from miles around to the central points of rallies such as the Lincoln-Douglas debates likewise suggest political motivation, as do occasional monster rallies in cities from the "log cabin" election of 1840 through the "sound money" and other parades of 1896. Moreover, the relative richness of content in written and oral political communication presupposes a judgment by the politicians of the time that the voter was both able and likely to pay considerable and sustained attention to the communication. Finally, it goes without saying that the political communications of the 1850s concerned political issues of transcendent importance.

All of this is compatible with an electorate having certain properties which no longer exist. First, voters were, for the most part, immensely dispersed across a continental countryside. Second, most of the voters lived on the land, or were at most a generation removed. Third, the primitive communications technology of the time sharply limited the voters' options for entertainment and diversion. This would help to ensure that politics would absorb a much greater share of the interested individual's attention than is now the case. Fourth, religion was primordially important in social and individual life. It now seems extremely likely that the clash between what Jensen describes as "pietist" and "liturgical" ethnoreligious subcultures was basic to the formation, maintenance, and transformation of partisan coalitions, at least in the Midwest and throughout the nineteenth century.[35] The mutual reinforcement of religion and politics which appears to have existed throughout the nineteenth-century political system was undoubtedly promoted by the extent to which Americans tended to settle in "island communities"[36] composed largely of people like themselves. Thus a visit to a Congregational or Methodist church on Sunday and a Republican rally on Monday in the Ohio Western Reserve a century or more ago would tend to produce cognitive consonance and partisan reinforcement. It would also be likely to produce intense party identification, highly stable party voting, and very high levels of political participation—at least so long as the military model of political campaigning was ascendant.

One very brief and partial, but instructive, example of the relation between religion, party identification, and party voting may be taken from a survey of Hendricks County, Indiana, which was made in 1874 and coded by Jensen a century later.[37] This was a county in which the pietist-liturgical ratio was lopsided in favor of the former, and one in which non-Protestant European immigration was extremely limited. It was correspondingly much more Republican than the state as a whole. Reworking the Jensen data and joining the global results of the 1874 survey with presidential voting behavior thereafter, we find the pattern described in Table 10. A comparison of this 1874 sample's aggregate of individual party preferences with the voting behavior of the county's entire electorate in the next five presidential elections reveals a correspondence so close that statistical demonstration is unnecessary. The mean Democratic percentage of the 1876–92 sample was 1 percent less than the 38.5 percent of 1874 preferences: the mean Republican voting percentage was 0.6 percent more. The central relevance of religious *tendence* in the largely native-stock population of this county seems clearly established here, as elsewhere in the Midwest where Jensen could generate similar microlevel data.[38] In this respect the nearly equal division of the large nondenominational vote between the major parties suggests nothing quite so much as the *absence* of a profoundly constraining group characteristic, a kind of aggregate "cross-pressure" situation.

I have argued elsewhere that a crucially relevant variable for the analysis

Table 10. Religious attachments, partisan preference, and voting behavior: The case of Hendricks County 1874–1908

A. Party Preference by Broad Religious Classification, 1874 Survey

Religious *Tendence*	% D	% R	% Other	% of total sample in category
Pietistic	23.8	72.3	3.9	49.5
No denomination	48.4	47.1	4.5	43.2
Liturgical	79.7	14.4	5.9	7.3
Total county	38.5	57.2	4.3	

B. Party Preference and Aggregate Voting Behavior, 1876–1908*

Year	% D	% R	% Other	N
1874 survey	38.5	57.2	4.3	1618
1876	37.1	58.4	4.5	5156
1880	36.9	59.1	4.0	5408
1884	38.9	56.4	4.7	5322
1888	37.0	58.6	4.4	5624
1892	37.8	56.4	5.8	5359
1896	40.3	58.1	1.6	5872
1900	39.7	57.6	2.7	5943
1904	37.1	58.8	4.1	5838
1908	43.2	54.3	2.5	5953

*Presidential elections.

of European political behavior may be the presence or absence in specific social strata of *political confessionalism*.[39] Such confessionalism, in European contexts, has tended to be closely associated with the development of specific religious-clientele parties (such as the *Zentrumspartei* in Germany between the 1870s and 1933), and also with the rise during industrialization of Marxist working-class parties with explicit ideological commitments which stand more or less sharply in opposition to the existing political and economic order. Strong confessional identities of this kind appear both to stimulate participation among the relevant sociological groups in the larger population, and to insulate them more or less thoroughly from electoral appeals arising from their traditional political oppositions on one hand or new and often extremist political movements on the other. In such cases, parties themselves become "churches," as it were, or are secular components of formally religious organizations.

We are used to thinking of such categories in terms of the class struggle in modern politics, or the epic conflicts between Catholic and Protestant in Ireland, but the work of a new crop of American political historians is now making it increasingly clear that American nineteenth-century electorates mediated their social experience and political goals through an intense, if typically American, form of political confessionalism. This confessionalism involved sharply antagonistic conflicts over right belief and right behavior, informed by highly divergent and intense polarizations over the "proper" form of Christian relatedness to society and politics. Religion and politics tended to constitute an extremely large part of the total formal communication likely to be received by any individual in that vanished political universe, and such communications heavily reinforced each other. The basis was thus laid for the mass participation and the political enthusiasm which defined nineteenth-century electoral politics, particularly during periods of critical realignment. Such linkages provided both ideologies and a coherent vehicle through which the individual could make sense of his world and his proper place in it. They likewise tended to produce an electoral structure whose thorough mobilization, extreme long-term stability, and other characteristics resemble nothing so much as the behavior of twentieth-century European confessional electorates. Finally, they provided the basis for a moral energy without which the great Republican revolution of the 1850s and 1860s could have been neither launched nor completed.

Indeed, elite strategies of the 1850–90 period make no sense except in the context of a high level of something very like political confessionalism at the mass base: the rallies, the dense packing of political information into party newspapers and broadsides, the staging and content of performances like the Lincoln-Douglas debates. These and other elite responses to politics bespeak an electorate which was intensely politicized, which had the opportunity, the capacity, and the motivation to pay relatively close attention to political argument, which participated—not only by voting—at rates phenomenal by contemporary standards, and which adhered tenaciously to partisan commitments once forged in the heat of realignment.

Thus our explanation is reduced to what would be a truism in any other context than this. High levels of political involvement existed in nineteenth-century America because the historical and sociological conditions of that time were at an optimum for such involvement. If modern religious or Marxist political confessionalism did not exist in this electorate, something very analogous in its political effects almost certainly did. It is hardly surprising that the grass-roots organizational effectiveness of the parties reached its peak in the last half of the nineteenth-century, or that the existence of a Republican-Union *party* organizational structure in the Northern states was so profoundly important to the success of the Union cause during the Civil War.[40]

A crucial measure of the comparative difference between electoral change in the United States and party development elsewhere in the Western world is the fact that this primordial quasi-confessionalism was progressively dissolved after the turn of this century *without being replaced by any other variety*. As the old cosmos of "island communities" and pietist-liturgical cleavages dissolved under the pressure of transition to urbanization and industrial capitalism, the modern shape of the American electoral universe emerged—a shape in which merchandising techniques tended to replace the old military campaign style, in which masses of Americans were selectively excluded, or excluded themselves, from participation in politics at any level, and in which party linkages between voters and their decisions at the polls suffered severe erosion.

If I were to make any comment on the original 1965 hypothesis which sought a coherent explanation of these remarkable developments, it would be to suggest that this process was largely unconscious and sociologically determined at all levels. It was certainly not a matter of conspiracy but of circumstance that things turned out as they did during and after the realignment of the 1890s: capitalist and party elites, like everyone else, were fumbling for the ball during that chaotic decade. Moreover, it would only be fair to emphasize how much of the demobilization and antipartisanship of the post-1900 era was achieved by a common mass consensus—at least among those who were still left in the shrinking electoral universe. As for the dissidents and dropouts, what viable alternatives did they have? Granted the manifest absence of conditions favoring class-solidaristic substitutes for the political confessionalism of the vanishing nineteenth-century religious cosmos, what other vision of the commonwealth and human liberty than one or another variant of liberal capitalism could survive the transition?

But when all is said and done, we would return to two basic points made in 1965. First, American electoral politics since 1900 has pursued a course which is retrograde to change elsewhere in the developed Western world, and the cumulative effect has been profoundly elitist. Second, the American electorates of the nineteenth century were *in fact*—and not in artifact—extensively different from contemporary American electorates, though perhaps similar in some of their attributes to some contemporary European ones. If this is indeed the case, then it is necessary to say once again that contemporary survey-research models of the American electorate should be used to explain voting behavior only at the times and places for which supporting survey evidence is available. They should be extended elsewhere only with great care. This may disturb paradigms, but it will also yield great benefits, for it will permit us to recover the historical and sociological dimensions of American electoral politics. These, in turn, will make it possible for our discipline to develop far better, far more complete empirical theories of democratic linkage and of American politics than any which we now have.[41]

NOTES

1. Thomas S. Kuhn, *The Structure of Scientific Revolutions,* 2d ed. (Chicago: University of Chicago Press, 1970).
2. Ibid., pp. 52–65.
3. *American Political Science Review* 59 (March 1965): 7–28. [This essay appears as Chapter One in the present volume.]
4. It is becoming increasingly apparent that the survey-research profile of individual voters laid down by the Michigan team—Angus Campbell, Philip E. Converse, Warren E. Miller, and Donald G. Stokes—in *The American Voter* (New York: Wiley, 1960), and in their *Elections*

and the Political Order (New York: Wiley, 1966), must be significantly modified for the post-1964 period—and on the basis of survey analysis itself. See, for example, Gerald Pomper's two articles, "Toward a More Responsible Two-Party System? What, Again?" *Journal of Politics* 33 (November 1961): 916–40; and "From Confusion to Clarity: Issues and American Voters, 1956–1968," *American Political Science Review* 66 (June 1972): 415–28; also, Norman Nie, "Mass Belief Systems Revisited: Political Change and Attitude Structure" (mimeo draft, August 1972); and especially Arthur H. Miller, Warren E. Miller, Alden S. Raine, and Thad A. Brown, "A Majority Party in Disarray: Social and Political Conflict in the 1972 Election" (Paper presented at the 1973 meeting of the American Political Science Association). In view of these findings, which are of course congenial to the arguments made in this paper, it seems about time that we had a complete retesting or replication of *The American Voter,* and one which is relevant to the contemporary electoral scene.

5. Philip E. Converse, "Change in the American Electorate," in *The Human Meaning of Social Change,* ed. Angus Campbell and Philip E. Converse (New York: Sage, 1972), ch. 8 (pp. 263–337). The controversy at issue here is centered in the first half of Converse's essay, pp. 263–301.

6. Jerrold G. Rusk, "The Effect of the Australian Ballot Reform on Split-Ticket Voting: 1876–1908," *American Political Science Review* 64 (December 1970): 1220–38. Also see the exchange of views between Rusk and Burnham in *American Political Science Review* 65 (December 1971): 1149–57.

7. Converse, "Change in the American Electorate," pp. 287–288.

8. To take but two *contemporary* examples of this problem: As of the 1970 election, there were no personal-registration requirements at all in thirty-seven (or 42.0 percent) of Ohio's eighty-eight counties, partial registration for small cities in another ten (11.4 percent), and countywide registration in the remaining forty-one (46.6 percent). In such a case, of course, no uniform standard exists even at the level of enumeration of electors provided by registration lists. Moreover, while Ohio gives the total number of ballots cast for its cities, it provides no such information for its county returns. In Massachusetts, on the other hand, complete information concerning the numbers of registered voters is published for each town and precinct; and the total number of votes cast, including blanks, is reported by town along with the vote for each candidate. Even in this case, of course, there is no indication of the number of potential voters—those who met legal requirements for the suffrage, but who did not register in the first place. In most European elections and in Canadian elections as well, it is possible to identify: (a) the total eligible electorate; (b) the total vote cast; (c) the number of invalid ballots cast; (d) the total valid vote cast; (e) the partisan distribution of that vote. For a more detailed discussion of heterogeneity and estimation problems in American electoral reporting, see the author's "Estimates of Potential Electorate," Introduction for section Y, 1974 edition of U.S. Census Bureau, *Historical Statistics of the United States.*

9. This urban-nonurban stratification has remained quite widespread down to the present day, particularly in Midwestern states with many low-density rural counties. In addition to the Ohio case mentioned in note 7, it is worth noting that as late as 1960, 87 of Missouri's 115 counties had no personal-registration requirements. As a group, these counties had turnout rates about 12 percent higher than those of the state's two major metropolitan centers. In New York, as of 1949, ten counties had no personal-registration coverage at all, while a majority of election districts in thirty-five additional counties also had no such coverage. In 1967 a permanent personal-registration statute was adopted for the entire state.

10. Pennsylvania is a typical example of the personal-registration transition which is occurring gradually in Ohio and has been completed in New York. The 1906 personal-registration statute applied these requirements to cities of the third class and larger, and was replaced by a statewide registration requirement in 1937. For text of the 1906 law, see any volume of *The Pennsylvania (Smull's) Manual* between 1907 and 1935. For text of the subsequent statute, see the 1937 or succeeding volumes of this publication.

11. The effectiveness of machine manipulation of the vote in Philadelphia can best be appreci-
    ated by examining turnout and voting in two wards: the 12th, a river ward populated heavily
    by immigrants and solidly controlled by the city Republican Organization, and the 21st, a
    middle-to-upper-class and politically independent suburban ward.

Turnout and machine control in Philadelphia, 1922–1930

Year and office	Ward 12 turnout	% D	Ward 21 turnout	% D	City turnout	% D
1922 Gov.	61.8	6.6	30.7	28.8	31.7	20.5
1926 Pres.	74.2	3.8	47.3	15.7	43.0	12.1
1926 Sen.	78.2	1.7	37.4	29.4	35.9	19.1
1928 Pres.	63.9	28.3	78.6	35.8	63.2	39.5
1930 Gov.	61.6	96.5	49.9	60.1	41.2	75.6
1930 Sen.	61.6	6.8	49.9	23.6	41.2	21.0

Statewide turnout rates during this period were 32.5 percent in 1922, 45.8 percent in 1924,
30.9 percent in 1926, 62.6 percent in 1928, and 40.3 percent in 1930. The inversion of usual
criteria of turnout—especially extreme nationally in the 1920s—is as remarkable in the contrast between lower-class Ward 12 and upper-middle-class Ward 21 as in the extraordinary
control of partisan outcomes in the former. Without further data presentation, it is also
possible to make one more assertion: the decline in turnout among middle-class "independent" wards such as 21 from 1890 through 1930 was vastly heavier than in the lower-class
wards which the machine controlled, in the former case approximating the rate of turnout
decay in the state's thirty-seven nonregistration counties. If the 1906 statute was designed to
protect middle-class and native-stock interests against the big-city machines and their ethnic
clienteles, it quite failed to achieve its purpose. In any event, there is a striking parallelism
between post-1890 declines in most of those wards of Philadelphia where antiorganization
sentiment was strongest and similar behavior in the nonregistration territories of the state
at large. Once again—and forcefully—it suggests the intervention of a behavioral variable
which can plausibly be linked to political alienation.

12. One notes the relevant parts of the Keystone (Progressive-Republican) party platform of
    1910: "A political trust, managed by cunning politicians, threatens the Commonwealth.
    Some of the conspirators are labeled Republicans and some Democrats, but they are all in
    league against the people and act in harmony with one treasonable purpose and under the
    orders of one head. . . . Both of the tickets nominated and both of the platforms adopted, it
    is well understood, were dictated by the same authority and were intended for the delusion
    of the voters and the further confirmation of the power of the political machine." *Pennsyl-
    vania (Smull's) Manual,* 1911, p. 520. It is perhaps also worth noting that the Socialist party
    reached its off-year apogee of 5.3 percent of the total vote in this election, with the Keystone
    candidate winning 38.3 percent, the Democrat 13.0 percent, and the Republican Organiza-
    tion's candidate 41.6 percent.

13. It is obvious that this metric is based upon an assumption of intercategory homogeneity in
    every respect other than that of the personal-registration variable. Such an assumption may
    or may not be partially or wholly valid. More elegant techniques of decomposition are
    undoubtedly called for. Even so, this is a possibly useful effort at *first* approximation.

14. Burnham, "Changing Shape." See also Walter Dean Burnham, *Critical Elections and the
    Mainsprings of American Politics* (New York: Norton, 1970), pp. 71–90; and E. E.
    Schattschneider, *The Semi-Sovereign People* (New York: Holt, Rinehart & Winston, 1960),
    pp. 78–96.

15. The best discussion still remains that of V. O. Key, Jr., *Southern Politics* (New York: Knopf, 1949). See also Paul Lewinson, *Race, Class and Party* (New York: Grosset & Dunlap, 1965).

16. It would be closer still if the partisan means included the Democratic percentages of the total vote cast for all statewide officers in the two periods.

17. Thomas A. Flinn, "Continuity and Change in Ohio Politics," *Journal of Politics* 24 (August 1962): 521–44; John H. Fenton, *Midwest Politics* (New York: Holt, Rinehart & Winston, 1966), pp. 117–54.

18. Burnham, *Critical Elections,* pp. 91–134 and appendix tables I–V.

19. For a classic presentation of this point for mostly European data before 1937, see Herbert Tingsten, *Political Behavior* (London: King, 1937), pp. 10–78; see also Angus Campbell, Philip E. Converse, Warren E. Miller, and Donald E. Stokes, *The American Voter* (New York: Wiley, 1960), pp. 483–93; and Stein Rokkan, *Citizens, Elections, Parties* (New York: David McKay, 1970), pp. 385–94.

20. Converse, "Change in the American Electorate," pp. 269–70.

21. Kentucky's 1920 turnout was 71.8 percent, more than 20 points above the national average, and falling midway between a mean of 81.2 percent for the 1900–1916 period and a mean of 63.1 percent for the period 1924–40. It is perhaps worth noting that the 1920 presidential election was closer in Kentucky than was any subsequent election except for 1952.

22. The sex-stratified data—available down to the precinct level in the city of Chicago—are found in the 1916, 1917, 1920, and 1921 volumes of the *Chicago Daily News Almanac.*

23. Rusk, "Effect of the Australian Ballot Reform."

24. Except with regard to roll-off, or ballot fatigue, which in Massachusetts underwent a sudden and substantial increase with the introduction of ballot reform. The failure of the partisan indicators to undergo any major change in the same reform-era period suggests the probability that ballot fatigue affected the followings of both parties about equally.

25. Burnham, *Critical Elections,* p. 115.

26. Ibid., especially pp. 52 and 109–10.

27. Richard Jensen, *The Winning of the Midwest, 1888–1896* (Chicago: University of Chicago Press, 1971), especially pp. xii–xiv. In the present author's view, this seminal work is "must" reading for anyone interested in diachronic analysis of American electoral politics.

28. Converse is particularly concerned to create the impression of endemic and massive voter frauds in nineteenth-century rural areas—not surprisingly, in view of his basic hypotheses—though part of the discussion of the fraud problem is based on Progressive-era formulations by Joseph Harris and others whose focus was urban rather than rural. Converse, pp. 288–93, 300.

29. Jensen, *Winning of the Midwest,* pp. 34–35.

30. Converse, "Change in the American Electorate," p. 287.

31. Louis Hartz, *The Liberal Tradition in America* (New York: Harcourt, Brace, 1955); see also Louis Hartz, ed., *The Founding of New Societies* (New York: Harcourt, Brace & World, 1964), especially pp. 1–22.

32. Converse, "Change in the American Electorate," p. 289.

33. For an extremely valuable discussion of the intellectual background of American political science generally and of voting analysis in particular, see Richard Jensen's two articles, "History and the Political Scientist" and "American Election Analysis: A Case History of Methodological Innovation and Diffusion," in *Politics and the Social Sciences,* ed. Seymour Martin Lipset (New York: Oxford University Press, 1969), pp. 1–28 and 226–43.

34. One perhaps suggestive quantitative characteristic of American electoral history has been the close association between periods of critical realignment such as those of 1856–60, 1894–96, and 1928–36 and waves of electoral mobilization or participation increases. Analysis of extreme-polarization situations in current American electoral politics likewise suggests inversions of "conventional wisdom" when racial polarizations are involved: for example, in the Cleveland mayoral election of 1967 the mean turnout in wards more than two-thirds

nonwhite was 69.2 percent, compared with a citywide average of 54.8 percent; and the Republican (white) candidate received about 80 percent of a working-class white vote which has typically been Democratic by margins of 2 or 3 to 1.

35. In addition to Jensen, see also Paul J. Kleppner, *The Cross of Culture: A Social Analysis of Midwestern Politics, 1850–1900* (New York: Free Press, 1970). For a review of this and two other recent works in the "new history," see Walter Dean Burnham, "Quantitative History: Beyond the Correlation Coefficient," *Historical Methods Newsletter* 4 (March 1971): 62–66.

36. The term was developed in detail by Robert Wiebe, *The Search for Order, 1877–1920* (New York: Hill & Wang, 1967), a nonquantitative but extremely useful reference for this critical transition period.

37. Jensen, *Winning of the Midwest,* pp. 59–60.

38. Ibid., pp. 61–62, 310–14.

39. Walter Dean Burnham, "Political Confessionalism and Political Immunization," *Journal of Interdisciplinary History* 3 (Summer 1972): 1–30.

40. For a lucid and illuminating discussion of this party contribution to Union victory, see Eric McKitrick, "Party Politics and the Union and Confederate War Efforts," in *The American Party Systems,* ed. William N. Chambers and Walter Dean Burnham (New York: Oxford University Press, 1967), pp. 117–51.

41. To a significant extent, this effort may require pursuit of exemplary "case studies" down to a very microscopic, "local history" level. A good case in point is that of corruption in rural voting among the Northern states. Here we may choose the case of Adams County, Ohio, which received extensive discussion in Jensen, *Winning of the Midwest* (see also sources cited there). One may summarize this case concisely. (1) Between about 1870 and 1911 the county was the scene of "undoubtedly the worst example of the corruption of the ballot ever known in American history . . ." (p. 38). (2) About 90 percent of the county's voters came to be involved, with each receiving on the average of $8.75. (3) The parties evidently wanted to curb the practice, but the voters would not let them. (4) In 1911, 1,690 men were finally disfranchised for five years by a judge for illegal vote selling (about one-fifth of the electorate). (5) At *no* time did this lead to turnout figures in excess of 100 percent, or to landslides for either party. As to the former, it may be inferred that the local (corrupt) voters had no incentive to bring in outsiders to share the spoils; there were very possibly normative constraints on "repeating" by local voters as well.

This is a curious case indeed; but system-wide inferences about "corruption" in earlier periods must be subjected to such micro-examination. Between 1872 and 1908, the mean turnout rate in Adams County was 93.1 percent of the estimated potential electorate, only 2.9 percent higher than the statewide mean turnout rate. Moreover, in 1916—after elections had supposedly been "purified" but the 1,690 disfranchised culprits of 1911 had been restored to their voting rights—the county's turnout rate was 87.3 percent, now more than 10 percent above the state's participation rate, but less than 6 percent below its corrupt pre-1911 mean.

It should also not go without remark that—as is well known—notorious examples of electoral corruption were widely used by the Progressives to obtain enactment of personal-registration statutes. But this *rural* county, perhaps proportionately the most corrupt of all, was entirely uncovered by such requirements until 1962, and a very large part of it is uncovered still.

3

Party Systems
and the Political Process
1967

I

The more deeply one reflects on the characteristic properties of political parties in the American party system of today, the more exceptional these parties seem to become by comparative standards. American mass mobilization parties are the oldest such phenomena in the modern world. After an impressive trial run which extended from 1790s to about 1820, they emerged as full-grown, recognizably modern structures by 1840—at least thirty years before the development of the first stable mass parties in Britain. During the enormously creative period extending roughly from the establishment of state nominating conventions in the 1820s to the creation of the Democratic National Committee in 1848, the party system took on a recognizably modern shape. But once that era had passed, very few major changes in organization occurred until the replacement of the convention by the direct primary during the years 1900–1915, and very few have taken place since.

No "branch" or "indirect caucus" types of party organization—still less "cell" or "militia" types—have been able, for example, to establish themselves in the American context.[1] Recruitment of elective elites remains closely associated, especially for the more important offices and in the larger states, with the candidates' wealth or access to large campaign contributions.[2] To this day the financing of the parties themselves remains to a substantial degree in the hands of those relatively few individuals and groups who make large donations; it does not rest on any stable mass base. This is, of course, but one reflection of the failure of European concepts or practices of party membership or organization to develop with democratization in the United States. Typically the American party is composed of an inner circle

From *The American Party Systems: Stages of Political Development,* edited by William Nisbet Chambers and Walter Dean Burnham. Copyright © 1967 by Oxford University Press, Inc.

of office-holding and office-seeking cadres together with their personal supporters and a limited number of professional party workers. This structure
has been decentralized but probably not significantly broadened under the
direct primary system. Outside this circle lies the mass of party identifiers,
only a tiny fraction of whom are involved in any more extensive partisan
activity than voting every two or four years. Still further outside is the
extraordinarily large part of the potential electorate which is not politically
involved at all.[3]

All of the available evidence suggests that the American party systems,
viewed comparatively, have exhibited an arrested development which stands
in particularly striking contrast to the extraordinary dynamism of the
nation's socioeconomic system.[4] This arrested development is visible both
organizationally and functionally. There are at least four broad functions
which are performed by fully developed democratic parties. The first of these
is a nation-building, integrative, or, in Lowi's term, "constituent" function.
This involves the regularized amalgamation and priority ordering of conflict-
ing regional, ethnocultural, group, or class interests through the mechanisms
of the political party. In modern politics, broad acceptance of the legitimacy
of the political regime as a whole depends in large measure on the successful
performance of this function. Second, political parties carry out an office-
filling function; orderly and democratic procedures are prescribed for elite
recruitment to a limited number of elective and appointive positions. Third,
parties perform a function of political education or political socialization for
their mass clienteles. In the most developed form, this function is carried out
through a network of party-controlled structures of education and media,
with the intent of producing a more or less coherent frame of political reference among those clienteles. Fourth, major parties may perform a policymaking function. This involves not only the capture of office, but the capture
of the policymaking machinery of government for programmatic purposes.

While the American parties have historically engaged in political education to some extent, and have occasionally been genuine policymaking vehicles, their involvement with such functions has been limited. Particularly in
our own century, American political parties have been largely restricted in
functional scope to the realm of the constituent and to the tasks of filling
political offices. So far as the function of political education is concerned,
indeed, there is evidence that during the nineteenth century the parties were
engaged in propaganda and political-socialization activities on a scale which
knows no parallel today. This intense activity seems to have been closely
related to the quasi-monopoly which election campaigns and the partisan
press of that period had on entertainment prior to the development of other
mass media, and also to the relatively extreme frequency of elections and
variability in election dates which existed prior to about 1880.[5] But as the
party press faded out of the picture toward the end of the nineteenth century,
as elections became concentrated in November and terms of office were

lengthened, and as nonpolitical mass media undermined the salience of the political campaign as a species of entertainment, the educative functions performed earlier by the parties tended to atrophy. American parties during the twentieth century have not been organized to provide political education or indoctrination for their clienteles on a month-in-month-out basis. Such an effort would require, in all probability, stable dues-paying memberships and the development of permanent ancillary party agencies for reaching a mass clientele whether or not an election was pending.

Similarly, the norm in American politics has been that the major parties lack the internal cohesion and organizational capacity to perform policymaking functions systematically or over a long term. In Great Britain, down through Lord Palmerston's time, majorities in the House of Commons—like those of the House of Representatives today—tended to be complex mixes of party and coalitional elements. Only after the sequential development of mass party organization following the Reform Bill of 1867 did the party whips in Britain become more than intelligence agents and persuaders for the leadership. In the United States, on the other hand, the development of mass parties in the 1830s destroyed the basis for consolidated party leadership which had existed during the time of the first party system, permanently activated the cleavage between executive and legislature, and gave an enormous impetus to the decentralizing influence of federalism.[6] Since then, American parties have performed the policymaking function only very infrequently— as a rule, only in situations marked by unusual social and political tension. Indeed, Dawson's study for this volume strongly implies that the link between party and policy outputs at the state level has been so ambiguous during this century as to generate doubt as to the weight of party influence upon those outputs.

The history of American party development, taken as a whole, suggests that every major upsurge of democratization has led to a dispersion rather than a concentration of political power in the policymaking arena. In this the United States has been quite exceptional. Students of comparative party organizations and their development have often argued that mass-based parties develop in an industrial era in order to array on a broad scale the power of numbers against the power of wealth, interest, and social position enjoyed by existing elites. Concentration of organization and decision-making authority is essential to this end. Were this argument to be accepted as a binding generalization about party development in advanced societies, we should be confronted with an intriguing paradox suggested by the late Morton Grodzins: that American parties may actually be antiparties.[7]

II

But this is not the end of analysis; it is merely a beginning. To make such an assumption is not only to take sides immediately in a substantive dispute with strong normative overtones, it is to plunge into a sea of analytical difficulties. It presupposes that there is, or more precisely ought to be, a clear-cut European confrontation between left and right in this country. It presupposes, moreover, that the cultural differences between the middle and working classes must necessarily dominate our domestic politics, and thus receive concrete expression in terms of party organization and evolution. But why should these differences receive such expression if they have not been dominant in the political culture? Nothing, indeed, is clearer than that the two major American parties are and always have been overwhelmingly middle-class in organization, values, and goals. From a European point of view they can hardly fail to appear as exquisitely old-fashioned, as nineteenth-century anachronisms. Yet they have not only survived, they have totally dominated the American electoral scene—and at no time more completely than during the past generation. If parties seeking to organize the lower strata of American society along collectivist lines have been so evanescent, if the two major components of the system have so feebly performed the educative and policymaking functions, the reason for this must lie not only in our dispersive political institutions but more fundamentally within the American political culture itself.

Of all these regarding that culture, the one which has the greatest explanatory power is the theory of the liberal tradition in America which has been developed by Louis Hartz.[8] According to this analysis, the United States is a fragment of Europe which was detached at a bourgeois point in history and subsequently developed a nearly monolithic, absolutist commitment to the values of individualist liberalism. These values, for most Americans at most times, have been so pervasive and have been so feebly challenged from within that they have tended to assume some of the properties of natural law in the American mind. Absolutist individualism has, of course, been buttressed both by the frontier experience and by the enormous relative affluence which most of the population has enjoyed throughout the country's history.[9] But the quite differing political experiences of other societies which have been relatively prosperous, or which have passed through a frontier stage, suggest the importance of the liberal cultural monolith in shaping the course of American political development.

Economic and social development consequently took place under very different political conditions than in Europe, other fragment cultures, or the modernizing states of the Third World. During an initial period in which there was inadequate capital which was both private and internally generated, governments at all levels played a major promotional role, especially but by no means exclusively in the fields of transportation and finance. This period was comparatively short, however, and was supplanted by an era in which an adequate supply of domestic takeoff capital in private hands and a widely accepted economic theory of laissez faire combined to restrict governmental intervention in the private sector. This arrangement, in its turn, survived with little effective challenge until the corporate world of private enterprise had clearly lost its capacity to regulate the development of the American economy without incurring ruinous social costs. But from beginning to end—from the age of neomercantilism down to the modern welfare-warfare state—the overwhelming value consensus in this field has been in favor of maximum private enterprise and minimum intervention by public authority. Neither the New Deal nor the post–New Deal periods can be adequately understood unless it is appreciated that this has remained official doctrine. Controversies over public involvement in the American economy have essentially tended to be differences, based on interest and ideology, over what "minimum intervention" ought to be at any given time.[10] Under such conditions there is hardly room for political parties to undertake sustained policymaking or government-control functions even if their leaders were for some reason inclined to make the attempt. Nor would such an attempt be likely to take place, since such control functions have not been considered necessary by any significant part of the population.

But this hardly means that there have been no political conflicts—or even that there have been no significant elements of class conflicts—in our history. Still less does it mean that the party system itself has played an insignificant role in that history. To be sure, it may be argued that its role has been peripheral and institutionally supportive of the American consensus concerning the primacy of the private sector in economic affairs. But there are other, and extremely important, dimensions of conflict which must be considered. Lee Benson is clearly right in his argument that the very consensus on economic and other fundamentals in the United States has produced not a lack of political cleavages, but a striking variety of them.[11] If the political culture and the party system have given only limited expression to vertical divisions within American society, they have accommodated an enormous variety of horizontal cleavages. Of these, three have been of paramount importance in our political history and, as the 1960 and 1964 elections have demonstrated, renain vital factors in American party politics to the present day.

The Clash of Sectional Subcultures. The literature which emphasizes and documents the vital historical importance of cultural and economic divergences along sectional lines is enormous. It may be enough here to mark the role of mass party politics in contributing to the full articulation of these cleavages. A student of the Missouri Compromise has pointed out that if the intense struggle in Congress over its terms had resulted in civil war, the war would have been fought out in the halls of Congress itself, with the rest of the country looking on in some amazement.[12] If a vastly different public reaction to its repeal was manifested a generation later, one can hardly doubt that the development of a mass-based party system, and the mobilization of mass opinion which accompanied it, bore much of the responsibility for the difference. In particular, the more Americans of the New England and Southern subcultures came to learn about each other's social values and political goals, the more pronounced their hostility toward each other grew. A strong argument can be made for the proposition that the Civil War was far more directly the product of the expansion of public political consciousness which is inherent in the dynamics of democratic development than of any economic causes.

The twofold decision of the period from 1861 to 1877—first to keep the South in the Union by force, and then to permit it to reassert substantially complete autonomy in its political affairs—enormously increased the need for the components of the party system to place major emphasis on integrative activities and functions. It is enough to recall that the influence of sectional cleavages was overwhelmingly dominant in American political history from 1854 to 1932, or to note the evidences of the age-old antipathy between the New England and white Southern subcultures which emerged in the massive voter shifts of the 1964 election. The South over the past century has influenced American politics in ways not dissimilar to the influence of Quebec in Canada or of Ireland in the United Kingdom in the years from 1870 to 1922. The massive influence of that deviant subculture upon national party politics and policy outputs has scarcely run its course a century after Appomattox.

The Clash Between "Community" and "Society." This cleavage, like the sectionalism with which it has some relationship on the national level, can in a sense be traced back to the controversies between the Hamiltonians and the Virginia agrarian school in the infancy of the republic. But it becomes particularly significant in the industrializing and postindustrializing eras of our history. As Hays points out in his illuminating study in this volume, between 1880 and 1920 a substantial portion of our industrializing elites and their supporters were pittted against the parties themselves. Whatever may have been the parties' contributions to centralized policy formation prior to the industrial takeover, after the end of Reconstruction the parties became bulwarks of localist resistance to the forced-draft change initiated by cosmopolitan elites. Thus was born a tradition of hostility to the political party as such, with an associated theme which emphasized that political centralization at any level in the system required the displacement rather than the further development of the political party as an instrument of popular government. This antipartisan tradition of the Progressive era, and the power shifts which developed from it at the local level, were both in profound harmony with the dominant function of the alignment system of 1896 to 1932: the insulation of industrial elites from the threat of effective, popularly based "counterrevolution." The persistence of this tradition and its consequences has undoubtedly made a significant contribution to the truncating of American party development during this century.

Viewed somewhat more broadly, this local-cosmopolitan cleavage may also be related to a larger regional phenomenon in post–Civil War America. This is the ongoing conflict between the Northeastern metropole on one hand and the Southern and Western regions, which stood in a quasi-colonial relationship to the metropole, on the other.[13] William Jennings Bryan, as H. L. Mencken sarcastically noted, was the political evangelist of white Anglo-Saxon provincial "locals" who were in revolt against "their betters"—by which Mencken meant metropole cosmopolitans like himself.[14] After the shattering blow which fell upon the Democratic party in the Northeast during the 1890s, the Democratic party became to a substantial extent the party of the "locals" or the community-oriented, while the GOP—especially in its Bull Moose wing—tended to become the dominant partisan vehicle of the "cosmopolitans" or the society-oriented.[15] This helped, no doubt, to reinforce the persistent Republican charge, which was reiterated down through the 1930s, that the Democratic party was incompetent to govern in an industrial society. Many of the urban "locals"—the working-class elements who were deliberately bypassed by Progressive-era reform in local government—supported this Republican position, at least at the presidential level, by their votes. That they did so probably reflects their memory of the "Democratic depression" of 1893–97, as well as their ethnocultural hostility to the Bryan crusade and its clienteles. It is possible that this support may also have been a manifestation of "working-class imperialism" akin to the support which Disraeli and his Conservative party successors enjoyed in England during the last third of the nineteenth century. By the beginning decades of the twentieth century the differential in economic development between the Northeast and the rest of the country had become enormous; to a greater or lesser extent most politically active groups in the metropole shared in its relative affluence.

Nor has intersectional antagonism disappeared as a force in contemporary American politics since, even though regional differentials in economic development have been severely eroded since World War II. While the terms of the political discourse involved have shifted beyond recognition, at the bottom of this antagonism may well be an ongoing "community" resistance—which still tends to be geographically concentrated—to massive, centrally directed social change. It seems reasonable to suppose that both the growth of a doctrinaire right and the contemporary conflicts between the executive and Congress can be at least partially explained in terms of this community-society continuum.

The Clash of Ethnocultural Groups. It is by now axiomatic that the diversity of ethnic origins in the American melting pot has been basic to American political alignments since the creation of the first party system in the 1790s. There appears to be a curiously circular relationship between ethnic diversity and the limited role which class alignments have played in the history of American politics. On one hand, as Benson argues, the American liberal value consensus on such fundamentals as economic and church-state relationships helped make possible a full articulation of ethnocultural antagonisms in party politics at a remarkably early date.[16] On the other hand, however, the very existence of ethnic fragmentation—especially after the "new immigration" of 1882–1914 added novel elements to the American melting pot—severely inhibited the formation even of mass trade unions, and was instrumental in preventing the emergence of socialism as a major political force.

In the Northeast, for example, lines of class and ethnic division tended to coincide during and after the era of industrialization. Most of the middle classes in such centers as New York, Boston, and Philadelphia were of old-stock Protestant background, while many of the working-class elements were relatively recent arrivals with wholly different cultural traditions. But if class and ethnic lines tended to coincide in the conurbations, the lower classes were also enormously internally fragmented along ethnic lines. Some groups, such as the Jews of Manhattan and the Germans of Milwaukee, had cultural traditions which favored the growth of socialist movements among them.[17] Groups such as these, coupled with certain radical native-stock farm elements in the colonial regions of the country, gave the American Socialist party most of such vitality as it possessed at its apogee around 1912. But most of the "new immigrant" groups came from a European stratum, the peasantry, which had no counterpart in the social experience of most other Americans. Such groups required the most extensive acculturation simply to come to terms with urban-industrial existence as such, much less to enter the party system as relatively independent actors.

It is hardly surprising, consequently, that the typical American form of urban political organization during and after the industrializing era was the nonprogrammatic, patronage-fueled urban machine rather than the disciplined, programmatic leftist party of the type which was emerging in European conurbations during the same historical period. As Robert K. Merton's pioneering analysis has made clear, the urban machine developed as an unofficial means of concentrating political power in an official context of power dispersion.[18] Its functions, at least on the latent level, were overwhelmingly integrative. For the lower strata, in return for their votes, it provided a considerable measure of primitive welfare functions, personalized help for individuals caught up in the toils of the law, and political socialization. For business elements it provided a helpful structure of centralized decision-making in areas of vital concern to them. Of the classical urban political machine it may be said, first, that it provided services for which there was a persistent social demand and which no other institution was capable of providing; and second, that the extraordinary fragmentation and lack of acculturation among its mass clientele ensured its preoccupation with ethnic-coalition politics to the virtual exclusion of class politics.

Here, as elsewhere in the study of American political cleavages and organizational structures, there appears to be an inverse relationship between the integrative or "constituent" functions of parties and their policymaking or government-control functions. If the social context in which a two-party system operates is extensively fragmented along regional, ethnic, and other lines, its major components will tend to be overwhelmingly concerned with coalition building and internal conflict management. The need to unite for electoral purposes presupposes a corresponding need to generate consensus at whatever level such consensus can be found. Surely at the promordial level the American liberal consensus in its various nuances provides a framework for a party action. Yet this essentially middle-class frame of reference and the feebleness of social solidarities above the level of the primordial have worked together throughout our history to exclude programmatic politics—or programmatic parties—from the mainstream of American political development. More than a decade ago Samuel Lubell argued that the major task of American politics was that of building a nation out of a congeries of regions and peoples.[19] So long as that task remains unfinished, there is little reason to suppose that our major parties will break out of the mold in which they were cast during the generative years of the second party system.

III

A study of the history of American voting alignments, paradoxically enough, reveals that, notwithstanding the substantial exclusion of the parties from policymaking and educative functions, the voting public has made vitally important contributions to American political development approximately once in a generation. Studies of American elections, especially in the past decade, have uncovered a remarkably stable pattern involving two broad types of elections which differ from one another not in degree but in kind. Most American elections, most of the time, are relatively low-pressure events. In such elections the voting decision seems to be a mix of traditional party identifications and short-term, "surge" factors associated with specific candidate or issue appeals.[20] Taken in the aggregate, such elections—whether "maintaining," "deviating," or "reinstating"—are part of a broad pattern of system maintenance.[21] They constitute reaffirmation of a "standing decision," even though the parties themselves may alternate in power as they did in the 1880s or the 1950s.[22]

Approximately once every thirty years, however, an entirely different cycle of elections emerges—a realignment cycle which precipitates massive grass-roots changes in voting behavior and results in a new coalitional pattern for each of the parties. Each of these critical realignments has been associated with a major turning point in the development of the American political system as a whole. There appears to be a typical pattern in the alignment cycle, although not all of its stages have clearly been followed in each historical realignment. After a more or less extended period of stability, broadly based discontent with the existing political order begins to emerge and then to crystallize. At a certain point the intrusion of a proximate tension-producing event, in a context of growing discontent, triggers either the creation of new major-party organizations or the capture of one of the older parties by insurgents against the political status quo. This proximate event may be economic, as were the depressions of 1893 and 1929, or political, as were the events leading from the Kansas-Nebraska Act of 1854 to the election of Lincoln; usually it has been a mixture of both elements. In the campaign or campaigns which follow this breakthrough, the insurgents' political style is exceptionally ideological by American standards; this in turn produces a sense of grave threat among defenders of the established order, who in turn develop opposing ideological positions. The elections which follow produce massive realignments of voters, and usually result in a stable majority for one of the parties which endures until the beginning of the next realignment phase a generation later.[23]

This cyclical pattern—long-term continuity abruptly displaced by an explosive but short period of change—seems not only to reflect the constituent function of the party systems and the electoral system in the United States, but to be a prime manifestation of the dominance of that function in our politics. The critical realignment, to be sure, drastically reshuffles the coalitional bases of the two parties, but it does far more than this. It constitutes a political decision of the first magnitude and a turning point in the mainstream of national policy formation. Characteristically, the relationships among policymaking institutions, their relative power and decision-making capacity, and the policy outputs they produce are profoundly affected by critical realignments. It was far from coincidence, for example, that the Supreme Court reached its height as an economic policymaker in a period—1890–1937—which almost precisely covers the period of the partisan alignment of 1894–1932, or that this role became untenable after the next realignment. With characteristic properties such as these, the critical realignment may well be regarded as America's surrogate for revolution. One of these experiences led directly to the outbreak of civil war, and every one of the others has been marked by acute political tension.[24]

The existence and significance of the critical realignment in American political history provides an excellent point of departure for analyzing our political development in terms of the constituent decisions and institutional modifications which have been associated with it. Examination of American party politics over time reveals the existence of not less than five national party systems. Each of these, to be sure, has constituted a link in a chain of development within the same polity and thus has numerous properties which it shares in common with the others. But to a marked degree each is also a discrete entity, with characteristic patterns of voting behavior, of elite and institutional relationships, and of broad system-dominant decisions. While a full exploration of these patterns in all their subtlety is the proper subject of a much larger and more detailed study than this, their broad contours can be briefly outlined here.

The Experimental System, 1789–1820. In a real sense, the first American party system was a bridge between a preparty phase in American political development and the recognizably modern parties found in the second and succeeding party systems. All of American political life was experimental and to a degree tentative during this period of nation-building, and the Federalist and Republican parties, as they developed, shared this experimental quality. In a developing society which was overwhelmingly agrarian and spread out along two thousand miles of coastline, a number of fundamental problems had to be faced almost immediately after the establishment of government under the Constitution. Full national independence from Europe—economic and psychological as well as political—remained to be realized. Both living and institutional symbols of a common American nationality had to be forged. In pursuit of the goals of nationality and full independence, it was essential to provide a political framework for "takeoff" into sustained economic growth under internally generated capitalist auspices. Finally, certain political issues required authoritative disposition. To what extent, if at all, should public deference to elite rule, characteristic of eighteenth-century British and colonial politics, be continued as a mainstay of republican institutions? To what extent, if at all, was partisan conflict, both over office and over policy goals, legitimate in itself?

The party system which developed in the 1790s exhibited peculiarities which were intimately associated with the attempt to find solutions to these problems. Both because of the narrow base of the active public in the initial phases of development and because of the primitive communication and transportation facilities of the time, both of the opposing coalitions were organized from the center outward toward the periphery.[25] While the importance of state politics and the significance of regional bastions of support for the Federalist and Republican parties should not be overlooked, these parties apparently were loose amalgamations of state organizations to a smaller extent than were their successors. Second, because of the heavy involvement of coastal America with Europe during the era of the French Revolution, foreign-policy controversy played an enormously salient role in the structuring of party conflict. Third, the deferential tradition was paralleled by a nearly complete identification by each rival group of the national good with their own partisan views; and this significantly affected the behavior of both parties during this period. While this identification with deference and the universal validity of their own policy views affected the Federalists especially adversely by rendering them inflexible and resistant to change, many Jeffersonians also tended to regard their opponents as a "disloyal opposition," and their leadership was hardly free of elitist bias.[26]

It would be a distortion to argue that domestic partisan conflict in this period was confined to the problem of defining how broad the ruling elite should be or what regional and economic interests should be included in it. But in practice it can hardly be argued that the election of 1800 produced any revolutionary change in the foundations of national economic policy-making, however important its political effects were. To some degree after 1800, the Jeffersonians were impelled toward active intervention in the economy by the exigencies of foreign affairs, as in the Embargo of 1808. But the rechartering of the Bank of the United States in 1816 and the harmony which developed over time between John Marshall's Supreme Court and the other branches of the federal government are sufficient indications that neomercantilism remained on the policy agenda after as well as before 1800. As the democratization of politics in the Jacksonian era was to reveal, such policy could be sustained in an overwhelmingly rural society only so long as systematic partisan mobilization of the vast majority had not yet occurred, and only so long as deference politics continued to display some vitality. In all probability the contribution of greatest lasting significance which the realignment of 1800 made to American political development was the precedent of a peaceful turnover of political power by a Federalist coalition which professed to regard its rival as subversive of the republic. Similarly, the greatest contribution of the parties themselves was to establish the tradition of partisan competition itself, as well as to supply practical working knowledge of such competition to a whole generation of Americans.

After 1800 as well as before, the first party system displayed certain properties which were unique to it and were not clearly transmitted to its successors. First, while contests for the presidency were of major importance in crystallizing party competition nationally, neither the office nor the modes of election involving it were as yet wholly democratized. Legislative choice of presidential electors, like such choice of governors in some states, survived in many states throughout this period. This, coupled with the "inner circle" characteristics of the Republican congressional caucus, permitted a semifusion of powers at the federal level which has known no counterpart since the election of Andrew Jackson.[27] Moreover, this system was very incompletely developed. Partisan competition did not spread throughout the country; particularly in the Southern and frontier regions, what weak foothold the Federalists had had down to 1800 virtually evaporated thereafter.[28] The forces which have operated over the past century to restore party competition were clearly inoperative after 1800. In the period from 1802 to 1822 (excluding the unopposed presidential election of 1820), Jeffersonian Republicans won three-quarters of all contests for presidential electors and congressmen, and four-fifths of all senatorial contests.[29] Finally, of course, the system evaporated in a decade-long nonpartisan "era of good feelings," for which there is no subsequent parallel in American history.

While all party systems can in a sense be regarded as artificial, the extreme imbalance between the components of this first system after 1800 suggests rather strongly that party competition under the first system had accomplished its dominant and relatively narrow purposes by that date. Were it not for disruptive internal pressures generated by repeated collisions with the major powers involved in a European "world war" which lasted until 1815, it seems probable that a nonpartisan interregnum would have emerged at least a decade before it did. Once the struggle between uncompromising and moderate elitism had been clearly settled in favor of the latter, there were evidently not enough points of internal conflict at the national level to sustain truly competitive party politics. In this sense, as in others, the first party system left no successors.

The Democratizing System, 1828–60. Of all of the five American party systems, the second was incomparably the most creative from an organizational point of view. The development of national two-party competition centering on a democratized presidency, and growing out of a host of local political alignments, was so massive an undertaking that it required more than a decade to complete after the critical election of 1828. There appear to have been three major stages through which the second party system passed before its dissolution. The first was the period of intense partisan organization between the election of John Quincy Adams in 1825 and his defeat at the polls in 1828. In this phase an extremely heterogeneous opposition to "insider" politics was mobilized around Andrew Jackson; this "outsider" was virtually compelled to raise the standard of popular revolt in order to unseat the incumbent political elite.[30] The second phase of development, by which time about half of the potential electorate had been mobilized, was the Jacksonian phase proper. This phase was characterized by the emergence of the convention system as a device for presidential nominations independent of Congress or the state legislatures, by an extreme sectionalism in voting patterns, and by the new separation-of-powers conflicts between president and Congress. This phase was also marked by a heavy majority for Jackson's supporters at the polls and a new issue-oriented politics which resulted in a resounding confirmation in 1832 of the "decision of 1828." The third, or mature, phase was inaugurated in 1834 by the founding of the Whig party, and was completed around 1840. The origins of the Whigs involved not only the definitive collapse of Jackson's "solid South," but by the recognition by Jackson's opponents that effective opposition depended on acceptance of both the policy and organizational implications of democratization. Thereafter, as McCormick observes, the second party system was notable for the extreme closeness of the party balance throughout the country and for exceptionally high rates of voter turnout.[31]

The "decision of 1828" generated a number of fundamental changes in institutional relationships and policy outputs. Broadly, it was a decision to democratize political opportunity and—at least rhetorically—to eliminate the last vestiges of elitism and deference politics from the American scene. Democratization in the context of the middle period came to involve a dismantling of neomercantilism on the federal level and a general recession of the federal government to an extremely low level of positive activity. During the first half of the lifetime of the second party system, this was to a substantial degree the expression of the agrarian yeoman's political style and political goals, just as Hamilton had feared half a century earlier. Thereafter, the emergence of sectional conflict reinforced the weakness of the federal government—as has been true in some contemporary societies with problems of regional integration—since systematic federal pursuit of any positive domestic policies gravely threatened the increasingly tenuous union between North and South.[32] In this context the presidential nominating convention produced candidates whose chief virtue was their "availability," and whose tenure was exceptionally short. The Senate's role during this period was that of a congress of ambassadors concerned with working out the terms of intersectional compromise. As for the parties themselves, the democratized political atmosphere, the increase in the number and variety of elective offices, and the perpetual mobilization campaigns necessitated the emergence of the plebeian electoral machine staffed by professionals who had to be paid for their services.[33]

The dramatic collapse of the second party system in the period from the mid-1850s to 1860 disclosed its essential fragility. Each party had been put together piecemeal from a bewildering variety of local cleavages and ethnocultural hostilities. On the national level, each was an electoral machine which sought to make voting capital out of these local antagonisms and the national symbolic rhetoric of the democratic "revolution." But precisely because the two parties were both so nationwide in their coalitional base, they found it increasingly difficult to accommodate sectionally divergent interests among their elites and mass followings.

The weakest link in the system was clearly the Whig party. In retrospect, it seems quite strange that a general sentiment arose in the aftermath of the 1852 presidential election that the party was already moribund, for even in defeat it had received 46.4 percent of the two-party vote.[34] Yet two major disruptive forces were at work. In the first place, the Whig party below the presidential level was a good deal weaker than it seemed, and this weakness tended to accelerate in every election after 1848. This weakness, and particularly its tendency to increase, probably helped certain Whig elites—especially in the North—to turn their attention to other political combinations which would be more profitable to themselves and the interests which they supported. Second, the structure of opinion on the slavery question and related issues was clearly sharply different among the mass bases of each party. In at least a number of Northern states, the Whigs were the party of the positive liberal state, and this in substantial measure reflected the presence of the large New England subcultural component in their mass base.[35] In the Deep South, on the other hand, there was a significant correlation in many areas between wealth—including slaveholding wealth—and Whig strength in given county units.[36]

Table 1. Partisan strength in the second party system, 1834–53

Office	% Democrat	% Whig	% Other
Presidential electors	53.9	46.1	—
U.S. representatives	54.9	42.6	2.5
U.S. senators	54.8	42.5	2.7
Governors	58.7	39.7	1.6

It is unlikely that this tendency toward a bimodal opinion structure among the Whigs was duplicated in anything like the same degree in the Democratic party. A posteriori, indeed, it can be noted that the swift and near-total collapse of the Whig party's mass support in the North after the Kansas-Nebraska Act had no counterpart among the Democratic following. Moreover, as elections after 1860 were to demonstrate, the critical split which did come to the Democrats that year was significant as an organizational rupture, but it did not result in permanent mass defections from the Democratic banner.

The Civil War System, 1860–93. The major "decision" associated with the realignment which culminated in the 1860 election was, of course, the reorganization of the party system and of institutional relationships and policy outputs along explicitly sectional lines. This was the only possible restructuring which could lead to the definitive containment and eventual extinction of slavery and the economic-cultural regime built upon it. Since this intersectional issue was neither "negotiable" nor one which the losing elite groups could permit to be resolved through the electoral process, the inevitable result of organizing the party system in this way was civil war. Put another way around, the only possible way to avert a breakup of the antebellum Union and the violence which followed was—precisely as the conservative compromisers of 1850 had always argued—to declare the entire question of slavery off limits and thus to prolong the life of the second party system indefinitely. But such an attempt to halt further political development in a broader system undergoing the most dynamic change seems to have been foredoomed to failure. It reckoned without the implications of accelerating cultural, demographic, and economic divergences along regional lines. In fact a classic prerevolutionary situation had developed by the 1850s. The system as a whole tended to be dominated by political elites who represented a declining sector of the national socioeconomic system. Elites who represented the values and interests of dynamically evolving sectors at first resented, and later rose in rebellion against, that traditional dominance.

It may be argued with great plausibility that the American Civil War and its aftermath constituted the only genuine revolution in the history of the country.[37] Certainly during the first half of the Civil War the party system was replete with characteristic deviations from normal patterns, deviations which could be expected in an era of violent transition. With the exclusion of an entire region from access to—much less control over—national policymaking institutions, a radical shift in policy outputs occurred. Not only was slavery given a violent *coup de grace,* but an integrated program of positive federal involvement in the fields of banking and currency, transportation, the tariff, and land grants to smallholders was inaugurated. While on some issues the majority coalition was fragmented, the central policy issues of the 1860s were closely integrated. During this era the dominant Republican party was genuinely, if unusually, a policymaking party. On the mass level, the partisan loyalties which were forged during this revolutionary era survived almost unchanged until the 1890s, and in some areas left traces which are still visible today.

It is usual to define the end of Reconstruction as occurring at the time of the famous bargain of 1877 between Northern Republicans and Southern Democrats. In reality, however, the revolutionary phase of this party system had largely run its course by 1870.[38] The administration of Ulysses S. Grant can probably best be regarded as a bridge between the era of convulsive revolution which preceded it and the era of industrializing-elite dominance which followed. Into the 1870s, Republican leaders were preoccupied with the danger that a Southern reentry into the political system might produce an overthrow of their coalition at the polls and a restoration of the Jacksonian coalition to its former dominance. Nor was this a chimera: the success of the Republican revolution in national policymaking had been predicated upon enormous artificial majorities that were produced in a Congress in which the Southern states were not represented. Indeed, the Republican fears were partially realized after 1872. Southern "Redemption" and the persistence of traditional Northern support for the Democrats resulted in a unique period of partisan deadlock which lasted from 1874 until Republican capture of all branches of the federal government in 1896.

Table 2. Partisan strength in the third party system 1854–92

Office	% D	% R	% Other
	A. 1854–73		
Presidential electors	29.4	70.0	0.6
U.S. representatives	34.4	58.4	7.2
U.S. senators	32.7	60.7	6.6
Governors	25.6	69.2	5.2
	B. 1874–92		
Office	% D	% R	% Other
Presidential electors	50.6	48.3	1.1
U.S. representatives	55.9	41.7	2.4
U.S. senators	47.5	51.0	1.5
Governors	48.9	49.5	1.6

The accelerating influence of industrial capitalism produced results, however, which differed sharply in many respects from those which had been feared by Republican leaders in the Reconstruction era. Both political parties fell substantially under the control of elites who favored industrial development and private enterprise. The Southern Redeemers were not Jacksonians *redivivi;* most of them were upper-class gentlemen who adapted their goals and styles quite well to the new industrial dispensation.[39] Paradoxically, the freezing of alignments along Civil War lines at the mass level gave maximum political latitude to the industrial elites and their partisan assistants to develop the economy on their own terms. As for the Negro, his interests were abandoned by the Republican leadership, in substance if not in rhetoric, as an essential part of the sectional bargain on which the stable deadlock rested. As the nature of the Southern Democratic leadership changed somewhat around 1890 by becoming rather more plebeian, it drew the logical consequences implicit in this bargain and, against ineffective Republican resistance on the national level, formally expelled the Negro from the Southern polity. During the 1880s and the 1890s this solution came tacitly to be accepted by the Republicans, by the Supreme Court, and by white public opinion.[40] The modern "solid South" thus came into being, and, as a political necessity in a one-party regime, so did the direct primary.

The Industrialist System, 1894–1932. The deadlock of 1874–92, however, was as unstable as the national bipartisan balance which had existed in the 1840s. In both cases a party system whose components were locked in an obsolescent pattern of alignments and partisan ideologies tended increasingly to underrepresent significant disadvantaged elements in the electorate. The processes of industrialization after the Civil War had produced two major strata of the disadvantaged: the farmers, especially in the cash-crop colonial areas of the country, and the growing, ethnically fragmented urban proletariat. As the history of the period from 1877 to 1896 so strikingly reveals, both groups became progressively more alienated from the established order as the Civil War system drew to its close. Suffering from the effects of a long-term crisis in agriculture, the cash-crop farmers in the Plains states and the South were already in active rebellion by 1890. Almost immediately thereafter, the second worst industrial depression in American history struck the urban centers. The almost instantaneous result, with a conservative Democratic administration in power, was the collapse of the Democratic party throughout the urban metropole in 1894. As many members of the Northeastern elite feared, the stage was thus set for a political coalition of both disadvantaged elements, with the objective of overthrowing industrial-elite rule. The fact that full democratization of politics had uniquely occurred in the United States before the onset of industrialization—and thus that such a mass assault against industrial elites could be conducted with constitutional legitimacy—undoubtedly increased the latters' anxiety. While the Supreme Court did what it could in its classic decisions of 1894–95 to undermine that legitimacy, only a critical realignment could dispose of the issues raised with any finality.[41]

In the event, the insurrection under William Jennings Bryan's leadership proved abortive. Among the factors leading to his defeat, several appear decisive. First, the urban working class was too immature and fragmented internally to work effectively with the agrarian rebels. But more than this, there appears every evidence that the combination of the "Democratic depression" of 1893 and severe ethnocultural hostilities between new-immigrant workers and old-stock agrarians created an urban revulsion against the Democrats which lasted into the late 1920s.[42] Moreover, the issues which appealed to the dominantly colonial-agrarian clientele of the Bryanites—especially currency inflation—meant nothing, and perhaps less than nothing, to workers whose wages were all too obviously at the mercy of employers and economic conditions. Finally, the essentially nostalgic and colonial character of the insurgents' appeal produced a violently sectional reaction throughout the metropole; the Democratic party in that region sank into an impotence which, save for a limited upswing between 1910 and 1916, lasted for a generation.

The alignment system which was set up during the 1890s marks the point at which American party development began clearly to diverge from developmental patterns in other industrial societies. This system was unique among the five under discussion: It was structured not around competition between the parties, but around the elimination of such competition both on the national level and in a large majority of the states. The alignment pattern was broadly composed of three subsystems: a solidly Democratic South, an almost equally solid Republican bastion in the greater Northeast, and a quasi-colonial West from which protesting political movements were repeatedly launched against the dominant components of the system.[43] The extreme sectionalism of this system can be measured by virtually any yardstick. For example, excluding the special case of 1912, 84.5 percent of the total electoral vote for Democratic presidential candidates between 1896 and 1928 was cast in the Southern and Border states. Gubernatorial contests during the 1894–1930 period, while showing somewhat greater dispersion of partisan strength, also demonstrate this sectional pattern.

A number of major consequences followed from this pattern of politics. With general elections reduced to formalities in most jurisdictions, the direct primary was developed as an imperfect and ambiguous alternative to party competition. Election turnout dropped precipitately from levels comparable with those of present-day Europe: by the 1920s national turnout ranged from less than one-third in off years to little more than two-fifths in presidential elections. Viewed in terms of the broader political decision-making system as a whole, the substantial disappearance of party competition, the discrediting of party itself as an instrument of government, the progressive fragmentation of Congress during this period, and the large but negative policy

Table 3. Sectionalism and gubernatorial elections, 1894–1931

Region	Percentage of governorships won by:		
	D	R	Other
South	96.8	2.6	0.6
Border	61.1	38.9	—
Midwest and West	31.0	67.2	1.8
Northeast	16.9	83.1	—
Total U.S.	43.0	56.0	1.0

Note: South—Eleven ex-Confederate states.
Border—Kentucky, Maryland, Missouri, Oklahoma, West Virginia.
Midwest and West—East North Central, West North Central (except for Missouri),
Mountain and Pacific census regions.
Northeast—New England and Middle Atlantic census regions plus Delaware.

role played by the Supreme Court all fitted admirably into the chief function of the fourth party system. That function was the substantially complete insulation of elites from attacks by the victims of the industrializing process, and a corresponding reinforcement of political conditions favoring an exclusively private exploitation of the industrial economy. One is indeed inclined to suspect that the large hole in voter participation which developed after 1900 roughly corresponds to the area in the electorate where a viable socialist movement "ought" to have developed but, for reasons discussed earlier, did not succeed in doing so.[44]

It can nevertheless be argued that the sectionalism of the 1896–1932 alignment significantly advanced the nationalization of American politics. First, the realignment of the 1890s destroyed or submerged a tangled network of diverse patterns of party allegiance which went back to the Civil War or earlier. It thus created both a severe loosening of the grip of traditional voting patterns and tended to establish broadly regional alignments in their place. Second, the apparently decisive rout of the Democratic party in the industrial-urban centers bore within it the seeds of the party's eventual regeneration. A power vacuum had been created in its state organizations in the Northeast by the desertion of the old, respectable Gold Democratic leadership. This vacuum was in time to be filled by representatives of the newer immigrants.[45] The stage was thus gradually set, via the Democratic convention of 1924 and the "Al Smith revolution" of 1928, for a transition from the old rural-colonial party of Bryan to the winning coalition of rural and urban underprivileged which the party was to become under Franklin Roosevelt.

The New Deal System, 1932–? The election of 1928, bringing as it did a huge block of new immigrant votes into the political system for the first time, has rightly been called the beginning of critical realignment in the Northeast.

Even so, it is doubtful that the extremely stable sectionalism of the fourth party system could have been destroyed by any force less profound than the Great Depression. The extended realignment of 1928–36, associated with that great shock and with the coming of age of the new immigrants, has rightly been called an event "very like the overthrow of a ruling class."[46] Permanent federal involvement in the mixed economy which arose from the ruins was substituted for a business rule which could no longer stay the course. The inevitable institutional modifications emerged; the presidency and its ancillary executive establishment moved into ascendancy as a center of policy planning and initiation, and the Supreme Court's veto over interventionist economic legislation was eliminated. The federal government also promoted the development of countervailing institutions of power in the larger society, especially in the labor field. As the Democratic party became the normal majority party, a substantial class cleavage was added to the traditional mix of voting alignments for the first time. Sectionalism was replaced piecemeal by the emergence of two-party competition where it had not existed for decades. The party system became nationalized, although the organizational structures and functions of the major parties themselves remained largely unchanged.

It is important to emphasize, however, that the nationalization of party organization and voting alignments which was inaugurated in the 1930s has been a gradual process which is still far from finished. On the state level, years had to pass before the conservative, old-line Democrats who were suddenly propelled to victory during the 1930s were replaced by leaders more in tune with the programs of the national party. Realignment of party organizations and followings along national lines did not spread into a number of states in the far North until the late 1940s and early 1950s.[47] Similar realignment in the South, at least below the presidential level, has begun to have statewide ramifications only since 1960.

Here, as elsewhere in the past, American party development has largely been derivative from major changes in the structure of the socioeconomic system. Urbanization and the development of autonomous sources of capital

Table 4. Gubernatorial elections in the fifth party system, 1932–66

Region	Percentage of governorships won by:		
	D	R	Other
South	98.4	1.6	—
Border	80.0	20.0	—
Midwest and West	44.4	53.7	1.9
Northeast	43.4	56.7	—
Total U.S.	57.4	41.6	1.0

in the former "colonial" areas, along with the enormous shift of middle-class populations southward and westward since the end of World War I, have resulted in a severe erosion of the formerly central distinction between economically developed and underdeveloped regions in the United States. Such postwar changes have helped to provide the social diversification essential to two-party competition. Moreover, the federal government during the past decade has conspicuously reversed the "decision" of the 1890s regarding the exclusion of the Negro from the Southern political system. This reversal is as characteristic of the fifth party system as the former decision was characteristic of the fourth. The original national acceptance of exclusion gave the cue for the organization and maintenance of the restrictive one-partyism which was essential to the classic Southern subsystem. The contemporary national insistence upon the inclusion of the Negro has destroyed most of the rationale for the preservation of that one-partyism.

There is much reason for subdividing the fifth party system into two parts at about the year 1950. The first period was that of the New Deal era proper. Broadly speaking, this was a period in which the major concerns of American politics centered on the domestic issues of prosperity and the full integration of the newer immigrants into American social and political life. As the sustained growth in the American economy after 1941 was still in its infancy, the dominant political themes at this time still turned on scarcity and the attendant problems of class relationships in American society. Despite the intrusion of World War II, foreign-policy issues remained on the whole distinctly peripheral to these concerns. The "American responsibility," about which so much has been heard of late, was barely in gestation among elite circles down until about 1950, and was virtually invisible to the country at large until the outbreak of the Korean War. The election of 1948 may well be remembered in retrospect as the last of the older-style elections: the last before the full emergence of television, the last to turn on explicit class appeals, the last in which the "farm vote" was considered a major factor in the outcome, the last in which foreign-policy issues were conspicuous by their absence from the major-party campaigns.

The period since 1950 may legitimately be described as one of great confusion in American party politics, a period in which the classic New Deal alignment seems to have evaporated without being replaced by an equally structured ordering of politics. The rapid development of public-relations techniques and the projection of candidate "images" have been accompanied at the mass level by a sharp decline in the salience of older-style class cleavages, and an equally significant erosion of party as a dominant factor in electoral decisions. The underlying partisan preferences of the electorate, as survey research has repeatedly demonstrated, have not significantly changed since at least the 1940s.[48] But the electorate since 1950 has displayed a willingness to engage in ticket splitting on an unprecedentedly massive scale. Probably as a consequence of image voting, the partial replacement of patronage politics by ideologically flavored politics, and the penetration of the mass media, short-term influences on voting have grown tremendously in recent years at the expense of long-term continuities.

For this reason, it remains impossible to say with any certainty whether or not the 1964 election was in fact a critical election, even though some of its behavioral properties would otherwise support this view. It is enough to point out that, as in 1964 the Republicans lost scores of counties in the Northeast which had been steadfast in their support since the Civil War, so in 1965 and 1966 Democratic candidates for major offices lost such once-impregnable party bastions as Boston, Manhattan, and Philadelphia to their Republican challengers. If a realignment is actually going on under these conditions, it is incomparably more subtle and diffuse than any of its predecessors.

IV

Underlying these developments, as Sorauf suggests, there may well be a party system which is undergoing extensive atrophy because so many of its functions are being performed by other agencies. At the bottom of such a development are certain rather obvious but momentous changes in American life. In the first place, a continually increasing majority of the active American electorate has moved above the poverty line. Most of this electorate is no longer bound to party through the time-honored links of patronage and the machine. Indeed, for a large number of people politics appears to have the character of an item of luxury consumption in competition with other such items, an indoor sport involving a host of discrete players rather than the teams of old. Connected with this tendency toward depoliticization is clearly a second factor: the extraordinarily complex and technical character of many key political issues as these become translated into concrete proposals for action and policy decisions. In such an intricate political environment, partisan cues to the public are perhaps less relevant than in an age of fewer and broader political issues. In a sense, this may be considered to involve an extension of the principles of civil service to the substance of policy: as there is said to be no Democratic or Republican way to run an agency, so it often appears—whether accurately or not—that there is no Democratic or Republican way to solve highly complex policy problems.

The third and most significant of all these changes has been the international emergence of the United States as an imperium in fact, if not in rhetoric or intent. This development has many profound implications for the subordinate role of the political party in the broader decision-making system. Two of these raise particularly serious questions of relevance. In the first place, as is well known, in the United States the making of foreign-policy and military decisions is the province of a bipartisan elite drawn from the executive establishment and from private industry. It is not the province of the political parties, and it is not normally capable of being structured into a cluster of issues on which the parties are opposed and between which the voters can choose. Since these decisions involve the expenditure of between three-fifths and three-quarters of the current federal administrative budget, depending upon which items are included, it is not inaccurate to say that most of the present-day activities of the federal government lie quite outside of areas in which parties can make any positive contributions to the political system. Second, the compartmentalization between foreign defense policies and domestic affairs which used to exist has increasingly dissolved since World War II. The nonpartisan military and foreign-affairs sector has come more and more to infiltrate the world of domestic politics and nominally domestic or private activities. Since this growing sector involves areas far removed indeed in complexity and remoteness from the lives of the voters, and consequently tends to escape even the forms of democratic control, its infiltration of the domestic scene suggests a growing restriction of the scope of effective party activity even there.

At the outset it was argued that this country has had a common middle-class, liberal political culture. Even though the absolutism of its assumptions has been brought into serious question in recent decades, the social center of gravity of the active electorate and the structure and financing of our parties have produced elective and appointive elites who have tended not to question the postulates of our traditional liberal absolutism. As the American cultural fragment has rejoined the world, it has been confronted by the most serious challenges to both the relevance and validity of the economic and political values which had so often been assumed to be "givens" in this country. What are the implications of this for the future of American party politics?

On the whole, the outlook is not reassuring. The socialist phase of historical development, and thus of party organization and development, is clearly a thing of the past all over the Western world. Whether for good or for ill, it was a phase through which American party organization never passed. Largely as a consequence of the failure of collectively based vehicles of mass political mobilization to develop here, American major parties have never performed most of the ranges of educative or policymaking functions in any sustained or systematic way. They are most unlikely to begin doing so now. Moreover, as we have indicated, there are large and rapidly growing areas of contemporary public policy from which they—and hence the public at large—have been effectively excluded.

It would clearly be a radical oversimplification, of course, to assert that the parties have been reduced solely to the office-filling function. There is, for example, no hard evidence that the parties have lost their constituent function; indeed, a study of Lowi's discussion might suggest a contrary view. Nor—as any examination of congressional roll calls in recent years makes clear—can it be said that policy differences between Republicans and Democrats are unimportant, much less that such differences do not exist. In the aftermath of the election of 1964, it would be particularly misleading to discount overmuch the very great salience which the political party continues to display in American political life. All that is suggested here is that the relative weight of nonpartisan factors at the top decision-making levels of the American political system has increased substantially over the past generation, and that, granted the massive scope of American involvement in the rest of the world alone, these factors will probably continue to grow in importance. Partisan politics, in such a case, will tend to be confined to a narrowing—though not necessarily unimportant—range of activity.

In politics, as in architecture, form follows function. In an affluent, corporate, technologically complex America, an America whose national policy processes seem likely to be increasingly dominated by its external commitments, it may be asked: what functions will be left for political parties to perform? The answer to this question is not at all clear. What does seem clear is that—in the future as in the recent past—the functions which the parties perform and the forms which they assume will be determined by the emergent needs of the broader social and political system, and not by the parties themselves.

NOTES

1. For a useful typology of modern party organizations, see Maurice Duverger, *Political Parties* (New York, 1954), pp. 1–60. A discussion in terms of the American party system of the present is found in Frank J. Sorauf, *Political Parties in the American System* (Boston, 1964), esp. pp. 38–58, 153–69.

2. This is especially the case at the crucial nominating stage. See Alexander Heard, *The Costs of Democracy* (Chapel Hill, 1960), pp. 34–35, 318–43.

3. Walter Dean Burnham, "The Changing Shape of the American Political Universe," *American Political Science Review* 59 (1965): 7–28 [Chapter One in the present collection].

4. This peculiarly nondevelopmental aspect of American party development has also been noted by Robert A. Dahl, ed., *Political Oppositions in Western Democracies* (New Haven, 1966), pp. 34–69.

5. An illuminating discussion of the exacerbating effects of perpetual electioneering on antebellum politics is found in Roy F. Nichols, *The Disruption of American Democracy* (New York, 1948), pp. 5–7.

6. Wilfred E. Binkley, *President and Congress* (New York, 1962), pp. 83–104. See also Theodore J. Lowi, "Party, Policy, and Constitution in America," in *The American Party Systems: Stages of Political Development*, ed. William Nisbet Chambers and Walter Dean Burnham (New York, 1967).

7. Morton Grodzins, "American Political Parties and the American System," *Western Political Quarterly* 13 (1960): 974–98.

8. Louis Hartz, *The Liberal Tradition in America* (New York, 1955); and idem, *The Founding of New Societies* (New York, 1964), pp. 1–122.

9. David M. Potter, *People of Plenty: Economic Abundance and the American Character* (Chicago, 1954).

10. Andrew Shonfield, *Modern Capitalism* (New York, 1965), pp. 298–329.

11. Lee Benson, *The Concept of Jacksonian Democracy* (Princeton, 1961), pp. 272–77.

12. Glover Moore, *The Missouri Controversy, 1819–21* (Lexington, Ky., 1953), p. 175.

13. An excellent statement of the case from a self-consciously colonial point of view is Walter Prescott Webb, *Divided We Stand* (New York, 1937).

14. H. L. Mencken, "In Memoriam: W. J. B.," in *The Vintage Mencken*, ed. Alistair Cooke (New York, 1959), pp. 161–67.

15. The *locus classicus* here, on the level of the Progressive intellectual, is surely Herbert Croly, *The Promise of American Life* (New York, 1909).

16. Benson, *Concept of Jacksonian Democracy*, pp. 165–207.

17. See, for example, Lawrence Fuchs, *The Political Behavior of American Jews* (Glencoe, Ill., 1956). To this day Jewish voters are very atypically Democratic (or "leftist") at levels of socioeconomic status where substantial Republican strength is normally encountered in other population groups. Angus Campbell et al., *The American Voter* (New York, 1960), pp. 302, 306.

18. Robert K. Merton, *Social Theory and Social Structure* (Glencoe, Ill., 1957), pp. 72–82.

19. Samuel Lubell, *The Future of American Politics* (New York, 1952), pp. 245–61; and idem, *White and Black* (New York, 1964), pp. 153–74.

20. Angus Campbell, "Surge and Decline: A Study of Electoral Change," *Public Opinion Quarterly 24 (1960): 397–418*.

21. Angus Campbell et al., *Elections and the Political Order* (New York, 1966), pp. 63–77.

22. For a useful discussion of the "standing decision," see V. O. Key, Jr., and Frank Munger, "Social Determinism and Electoral Decision: The Case of Indiana," in *American Voting Behavior*, ed. Eugene Burdick and Arthur J. Brodbeck, (Glencoe, Ill., 1959), pp. 281–99.

23. V. O. Key, Jr., "A Theory of Critical Elections," *Journal of Politics 18* (1955): 3–18.

24. See a preliminary discussion of these characteristics in Walter Dean Burnham, "The Alabama Senatorial Election of 1962: Return of Interparty Competition," *Journal of Politics 26 (1964): 798–829, esp. 822–9*.

25. William N. Chambers, *Political Parties in a New Nation* (New York, 1963), pp. 103–12. See also Noble E. Cunningham, Jr., *The Jeffersonian Republicans in Power, 1801–1809* (Chapel Hill, 1963), esp. pp. 299–305, and Richard P. McCormick, *The Second American Party System* (Chapel Hill, 1966), pp. 19–31.

26. Cunningham, *Jeffersonian Republicans in Power*, pp. 8–9, 303–4. For an interesting discussion of the ambiguous attitude of many Republican leaders concerning expansion of the suffrage, see Chilton Williamson, *American Suffrage from Property to Democracy, 1760–1860* (Princeton, 1960), pp. 138–64.

27. Binkley, *President and Congress*, pp. 60–82. See also Leonard D. White, *The Jeffersonians* (New York, 1951), pp. 29–59. White also duly notes that very little break in basic administrative patterns occurred after 1800. For a eulogy of Alexander Hamilton delivered by a Republican secretary of the treasury, see pp. 14–15 of White's work.

28. McCormick, *Second American Party System*, pp. 27–28.

29. For a statistical confirmation of the existence of forces tending to restore two-party competition, extending back to 1866, see Donald E. Stokes and G. R. Iversen, "On the Existence of Forces Restoring Party Competition," in Campbell et al., *Elections and the Political Order*, pp. 180–93.

30. For a lucid discussion of the organizational effort involved, and Martin Van Buren's key role in it, see Robert V. Remini, *The Election of Andrew Jackson* (Philadelphia, 1963), pp. 51–120.

31. McCormick, *Second American Party System*, pp. 329–56.

32. Studies of elections in certain developing nations can provide interesting comparative insights into the interrelationships between the integrative and policymaking functions under acute conditions of regional heterogenity. See K. W. J. Post, *The Nigerian Federal Election of 1959* (Oxford, England, 1963), esp. pp. 437–43; and also Leon D. Epstein, "A Comparative Study of Canadian Parties," *American Political Science Review 58 (1964): 46–59*.

33. Still extremely useful for a study of American party development with these particulars in mind is M. Ostrogorski, *Democracy and the Organization of Political Parties* (New York, 1902) 2:207–440.

34. Arthur C. Cole, *The Whig Party in the South* (Washington, 1913), pp. 274–76. It is particularly suggestive that Whig leadership opinion that the party was dead after the 1852 election was concentrated in the North, and that this conclusion tended to be resisted by Southern Whig leaders with the exception of the group around Alexander H. Stephens.

35. One notes this, for instance, in Benson's estimates of the demographic composition of the major parties in New York around 1844: Benson, *Concept of Jacksonian Democracy*, p. 185. The most predominantly Whig area in New York State was also known as the "Burned-over District," because of the plethora of social-reform and millenarian movements which flourished there during the middle period. See Whitney R. Cross, *The Burned-Over District* (Ithaca, N.Y., 1950).

36. Seymour Martin Lipset, *Political Man* (New York, 1960), pp. 344–54.

37. A most provocative recent discussion, which raises the largest issues of development, is found in Barrington Moore, Jr., *Social Origins of Dictatorship and Democracy* (Boston, 1966), pp. 111–55.

38. Kenneth J. Stampp, *The Era of Reconstruction, 1865–1877* (New York, 1965), pp. 186–215.

39. See, for instance, C. Vann Woodward, *Reunion and Reaction: The Compromise of 1877 and the End of Reconstruction* (Boston, 1951), pp. 23–53.

40. C. Vann Woodward, *The Burden of Southern History* (Baton Rouge, La., 1960), pp. 69–87. See also *Civil Rights Cases,* 109 U.S. 3 (1883); and *Plessy* v. *Ferguson,* 163 U.S. 537 (1896).

41. Among the best studies of this interaction are Arnold M. Paul, *Conservative Crisis and the Rule of Law* (Ithaca, N.Y., 1960), esp. pp. 131–235; and Alan F. Westin, "The Supreme Court, the Populist Movement and the Campaign of 1896," *Journal of Politics* 15 (1953): 3–42.

42. One of the many myths—or, more appropriately, half-truths—which are propagated among liberal historians is the notion that employer coercion or intimidation of urban workers was a dominant factor in the 1896 outcome. See, for example, Ray Ginger, *Altgeld's America* (New York, 1958), pp. 172–79. Quantitative analysis of urban election data—especially over time—indicates how dubious this explanation is. It fails to take into account, first, that the realignment of 1896 endured in the cities with very little change until 1908, and remained substantially dominant until the "Al Smith revolution" of 1928; and second, the existence even in 1896 of very considerable ethnic differentials in the impact of the campaigns and realignments of this era on the urban vote. For a study which remains a seminal contribution by a historian, see Lee Benson, "Research Problems in American Political Historiography," in *Common Frontiers of the Social Sciences*, ed. Mirra Komarovsky (Glencoe, Ill, 1957), pp. 113–83.

43. This discussion partly follows E. E. Schattschneider, *The Semi-Sovereign People* (New York, 1960), pp. 78–96. See also his "United States: The Functional Approach to Party Government," in *Modern Political Parties, ed. Sigmund Neumann* (Chicago, 1956), pp. 194–215.

44. For a more detailed discussion, see Burnham, "Changing Shape of the American Political Universe."

45. An excellent state case study of this process is J. Joseph Huthmacher, *Massachusetts People and Politics, 1919–1933* (Cambridge, Mass., 1959).

46. Schattschneider, *Semi-Sovereign People,* p. 86.

47. In New England and elsewhere, as Richard E. Dawson points out in "Social Development, Party Competition, and Policy," in Chambers and Burnham, *American Party Systems*, 1954 appears to have been a breakthrough year. See Duane Lockard, *New England State Politics* (Princeton, 1959), pp. 30, 45. In the upper Midwest, the breakthrough came around 1948. John H. Fenton, *Midwest Politics* (New York, 1966), pp. 18–29, 44–64, 87–100.

48. Philip E. Converse, "The Concept of a Normal Vote," in Campbell et al., *Elections and the Political Order,* pp. 9–39, esp. 13.

II

Who Votes? Who Doesn't? Does It Matter?

4

The Appearance and Disappearance
of the American Voter
1978

I. Introduction

This paper has been designed to deal with only a part of the larger problem of political participation and the state of health of democracy in the United States. The part of the task which it addresses is the tracing of a very peculiar historical pattern. This pattern may, very succinctly, be described as the appearance and disappearance of the American voter. But the larger task, even in this smaller part, requires not merely a documented pattern but a scheme of explanation.

We begin with several propositions, all of which will be supported by the discussion below. In the first place, mobilization of the mass electorate is and always has been contingent on the existence, competition, and organizational vitality of political parties. The two cannot be understood in isolation from each other. Second, one widely and rightly used criterion of democratization in a political system is the proportion of the potential electorate which actually votes. By this criterion the American political system is markedly less democratic today than it was in the last century. Moreover, it is significantly less democratic today than is any other Western political system which conducts free elections.[1]

Third, there are two features of American participation in elections which stand out as essentially unique: (1) the class skew in American voting participation is today greater by a very wide margin than it is in any other Western country with the possible exception of Switzerland; and (2) long-term trends in American participation run exactly counter to trends in these other countries. The long-term movement there has been toward substantially complete incorporation of the mass public into the political system during this century,

A paper delivered at the American Bar Association's symposium on "The Disappearance of the American Voter," June 23–24, 1978, Palo Alto, California.

121

while here it has since about 1900 been a move toward functional disfranchisement, especially of the lower classes.[2] Finally, turnout trends in the most
recent period—since 1960—show a renewed but very unevenly distributed
downward trend. Participation during this period has declined most rapidly,
both in absolute and in relative terms, among those groups of the population
(outside the South) which already had come to participate least. In other
words, the class skew in turnout is noticeably greater now than it was in the
early 1960s, and is in many metropolitan areas as great as it was in the previous trough, the 1920s.

An extensive polemical and analytical literature has grown up in the past
decade on this question of participation in American elections.[3] Much of this
literature has been preoccupied with the attempt to show that nineteenth-
century participation rates were grossly inflated by the kind of electoral corruption known as ballot-box stuffing.[4] Much of the rest of it has been aimed
at the identification of changes in the rules governing access to the ballot box
(particularly personal registration statutes) which occurred during the Progressive era just after the turn of the century.[5] My own research not only
accepts but can identify particular instances of ballot-box stuffing in national
elections (for example, New York City in 1868, Kansas City in 1936). It
rejects as unprovable, unfalsifiable, but also unreasonable a notion of "universal corruption," particularly in the rural areas of the North which cast
most of the votes in elections down until the 1880s. Likewise, it accepts and
can measure to some extent the effects of personal registration statutes and
other rules-of-game changes in depressing turnout after 1900. But it rejects
the view that *all* or even *most* of the decline outside the South between 1900
and 1920 can be accounted for by these rules changes. Moreover, it insists
that the changes themselves must be understood as the result of policies
which were designed in part to eliminate "undesirable" voters from the electorate (typically, in the South), or at least to dilute heavily their presence in
it (typically, in the North).

Certain other literature, especially during the 1950s and 1960s, accepts the
low participation rate and the heavy class skew, and indeed celebrates it as
part of a "politics of happiness" which bespeaks the middle-class content-
ment of our pragmatic, low-keyed politics.[6] Such views have tended to go
out of fashion after the convulsions of the late 1960s and early 1970s. They
were not supported by the seminal report of the President's Commission on
Registration and Voting Participation (1963). Nor have they been supported
by subsequent developments in public opinion, most especially the collapse
of public affective support for national political leaders and institutions since
the mid-1960s which Harris and others have reported.[7] In short, they have
not been supported by history. One may assume, if one wishes, that there is
really no problem here; in which case, it would be puzzling as to why this
conference was held at all. On the other hand, certain research suggests
rather forcefully that the presence of a large, politically inert mass of people

who have not been in effect politically socialized may also mean that they (and others) have not been politically immunized against antidemocratic demagogic mass appeals. This could spell real trouble for the system and its values at some future crisis point.[8]

In sum, then, it is the argument of this paper that the decay in participation during this century is real, that the high levels of democratization during the nineteenth century were real, and that the decay—for which no parallels can be found elsewhere in Western electoral history—is largely the result of *political* change. This political change, whatever it is, is obviously and tightly linked with the decay of political parties in the United States during the twentieth century. It has clearly worked to the comparative political disadvantage of the lower classes in American society, as recent comparative analysis is beginning to demonstrate.[9]

II. Comparisons with Other Countries Today

It is probable that other papers in this series will deal more fully with contemporary comparisons. The remarks here will necessarily be compressed, but a comparative view of where we are today seems essential for full understanding of American electoral development.

Comparative literature has emphasized the close links between the density of party and other social organization throughout the social structure on one hand, and the mobilization of voters on the other.[10] It is virtually axiomatic that such structures for penetrating and mobilizing the American working classes are much feebler than in most European cases. Gunnar Myrdal, indeed, has gone so far as to comment that the United States has the most disorganized social infrastructure (or, if one prefers, the largest *Lumpenproletariat*) to be found in any advanced industrial society.[11] A brief comparison of American with non-American turnout patterns stratified by social class and age in recent elections identifies the groups among which the "hole" in mass participation is concentrated.

The data in Table 1 are quite unambiguous. Turnout in Sweden is essentially invariant along class lines, and is extremely high in each category. In the United States there is a massive class skew in turnout. The Swedish turnout ratio between manual workers and propertied middle-class voters is 1.03 to 1 in favor of the latter; in the United States, this ratio was 1.42 to 1 in 1968 and 1.65 to 1 in 1976 in favor of the latter. Even more extreme American skews can be found when cross-tabulating educational and occupational levels. In 1972, where such cross-tabulation can be performed, the turnout rate among white male laborers with a grade-school education was 40.9 percent, compared with 86.6 percent among people of managerial occupation and a B.A. degree—a ratio of 2.12 to 1. Quite naturally this class skew also shows up in the 1976 survey when adults respond that they have *never* voted

124   WHO VOTES? WHO DOESN'T? DOES IT MATTER?

Table 1. Voting participation and occupation in Sweden, 1960, and in the United States, 1968 and 1976

Socio-occupational classification	% Voting, Sweden 1960	% Voting, U.S. 1968	% Voting, U.S. 1976	Difference between Sweden and U.S. 1968	Difference between Sweden and U.S. 1976
Propertied middle class (Professional, managerial, farm owners)	89.6	81.7	76.9	−7.9	−12.7
Nonpropertied ("white-collar") middle class*	90.4	77.4	70.4	−13.0	−20.0
Subtotal: middle class	90.0	79.9	73.6	−10.1	−16.4
Craftsmen, foremen**	93.3	66.9	58.0	−26.4	−35.3
Workers (including farm laborers)***	87.0	57.5	46.7	−29.5	−40.3
Working class	87.4	62.0	52.3	−25.4	−35.1
Total turnout from survey	86.6	67.8	59.2	−18.8	−27.4

*The occupational categories are not quite the same in the two countries. Nonpropertied "white collar" (salaried employees) includes two categories in the U.S. and three in Sweden.

**In the U.S. this has been expanded in this table to include also service employees. For the craftsman-foreman category alone in the U.S., the turnouts were 68.6 percent in 1968 and 56.1 percent in 1976, with differences with Sweden of −24.7 in 1968 and −37.2 in 1976.

***In the Swedish sample, this includes two categories: all noncraftsman blue-collar workers and farm laborers. In the American case, it includes three categories: operatives, laborers (except farm), and farm laborers.

Note: The Swedish survey contains a substantial category of people not in the labor force ("Independent, members of families"), with a turnout rate of 75.5 percent, reducing the overall to 86.6 percent.

Sources: Sweden, 1960—*Statistisk Arsbok for Sverige* (1964), p. 380. Based on a 1/30 sample of total voting-age population.

United States, 1968—U.S. Bureau of the Census, Current Population Reports, series P-20, no. 192 (December 2, 1969), *Voting and Registration in the Election of November 1968*, p. 22; 1976—Ibid., series P-20, no. 322 (March 1978), *Voting and Registration in the Election of November 1976*, pp. 63–64.

in elections. Among white males, 6.0 percent of the propertied middle classes reported that they never voted, compared with 13.5 percent of the subaltern middle classes ("white-collar" salaried workers), 20.2 percent of craftsmen and service workers, and fully 30.4 percent of people in lower working-class occupations.[12]

Two other points should be made about the data in Table 1. Turnout patterns in Sweden and a number of other continental European countries have been not only very high but largely invariant from election to election in recent decades. This corresponds not only to the capacity of Social Democrats and other parties and organizations to penetrate and mobilize the lower classes, but also to the existence of two sets of rules of the electoral game which facilitate mass participation in elections. The first of these is automatic state enrollment of voters, a practice universal in one variant or another in every Western democratic system except for the United States. The second is proportional representation, which essentially eliminates "wasted votes" and the potential turnout-depressing effects of "safe" seats in first-past-the-post representation systems like the American or the British. In the United States, on the other hand, the combination of personal registration and first-past-the-post is no doubt an important set of contributing factors in depressing the participation rate. But it should be noted that their contribution is marginal at best, as we shall see below. Moreover, there has been a notable relaxation of barriers to the franchise since the early 1960s, not in the Southern states alone but in the whole country. Yet this has not been associated with increased participation (except in the South), but on the contrary with major contemporary declines—declines which, as Table 1 reveals, are moreover concentrated among those lower classes who already participate least. What one finds in the contemporary United States, then, is turnout which is not only low but highly variable. The primary explanation for this must be found not in structural factors (which in any case are not politically uncaused causes) but in politics itself. What is it about? How relevant is electoral politics in the United States to the latent or manifest needs of the lower classes?

As with class, so with age. It is well known that younger voters in the United States participate much less than those of middle age. But this is not an inexorable law. A review of American turnout rates in the nineteenth century makes it clear that class differentials in participation then must have been much closer to the contemporary Swedish example than to the current American situation. The same considerations apply to age stratification in voting participation: if your authentic aggregate turnout rate is 90 percent or better, it is trivial to show that the participation inputs of the youngest age cohorts must be very close to those of middle-aged people. An obvious comparative case in point involves two recent elections held at the same time (1972) in the United States and West Germany, both of which had enfranchised 18-year-olds since their last general elections. The data in Table 2—even though the age brackets are not identical in the two countries—speak for themselves. In Germany, the ratio between the participation rate of the 18-to-20-year-old age group and the highest-turnout group (50–59) is 1.11 to 1. In the United States, the ratio between the participation of the newly enfranchised and the highest-participation group (45–54) is 1.47 to 1.

Here again we can observe a notable penetration-mobilization gap in American politics which does not exist on anything like a comparable scale elsewhere. With an aggregate turnout rate of 91 percent, the maximum age-cohort gap in German turnout is less than 10 percent. With an aggregate survey turnout rate of 63 percent (a 7 percent overreporting, it should be noted), the maximum age-cohort gap in American turnout is nearly 23 percent. This may be taken in conjunction with the age-stratified data in *The Changing American Voter,* which stresses the magnitude of rejection of political parties among the youngest age cohorts.[13] When it is, we conclude that in addition to a massive failure in political *incorporation* of the lower classes, there is an almost equally massive failure in political *socialization* of the young into participation in the American electoral process.

Table 2. Turnout rates by age: United States 1972, and West Germany, 1972

Age group	United States			Age group	West Germany		
	M	W	Total		M	W	Total
18–20	47.7	48.8	48.2	18–20	85.0	84.3	84.6
21–24	49.7	51.7	50.7	21–24	83.9	85.0	84.4
25–29	57.6	58.0	57.8	25–29	87.6	88.8	88.2
30–34	62.1	61.7	61.9	30–34	90.4	91.3	90.8
35–44	65.9	66.7	66.3	35–39	92.4	92.3	92.3
				40–44	93.4	92.8	93.1
45–54	72.0	69.9	70.9	45–49	94.5	93.5	93.9
				50–59	95.2	93.3	94.1
55–64	72.4	69.2	70.7	60–69	94.5	92.2	93.2
65–74	73.1	64.3	68.1				
75 and over	65.9	49.1	55.6	70 and over	90.2	83.3	85.9
Total	64.1	62.0	63.0	Total	91.4	90.2	90.8
Actual turnout (percent of potential electorate)			55.7				91.1

Sources: United States—U.S. Bureau of the Census, Current Population Reports, *Population Characteristics,* "Voting and Registration in the Election of November 1972," series P-20, no. 253 (October 1973).
West Germany—Statistiches Bundesamt, *Statistisches Jahrbuch 1975 für die Bundesrepublik Deutschland,* p. 143.

III. The Evolution of Voting Participation in the United States

The Nineteenth Century. In general, as we have suggested, there is a great and irreducible difference between American participation rates in the nineteenth and twentieth centuries. Samuel Huntington and J. R. Pôle, among others, have pointed out that the United States was the first major Western country to adopt substantially universal male suffrage.[14] A major comparative distinction exists between the situation in the United States in this period and in most other Western political systems. In the latter, it was not until the period 1885–1918 that full enfranchisement of the adult male population was accomplished. In all cases the struggle to expand or to contain the franchise was intensely, explicitly political. It was closely linked with the emergence of socialist mass movements whose first demand, naturally, was for class equality in political participation.

As the European franchise was broadened under this kind of pressure, the simple, often dualistic, competition between middle-class "parties of notabilities" was immensely complicated by the entry of socialist and in some cases—notably Ireland and Austria—nationalist mass movements. The transition to the European party system which we take for granted today was by no means a smooth or gradual process. In some countries (notably Germany, Italy, and Spain) it was derailed by fascist counterrevolutions. In Britain it led to a highly volatile multiparty system which came to pivot on the Irish question, and toward a civil war which was averted at the last moment by the international explosion of August 1914. This problem was settled (at least for the next fifty years) only by the separation of southern Ireland from the British polity in 1921.

In the United States, on the other hand, matters were far different. First, as Pole has demonstrated, the proportion of adults legally able to participate in electoral politics was generally very large even in the colonial era, and especially so by contemporary British standards.[15] Secondly, the period from 1818 to 1830 was marked by a very rapid elimination of old property barriers to universal white male suffrage in virtually all states where these had existed except for Louisiana, Rhode Island, and South Carolina. This "constitutional revolution" was not associated with the emergence of mass movements hostile to the foundations of the economy or the state. It occurred very early, at a time when four-fifths of the population lived in rural areas. Moreover, with the sole if important exception of the issue of black participation in the electoral system, the reform was not only speedily carried through but almost instantly won universal acceptance. Even in the antebellum South this was very largely the case. So long as the institution of slavery was not threatened by organized electoral movements elsewhere in the country, participation and the closeness of interparty competition reached heights fully comparable with those found elsewhere at the same time. If, after 1860 and 1877, Southern cultural norms turned away from democracy, it seems fair to say that this happened because there was no way to preserve it without accepting the presence of large numbers of blacks in the active electorate.

These differences are largely explicable by the difference between what Louis Hartz calls the American "fragment culture" and the much more complex European "matrix."[16] This "fragment culture" has enjoyed a consensus over the nature of the political economy, the organization of the political system, and the place of religion in public life which was and is quite absent in any European context, even the British. Put another way, the United States is the most conspicuous example of *uncontested hegemony*. If one were to find a single term which might describe this hegemony most concisely, it would be *liberal capitalism:* a consensual value system based on support for private property and its accumulation on one hand, and on an appropriate form of political democracy on the other. The primordial value which this consensus enshrines is that of *self-regulation* extended right down to the individual level; hence, for example, the enduring vitality of Western movies in the popular culture. In Europe, on the other hand, the hegemony of the middle classes (the bourgeoisie) has been anything but uncontested. They have had instead to fight a two-front war first against organized traditionalist social forces (the church, the nobility, to a large extent the peasantry), and subsequently against a socialism which seeks the speedy or eventual replacement of capitalism altogether.

These differences are fundamental and have enormous ramifications, to some of which we will return at the end of this paper. With regard to democratization and the suffrage, it is enough to make a few preliminary points. First, in the United States there was a paucity of traditionalist social forces which were threatened by, and resisted, mass enfranchisement. Second, unlike the situation in Europe, the propertied classes in the United States accommodated themselves both speedily and effectively to the democratic dispensation, since it rapidly became clear that the democracy which came into being offered no serious threat to themselves or their property. Third, substantially full democratization occurred—radically differently from Western Europe—well before the onset of urbanization and the concentrated-industrial historical stage of capitalist development. As we shall see, this was to create a sensation of acute vulnerability among the propertied classes of the industrial Northeast by the end of the century.

Fourth, as Hartz suggests, the absence of effective organized opposition to the liberal-capitalist consensus involved a significant blurring of ideas and of alternatives. So far as popular participation was concerned, we have already noted the relative speed and ease with which property-holding barriers to the ballot box were removed at the beginning of the century. By its end, the growth of huge ethnically polyglot urban "masses" led to the creation of personal registration legislation—ostensibly, at least, to combat corruption, i.e., stuffing the ballot box. This legislation was not only created, but varied extremely widely in its incidence and effects (like all other electoral legislation) from state to state and locality to locality. Moreover, it too was accepted consensually: it is very hard to find any critical literature of political science on this subject until as late as the 1960s.

In Europe, Canada, and other democratic political systems, on the other hand, the burden of registering electors is universally borne, in one way or another, by the state. There are two obvious reasons for this. First, the epic struggles over the state, lasting several centuries, had produced a "hard" or "internally sovereign" state with elaborate and prestigious bureaucracies. The sheer physical competence required for the state to enroll voters was in place. Second, these struggles among social groups (collectivities) produced great clarity about the political uses to which electoral legislation could be put by those in power, and an insistence that individuals (especially of lower and dependent classes) not be burdened by cumbersome procedures which might easily lead to their abstention. In the United States, on the other hand, it is very difficult to escape the impression that personal registration came into being as something beyond organized political criticism because it fit so well with the underlying, hegemonic ideology. The other side of the individual-self-regulation coin is individual responsibility for one's existence: if the individual wants to vote, it is up to him to prove to the electoral officials that he is legally qualified to do so.

Finally, it can be speculated that political parties could develop their organizational capacities to penetrate and mobilize an electorate in such a system only up to a point. Maurice Duverger points out that American parties have a "very archaic general structure" compared with newer types of mass parties in the European context.[17] If so, an essential reason for this is that these newer parties reflect a *collectivist* organizational capacity which cannot strike root in the United States so long as the consensual ideology of liberal capitalism retains its uncontested hegemonic position. So long as the demographic structure and the political economy were of American nineteenth-century type, the traditional major parties could and did mobilize very effectively indeed. In the radically different social and economic conditions of today, their mobilization capacity would almost inevitably be seriously impaired: neither their organizational form nor their leadership are geared to do so, and the cultural conditions for a serious effort along those lines do not exist.

Returning to our discussion of nineteenth-century participation, the most notable point about it is, in general, its rapid growth with the growth of organized, mass-based democratic politics in the 1830s. One case among many which could be chosen is Connecticut's presidential vote between 1820 and 1844 (Table 3). The process of growth in this period took extremely diverse forms from state to state; these matters need not detain us here. By the early 1840s a moderate to very full mobilization had occurred in most states. While the conduct of elections also varied to a considerable extent, it is possible to give a modal description. Essential to the new parties of this period was their use of a variety of devices to get their voters to the polls. Throughout most of the nineteenth century, party identification was obviously much more complete throughout the voting population, and much more intense, than it is today. This meant, typically, that parties could not expect very many conversions from already active opposition voters, and that they could not count upon a large pool of free-floating independent voters. Their propaganda, torchlight parades, and other activities were designed to

Table 3. Classic mobilization in the Jacksonian period: The case of Connecticut (Percentage of potential electorate)

Year	Voting	Nonvoting	Jacksonian (D)	Opposition	Other
1820	7.6	92.4	6.4 (Monroe)		1.2
1824	14.9	85.1	—	11.8 (Adams)	3.1 (Crawford)
1828	27.2	72.8	6.6	20.6 (Adams)	—
1832	46.0	54.0	16.0	25.2 (Clay)	4.7 (Wirt)
1836	52.3	47.7	26.5 D	25.8 W	—
1840	75.7	24.3	33.6	42.0	0.1
1844	80.0	20.0	36.9	39.1	2.4 (Liberty)

secure a maximum turnout of their own voters, since that was normally the way one won elections. This accounts, among other things, for the creation of Tammany's Naturalization Bureau in 1840, since it was already clear to this organization that most immigrants, if naturalized in time for the election as required by New York law, would vote the Democratic ticket. The resultant turnout rates in closely contested states like New York throughout this period, or Indiana from 1860 on, were truly awesome. Similarly, when (as in 1848 or 1872) turnout sharply dropped, there were usually clear-cut partisan-political reasons for the drop. These often involved a rupture of one of the parties into competing factions or rejection of a party's nominee by many rank-and-file supporters (the latter being particularly the fate of Horace Greeley among many Democratic voters in the 1872 election).

In this period there were, effectively, no personal registration statutes or other significant barriers to the franchise in most states. What amounted to freehold or property qualifications survived in Louisiana and South Carolina until the Civil War, and in Rhode Island until 1888. In a few states (Massachusetts, Connecticut, and Rhode Island), anti-Irish sentiment led by the American (Know-Nothing) Party resulted in the adoption of literacy qualifications in the mid-1850s; in Connecticut they do not appear to have been effective in curbing turnout.

Similarly, the conduct of elections was very largely a party matter. Printed tickets were typically produced by the parties and distributed to their voters—initially before the election, but toward the end of the century on election day near the polls. Such tickets, of course, immensely simplified the act of voting for the plethora of candidates and officers which are elective in our complex constitutional system. They could be and often were "scratched" by voters dissatisfied with one or more of the candidates on the printed ticket. It was possible in effect to cast a "split ticket" under this system, but in general it facilitated straight-ticket voting as a norm which reflected the intense partisanship of the age.

In a great many jurisdictions, the act of voting itself was at best only semisecret: often, separate ballot urns were maintained for the supporters of each candidate. Obviously, this kind of system could and did lend itself to abuse on occasion. But this typically occurred when a party organization was willing to terrorize or bribe officials belonging to the opposition (as well as the police). This sort of thing was episodic. Well-known cases include the Plaquemines Parish frauds of 1844, which helped Polk win the presidency, and the New York City frauds of 1868, when the Tweed machine produced turnouts of well over 100 percent in many election jurisdictions and helped Seymour carry New York State by exactly 10,000 votes—it was said at the time—in order for a leading city politician to win an election bet! That these cases were episodic rather than the norm is suggested not only by data analysis, but by the very fact that they were notorious "horrible examples" to contemporaries. Moreover, the Philadelphia machine, among others, was to

prove in the twentieth century, under the new reform dispensation, that it
was perfectly capable of producing occasional turnouts in excess of 100 percent in order to get its way. The Pendergast machine in Kansas City proved
the same point in the 1936 elections.[18] Normally, it would seem, the system
worked without excessive ballot-box stuffing because in the rural areas of the
North and West people knew each other and because in the cities the two
parties had every incentive to watch each other. Moreover, each organization's cadres had an extremely precise idea of just how many "troops" the
other "army" had.

In general, turnout rates reached their historic highs in the United States
during the last quarter of the nineteenth century. The broad reason for this
is pretty obvious. The most traumatic collective experience through which
Americans have ever passed was the Civil War (and, for Southerners, the
Reconstruction which followed the war). In many respects it was, as Barrington Moore has argued, the "last capitalist revolution" in modern world
history,[19] and it partook of many of the characteristics of revolution—not
only in the South, it may be added. This war produced a much higher tax of
men than did World War II. Using as a benchmark the total male population
21 and over, about two-fifths of that figure served in the Union army, and
slightly more than half of it in the Confederate army, compared with about
one-third of this age and sex group during World War II. Even more dramatic were the differentials in casualty rates: 18.0 percent of those who
served in the Union army died during the war, while another 13.8 percent
were wounded but apparently survived. The Confederate toll was still more
ghastly: evidently about 34 percent of those who served in that army died
and another 26 percent were wounded, an overall casualty rate of about
three-fifths. By comparison, only 1.1 percent of all those who served in
World War II died, and another 1.6 percent were wounded.

It seems quite accurate to say that the most important collective task ever
performed by our party system was, first, the creation of the political preconditions for this conflict and, second, its political management.[20] In view
of the magnitude of this trauma and the fundamental issues at stake in it, it
is hardly surprising that in the North the quality and quantity of party organizational effectiveness reached levels in the decades of the 1860s, 1870s, and
1880s never seen before or since, or that overall turnouts there also reached
heights never seen before or since. The trauma in the South was not only
vastly more massive in terms both of lives and property than in the North,
but involved the victors' requirement of incorporating the just-freed black
population into the electorate. By precisely the same token, but in this highly
specialized—almost "Third World"—context, its effects were to destroy any
long-term stable basis for electoral competition, and to promote ultimately
highly successful efforts by the surviving white population to get rid of the
region's black voters.

To this must be added another concomitant of the Civil War period: the shift toward urbanization and concentrated industrial capitalism which achieved a temporary maximum velocity during the 1880s. It is not possible in the confines of this paper to review in any detail the complex issues which this transformation in the foundations of human existence created. It is enough to sketch a few points. (1) Devotion by most Americans to the premises of our uncontested hegemony—self-regulation and "business" in the broader sense of the term—continued after the war as it did before. In politics it manifested itself in the increasingly machine-dominated character of party organizations, fueled both by rapid increase in the concentrated power of money and by the growth of dependent immigrant urban populations, which were used effectively as electoral cannon fodder. (2) The growth of the "new" capitalist and the "new" political boss, along with the "new," i.e., post-1882, immigration, led to a massive sense of loss—in cultural values, prestige, and status—among older-stock middle-class elements. The growth of the forces promoting this alienation eventually produced a reaction with major consequences for the electoral mechanism and for representation itself. (3) The shift toward new industrial-urban concentrations of power produced not only much the same "cultural" or "social issue" processes among farmers, but massive-economic dislocation. In very large part this was a by-product not only of the subordination of the country to the industrial-capitalist city, but quite directly of public policy. In addition to its other revolutionary aspects, the Civil War had left an enormous public debt behind, and this had largely been funded (as in all total wars) by inflation. The twin objectives of all Treasury policy from 1865 to the 1890s were to shrink the money supply and return to an implicit gold standard—achieved by 1879—and to retire the debt. The result was a generation-long deflation which particularly hurt all who were debtors and undercapitalized, especially the farmers of Western and Southern cash-crop areas. The stage was thus set for a massive agrarian insurrection against the new capitalist order—and it should be noted that an absolute majority of the work force was employed in agriculture until about 1878–79, and very close to a majority was so employed until after 1900.

Let us now examine the general pattern of electoral participation during this crucial era of socioeconomic transformation, and examine these patterns somewhat more microscopically, in New York State. Table 4 provides the mean presidential and off-year congressional turnouts by state for the periods 1874–92 and 1952–70.

Beginning first with an examination of the 1874–92 array, several things can be noted immediately. In the first place, the most densely populated states were on the whole those with the highest turnouts.[21] Among the fourteen whose turnout rates ranged from 80.1 to 92.7 percent (roughly equivalent to a range of participation from France to Italy in contemporary Europe), only two (Iowa and Kansas) could be described as overwhelmingly rural. Of the top ten cities in the United States as reported by the 1890 census, eight were located in states whose mean presidential participation was 80 percent or greater in this period. Second, the Southern shift toward very low rates of participation was clearly visible (six of the eleven states in the lowest quartile), but was not yet consolidated. Three of the ex-Confederate states (Florida, North Carolina and Virginia) had presidential participation rates in excess of 75 percent. Third, outside the South, as the "center" area of the Northeast and near Midwest had in general the highest volume of participation, so the "peripheries," and especially the frontier areas, tended to have the lowest turnout rates. Finally, the drop-off in this period between mean presidential and mean off-year congressional turnout rates was relatively tightly constrained, especially in the Northern and Western states,[22] one of many indicators which reveal the relatively intense partisanship and commitment to participation which existed in that period.

On the whole, then, the most socioeconomically advanced or "developed" parts of the United States—concentrated largely in the "metropole" of Northeast and near Midwest—showed the highest mobilization levels. Additionally, as has been suggested before, turnout rates which are in excess of 80 percent or so—and especially those in the high 80s and low 90s—constitute presumptive evidence that all social-occupational classes in a given area tended to participate at about equal (and high) rates; the same may be said so far as cohorts are concerned. It has long been observed that class cleavages in American elections, once one disentangles ethnic variables, probably did not autonomously exist prior to about 1934. Their emergence was one of the things the New Deal realignment was all about. Similarly, it seems correct to say that before the first decade of the twentieth century—and in some urban jurisdictions like Chicago and Philadelphia for a very long time thereafter—class differentials in voting participation tended to be either small or nonexistent.

Quite a different pattern emerges when we compare that mobilized universe with that of the 1952–70 period—the latter chosen to exclude the effects of the deep drop in participation which has gone on during the 1970s. The first global observation which can be made is that mean levels of presidential turnout in this contemporary period are below participation rates in off-year congressional elections during the 1874–92 period. In this connection, we note that drop-off rates between presidential and off-year elections are uniformly higher in every comparable state except for the very special case of Rhode Island. In some cases they are very much higher (especially in the Southern states), and this despite a major decline in the outer size of the active electorate as measured by residential election turnouts. If participation is much lower in this contemporary period's presidential elections than was the case before 1900, the decline in off-year turnouts is both absolutely and proportionately greater still. In view of the fact that the electoral linkage system has been still further fragmented during the past generation by a parade of states moving their gubernatorial elections to four-year terms *fall-*

Table 4. A mobilized electorate: Mean turnout in presidential and off-year congressional elections

Rank (1874–92)	State	1874–92			1952–70		
		President	Off-year congressional	Mean drop-off	President	Off-year congressional	Mean drop-off
1	Ind.	92.7	83.5	9.9	73.5	59.9	18.5
2	N.J.	92.2	76.6	16.9	69.6	49.7	28.5
3	Ohio	92.1	76.8	16.6	67.1	48.7	27.4
4	Iowa	91.8	73.4	20.0	72.9	49.0	32.8
5	N.H.	89.7	83.0	7.5	74.7	53.8	27.4
6	N.Y.	89.0	68.4	23.1	66.0	49.5	25.0
7	W.Va.	87.5	72.6	17.0	74.2	50.5	31.9
8	Ill.	86.1	67.4	21.7	73.4	54.2	26.2
9	N.C.	84.5	67.3	20.4	51.8	28.8	44.4
10	Conn.	82.8	69.7	15.8	74.8	61.2	18.2
11	Pa.	82.7	70.5	14.8	66.7	54.8	32.8
12	Md.	81.4	65.3	19.8	56.1	37.7	32.8
13	Wis.	81.3	64.4	20.8	69.6	50.1	28.0
14	Kans.	80.1	66.6	16.9	67.6	50.3	25.6
15	Mo.	78.2	64.9	17.0	68.7	42.3	38.4
16	Fla.	76.6	67.9	11.4	49.4	25.1	49.2
17	Va.	76.4	54.2	29.1	37.6	22.5	40.2
18	Ky.	76.3	45.6	40.2	56.8	30.7	46.3
19	Del.	76.1	65.8	13.5	72.7	56.1	22.8
20	Mont.	74.2	70.0	6.0	70.8	61.9	12.6
21	Nev.	73.9	73.4	0.7	62.5	50.9	18.6
22	Cal.	73.4	65.1	11.3	65.9	53.1	19.4
23	Vt.	73.3	56.2	23.3	67.2	53.5	20.4
24	Me.	73.2	70.1	4.2	66.1	49.2	25.6
25	Tenn.	72.7	54.9	24.5	49.1	28.0	43.0

26	Tex.	72.3	54.9	24.1	43.2	20.5	52.5
27	Mass.	71.8	58.2	18.9	71.7	54.7	23.7
28	Mich.	71.5	63.3	11.5	68.7	51.3	25.3
29	S.D.	70.7	80.4	13.7 Acc	75.3	61.5	18.3
30	Minn.	70.3	60.3	14.2	72.8	58.5	19.6
31	Wash.	67.3	48.2	28.4	69.4	50.2	27.7
32	Neb.	66.1	55.2	16.5	67.6	50.3	25.6
33	Ore.	64.9	70.2	8.2 Acc	69.0	55.0	20.3
34	Idaho	63.1	67.0	6.2 Acc	76.4	63.0	17.5
35	Ala.	62.2	47.0	24.4	34.4	25.4	26.2
36	Ark.	61.4	34.9	43.2	44.3	36.9	16.7
37	S.C.	58.2	54.8	5.8	34.0	19.9	41.5
38	N.D.	56.6	66.4	17.3 Acc	73.8	55.4	24.9
39	Col.	55.4	51.4	7.2	69.7	52.9	24.1
40	La.	54.6	49.7	9.0	44.7	17.6	60.6
41	R.I.	52.5	26.4	49.7	73.0	59.0	19.2
42	Ga.	48.9	30.6	37.4	37.2	19.5	47.6
43	Miss.	48.3	35.3	26.9	31.6	16.7	47.2
44	Wyo.	47.7	49.1	2.9 Acc	70.6	60.5	14.3
	Admitted since 1892:						
	Ariz.	—	—		52.6	41.1	21.9
	N.M.				60.9	50.9	16.4
	Okla.				63.3	40.9	21.2
	Utah				78.4	63.1	19.5
	Alaska				53.4	43.6	18.4
	Hawaii				55.0	47.4	13.8
	Subtotals						
	North and West	85.4	70.8	17.1	68.5	52.6	23.2
	Non-South	84.5	69.3	18.0	68.0	51.2	24.7
	Border	79.1	60.3	23.8	63.2	39.7	37.2
	South	65.6	49.9	23.9	42.6	23.3	45.3
	Total U.S.A.	78.5	64.7	17.6	62.0	44.5	28.2

ing in these off years (the most recent case being Illinois), this growing trough in off-year participation is not without political significance.

Another major change can be seen in the distribution of states in the top and bottom quartiles. In the 1952–70 period the bottom quartile is composed of a solid phalanx of the eleven ex-Confederate states plus Arizona and Alaska. While Southern turnout in this period was markedly higher than in the years intervening between 1900 and 1952,[23] the Southern consolidation which V. O. Key definitively analyzed in his *Southern Politics* remained.[24] The top quartile shows equally profound changes. Connecticut and Rhode Island give the only Northeastern representation; Illinois is on the margin of this quartile and if included gives the sole near-Midwest representation. All the rest are found in the Plains, Mountain, and (with Oregon) Pacific states. A useful way to look at the shift away from concentrated metropolitan populations toward sparsely populated and peripheral areas is to examine the number of representatives found in the top-quartile states in each era from 1874 to the present.

The Disappearing American Electorate, 1986–1930: Social Crisis and Political Response. The late-nineteenth-century electorate we have been describing was, as we have seen, quite fully mobilized. Its participation was, essentially, the fruit of several generations of effective work by America's traditional party organizations. But it presented dangers for elites, chronic and growing irritation among old-stock middle-class elements who saw themselves between the hammer of corporate giantism and the anvil of machine corruption, and serious impediments to the realization of technocratic and bureaucratic ambitions. The great socioeconomic crisis of the 1890s appeared to demonstrate that this electorate was not only large but might be a mortal danger to our emergent hegemonic interest.

The point has been made before that the United States was unique in that it had a fully operating set of mass democratic institutions and values before the onset of industrial-capitalist development.[25] In every other industrializing

Table 5. The long-term shift away from the metropolitan: Number of representatives in states in top turnout quartile, 1874–1976 (Presidential elections)

Period	Range of turnout, top quartile	N of representatives	% of total	Mean N of reps per state
1874–92	82.7–92.7	158	44.4	14.4
1894–1910	81.5–91.7	154	39.4	14.7
1912–30	67.7–77.8	120	27.6	10.0
1932–50	71.0–78.4	90	20.7	7.5
1952–70	72.8–78.4	77	17.7	6.4
1972–76	62.0–69.7	46	10.6	3.8

nation of the 1850–1950 period, "modernizing" elites were effectively insulated from mass pressures. In most of the West this insulation was accomplished by the *regime censitaire* or other formal devices for keeping lower
classes out of electoral politics, supplemented in some cases by ingenious
trasformismo-type manipulations and corruptions both of electorates and of
legislatures. In the Communist world, insulation involved dictatorship and
the use of elections as devices of acclamation for the regime rather than
choice. In today's so-called Third World, a mixture of the two—coupled
with the entry of the army where "necessary" as an ultimate insulation control—generally prevails.

Since the processes of capitalist development have always been exploitative, harsh, and even brutal, it is not surprising that so wide a range of devices
has been employed to prevent the "modernizing elites" overthrow by an out-
raged population, and that the quest for such insulation has been so universal.
This was a problem which had particular urgency in the American context,
precisely because of the relative breadth of popular participation at all times,
and because so many of the people participating during the country's first
century were agrarians. Alexander Hamilton's whole policy, from beginning
to end, betrays preoccupation with an overriding question: how can one find
a legitimate way to win mass acquiescence for measures designed to promote
the creation of what Madison, Monroe, and others quite correctly called
"empire" and to accumulate movable capital? It is hardly surprising that Jefferson suspected Hamilton of harboring dictatorial designs. It is still less surprising that the turn-of-century Progressives—particularly those of a techn-
ocratic-corporatist bent—revered Hamilton as their hero and rejected
Jefferson as a tribune of all the backward (agrarian) forces in American society.[26]

How, then, does one tame a fickle, numerous, and dangerous electorate
without outright overthrow of the engrained democratic tradition in American politics? It is more than doubtful that any of our corporate elites or the
organic intellectuals and politicians who supported them at the turn of the
century ever put the question in such nakedly self-conscious terms. Still more
doubtful is it that what happened to participation and the nature of the party
system in this period was the result of a diabolically clever conspiracy among
elites who knew just what they wanted and how to get it—though particular
or localized conspiracies abounded in this period, chiefly in the South, where
the stakes were higher because the social structure was so primitive. But this
was the underlying political problem nevertheless.

The questions to ask ourselves, then, are these: What was done? Against
whom was it done? What was the convergence of interests which made it
possible to do it at all? What were the effects of what was done? And finally,
why was what was done so rapidly and fully legitimated that all of it
remained beyond criticism for more than a half-century and much of it
remains with us still?

We may begin by identifying schematically the major "target" groups of legislative changes affecting elections and their opponents (see Table 6).

Whether or not elites "conspired," the patterns of political change which emerged during the first third of this century functioned as though its proponents and beneficiaries agreed on a number of points which can scarcely be called in accordance with participatory-democratic political theory.

(1) The greater the preliminary hurdles which a potential voter had to cross in order to vote, the fewer voters there would be.

(2) These hurdles would be easily cleared by people in the native-stock middle classes, would not exist at all for small towns and rural areas, would be serious impediments to lower-class voting, and would be virtually impassable for Southern blacks.

(3) Political parties were dangerous anachronisms when it came to giving cities efficient, expert corporatist government. They were to be abolished where possible (especially on the urban level), and were elsewhere to be heavily regulated by the state. In particular, their monopoly over nominations

Table 6. The sociopolitical thrust of electoral reform

Target groups	Opponents	Rationale	Sanctions
1. Urban machines	Middle-class WASPs (status anxiety) Technocratic progressives Upstate agrarians	Anticorruption Inefficiency "Sodom and Gomorrah"	Ballot reforms; special prosecutors; nonpartisanship malapportionment, discriminatory registration laws
2. Political parties	Progressives; many of the above	Unrepresentativeness; "old politics"; corruption	Detailed legislative regulation; direct primaries
3. Immigrants	Agrarians (upstate in NY); small-town and metropolitan middle-class WASPS	Not adequately culturally developed; "Romanism"; "Anarcho-Communism"; lazy, drunk, etc.	Elimination of alien voting (Midwest and West); personal periodic registration for cities only; literacy tests
4. Populist agrarians (North and West)	Corporate elites, urban/town middle classes	Danger of "revolution" by the "backward"	Electoral activity plus some corruption and "pressure" on dependent voting population
5. Populist agrarians (South)	Elites (regional and local), Democratic organization	Same as above	Same, but massive use of fraud and force, and sanctions below
6. Blacks (South)	Southern white progressives; gradually, ex-Populist white agrarians	Racism; "traditionalist Southernism"; anticorruption	Violence and terror; later largely replaced by poll tax, literacy tests, "white primaries," grandfather clause, etc.

would be stripped from them, so that "the people" (functionally between 5 and 35 percent of the potential electorate, as a rule) could choose candidates in direct primaries.

(4) Close partisan competition was generally undesirable, though for different reasons in the South (where a quasi-absolute prohibition of party competition developed) and in the North and West. Desirable or not, it was substantially abolished down through 1932 not only in the South but in vast reaches of the North as well, extending from Pennsylvania to California.

The basic legal devices which were adopted—particularly the device of personal registration—without question contributed to the massive decline in voter participation after 1900. In the extreme case of the South, the result was that by the mid-1920s presidential turnout had declined to about 18 percent of the potential electorate, while in the off-year congressional election of 1926 only 8.5 percent of the South's potential voters actually went to the polls. Participation fell in the North and West as well, but very unevenly. How much of this was attributable to these rules-of-game changes? In an earlier study I concluded that in the period prior to the enfranchisement of women (nationally adopted in 1920), not more than one-third of the decline could be attributed to these causes alone.[27] It is possible from recent Ohio data to give a more precise estimate of the effect of introducing personal-registration requirements on the participation rate.

Ohio is a state, like many others, where the basic electoral law prescribes personal registration in counties of a certain size, and allows local option as to whether to adopt it or not for all other counties. In some counties, registration applies only to one or two small towns, with the rural areas having no such requirement for voting. In the period from 1960 to 1976, a significant number of counties have shifted from nonregistration to registration status. It therefore becomes possible to derive a relatively good estimate of the effects of introduction by comparative analysis (Table 7).

Simple counterfactual manipulations point toward a fairly well-defined range. Taking the transitions in each group but subtracting the year-on-year

Table 7. Turnouts in Ohio counties, 1960–76

Registration began	N of counties	Turnout 1960	1964	1968	1972	1976
Before 1960	28	69.0	64.0	62.3	57.7	57.1
Before 1964	5	75.5	60.4	57.9	54.6	53.8
Before 1968	4	77.7	72.1	63.3	57.9	58.3
Before 1972	10	79.2	72.3	69.3	59.9	59.0
Before 1976	5	79.5	72.7	70.6	63.7	57.1
Never	23	79.8	73.0	70.0	65.8	71.4

Note: Does not include counties with partial registration coverage during this period.

turnout change for full-registration counties yields a differential of 6.1 per-cent. Another approach is a straightforward year-on-year comparison between those counties fully covered and those never covered across the period. Not surprisingly, this yields the highest general differential of 10.0 percent, though it provides no basis for estimating possible turnout differentials arising from the sheer rurality of the twenty-three counties which do not have registration as a requisite for voting. Still another way of analyzing the data in this table is to estimate the positive deviation in preregistration turnout in each category with "before" and "after" properties, utilizing the turnout data in each preceding group of full-registration counties. This produces a positive deviation of 7.8 percent, which—when one attempts such crude controls for the downward movement in turnout—seems perhaps to be the most accurate estimate of the three. It is worth noting that all of these measures except the simple comparison between full and nonregistration counties are at their maximum in 1960, the election with the highest overall turnout rate. Even more important, between 1960 and 1968 there was a substantial decline across the board, both in the consistently registered and non-registered counties, and in comparable intervening categories. In the former category this decline was 6.7 percent (or 9.7 percent of the 1960 base). In the categories "before 1972," "before 1976," and "never," this decline was 9.9 percent (12.5 percent of the 1960 base), 8.9 percent (11.2 percent of the 1960 base), and 9.8 percent (12.3 percent of the 1960 base) respectively; these, of course, were at the time all without the burden of personal registration, and prior to the enfranchisement of the 18–20 age group.

It would be reasonable to conclude that *ceteris paribus,* the creation of a fully automatic system for enrolling voters would serve to increase participation by between 7 and 9 percent. Applied to the North and West for the 1952–70 period, and correcting for the wide local diversity in registration and other statutory coverage, this might yield a turnout ranging between 72 and 74 percent instead of the 68.5 percent actually recorded. This still leaves us far short of the 85.4 percent turnout of the 1874–92 period, with between three-fifths and two-thirds of the difference left unaccounted for. Moreover, *ceteris* is anything but *paribus* in the real world. Despite the considerable liberalization of electoral law during the 1970s, the mean turnout rate for 1972–76 fell to 58.5 percent in the North and West, with many states showing turnout levels at or below the all-time national lows achieved in 1920 and 1924.

Viewed particularly over the long run, the same considerations apply very largely to expansions of the size of the electorate—notably the enfranchisement of women, which was legally completed in 1920. There is no doubt at all that the enfranchisement of women had profound effects on the turnout rate, but virtually no autonomous effect on the partisan distribution of outcomes. We may take one case in point for which data during the transition period was exceptionally precise: Chicago, in which women were first enfranchised in time for the mayoral election of 1915 (Table 8).

Table 8. Registration and voting by sex: The case of Chicago, 1892–1932
(Percentage of potential electorate)

Year and office	Registered Male	Registered Female	Voting Male	Voting Female
1892 Pres.	(Registration not adopted until		84.5	—
1896 Pres.	1902)		94.3	—
1900 Pres.			87.5	—
1904 Pres.	87.8	—	78.9	—
1908 Pres.	83.2	—	75.4	—
1912 Pres.	82.0	—	70.3	—
1915 Mayor	81.2	48.6	72.5	43.1
1916 Pres.	82.7	51.1	77.1	47.3
1919 Mayor	74.7	46.5	65.1	38.7
1920 Pres.	80.1	49.2	71.5	43.2
1920 Pres.	81.2	52.4	73.4	46.5
1928 Pres.	80.5	64.2	74.5	59.1
1932 Pres.	84.4	66.6	80.0	61.5

Note: Chicago elections, 1915–20, were unique in that the ballot boxes were segregated by sex. The basic voting data, so segregated, are reported in the relevant annual volume of *The Chicago Daily News Almanac*, 1916–21.

The initial gap between male and female participation rates stand out very clearly. It lasted for more than a decade. The female turnout of about 45 percent in the 1915–24 period in Chicago contrasts remarkably with some post-enfranchisement elections elsewhere. In his pioneering study *Political Behavior*,[28] H. L. A. Tingsten reports 61.8 percent turnout among women in Sweden in 1919 (−7.1 percent compared with men), 68.1 percent in New Zealand in 1896 (−7.8 percent) and 82.1 percent in Austria, 1919 (−4.9 percent).

Two other points are worth mentioning here. First, as is well known to students of American elections, if one goes to the trouble of registering, the probability is very high that one will vote. After 1900, as Table 8 indicates, the level of registration very largely determines the subsequent level of voter participation. Second, this table captures something of the relatively immense impetus to mobilization which the extreme ethnocultural polarizations of the 1928 campaign produced. This is a well-known theme, as is the obvious importance of further female mobilization to the subsequent New Deal critical-realignment sequence. It is enough to say here, as a note of warning, that the analytic issues involved in attempting to assess the extent to which mobilization of former nonvoters was the overwhelming "cause" of this realignment are much more complex than much of the literature would imply.[29] It is of course also obvious that low turnouts *after* the 1930s (e.g., 1942, 1946, 1948, 1974, 1976) can hardly be explained by the imperfect political socialization of women. Indeed, sex differences in the turnout rate of about 30 percent in cities like Chicago and Philadelphia in the 1920s had apparently declined to 10 percent or less in the early 1950s and virtually disappeared altogether by 1976.

All of the factors which we have discussed—the adaptation of the political culture to the norms of Progressivism and, more broadly, corporatism; the introduction or procedural barriers to the franchise for certain classes of voters; the suppression of party competition in the South and its massive displacement in much of the North and West; and the enfranchisement of women under substantially depoliticized conditions—all conduced to the creation of an essentially oligarchic electoral universe by the end of World War I. Table 9 provides a synoptic comparison between the decades of the 1880s and the 1920s for the North and West. It speaks for itself.

Essentially, a political system which was congruent with the hegemony of laissez-faire corporate capitalism over the society as a whole had come into being. The era of "normalcy" was one of two noncompetitive party hegemonies: the Democrats in the South and the Republicans throughout much of the North and West. Both implicitly rested on a huge mass of nonvoters.

This implies, of course, that a "modern" American electorate had come into being, i.e., one which was heavily class-skewed in participation *in the absence of an organized (or organizable) socialist mass movement which could mobilize lower-class voters in modern industrial-urban conditions.* This was obviously the case in general. But the individual-level data for the 1920s is, naturally, extremely fragmentary. Tingsten reports one study of 1925 which focused on the small city of Delaware, Ohio.[30] This study found

Table 9. The achievement of "order": Turnouts in the North and West, 1880–90 and 1920–30

Year	President	Congress	Drop-off
1880	87.6	86.8	
1882		71.4	18.5
1884	84.9	84.2	
1886		71.6	15.7
1888	86.4	87.7	
1890		71.3	17.5
1920	56.3	53.5	
1922		42.8	24.0
1924	57.4	52.4	
1926		39.7	30.8
1928	66.7	61.6	
1930		43.7	34.5

the expected gap between classes: the local upper class, composed of corporate and banking elites, had an 86 percent turnout rate, compared with 63 percent for the "mass of industrial workers." Since personal registration requirements were not imposed there until 1965, the gap is particularly suggestive of our argument that the interaction of social change and organizable political alternatives, rather than rules changes, was primarily responsible for what happened after 1900.

On the other hand, if America did not have a viable urban socialist movement in the 1920s, it did have political machines. There are a number of cases which suggest pretty clearly a marked inversion, caused by the activities of these machines, in the usual "modern" turnout pattern. Philadelphia—both before and after the electoral reform and personal registration requirements enacted in 1906—provides a classic case in point. Table 10 presents the basic participation data between 1896 and 1930 for two groups of eight Philadelphia wards each: machine-controlled wards, called "river wards" from their location near the Delaware River, and antimachine wards, located without exception in the far and essentially suburban reaches of the city. While time and space preclude a systematic statistical analysis of these two groups of wards, it is obvious that the river wards were densely populated slum areas with high proportions of foreign-born whites, while the semisuburban antimachine wards were much less densely settled, by a population with much larger proportions of middle-class and native-stock whites than the machine wards contained.

The Philadelphia Republican machine was without question one of the most awesome organizations of its kind ever constructed. Despite the arrival of the New Deal, it remained in power locally until as late as 1951. But its control of local politics and its share of statewide control were sometimes seriously threatened. The Organization's leadership was especially challenged statewide in 1902–4, 1910, and 1926.[31] When it became necessary to manufacture votes, the Organization was capable of doing so, as the extraordinary surpluses of 1902 and 1904 reveal. This sort of thing, of course, fanned the flames of reform, which led to the adoption by the state legislature of a more or less effectively enforced personal registration law in 1906. Even thereafter, however, the presence of a powerful incentive for stimulating voter turnout in the areas it controlled produced remarkable electoral performances for this period—conspicuously in the threat-filled years of 1910 and 1926, when the river wards' turnouts ran to 30 percent higher than the statewide average. Conversely, as a comparison of turnout differences between the top machine-controlled and antimachine wards shows, these surpluses disappeared in years—especially presidential years—where no serious Organization interests were at stake (e.g., 1896, 1900, 1908, 1912, 1916, 1920). Even then, however, the overall picture which emerges is that when such interests were not at stake, turnout was largely uniform across residential areas from the slums near the docks to the mansions in Mount Airy.

Table 10. Participation patterns in machine and antimachine wards of Philadelphia, 1896–1930

A. Long-term Flow

| | Turnout rates | | | Turnout differential | |
	Eight machine wards	Eight antimachine wards	PA	Machine-antimachine	Machine-PA
Year and office					
1896 Pres.	85.4	83.0	81.8	2.4	3.6
1898 Gov.	70.9	59.1	66.6	11.8	4.3
1900 Pres.	74.0	74.9	75.0	−0.9	−1.0
1902 Gov.	105.3	68.0	67.8	37.3	37.5
1904 Pres.	118.7	72.2	74.3	46.5	42.4

(Personal registration required in cities of the third class or larger, 1906)

1906 Gov.	66.3	63.5	58.7	2.8	7.6
1908 Pres.	77.9	74.1	71.8	3.8	6.1
1910 Gov.	84.8	63.3	54.6	21.5	30.2
1912 Pres.	68.8	69.3	64.4	−0.5	4.4
1914 Gov.	69.6	62.3	56.5	7.3	13.1
1916 Pres.	69.9	70.5	63.4	−0.6	6.5

(Enfranchisement of women, 1920)

1920 Pres.	45.4	46.9	42.8	−1.5	2.6
1922 Gov.	49.0	27.1	32.5	21.9	16.5
1924 Pres.	56.3	40.8	45.8	15.5	10.5
1926 Gov. (primary)	60.1	41.5	29.9	18.6	30.2
1926 Gov. (general)	57.4	29.6	31.0	27.8	26.4
1928 Pres.	54.6	65.7	62.7	−11.1	−8.1
1930 Gov.	48.8	37.3	40.6	11.5	8.2

B. Estimated Turnout by Sex, 1924–28

Year and sex	Eight machine wards	Eight anti-machine wards	Differential
1924 Pres.			
Male	71.9	59.0	12.9
Female	34.9	24.3	10.6
1926 Gov. (primary)			
Male	75.3	58.1	17.2
Female	37.4	26.4	11.0
1928 Pres.			
Male	65.5	76.0	−10.5
Female	39.2	56.4	−17.2

C. Support for Organization Candidates, 1910–1926

Type of ward	Wards included	% For Org., 1910	1926 primary
Machine	1–5, 11, 13, 14	85.4	96.7
Antimachine	21–23, 34, 37, 38, 42, 46	38.3	52.9

Estimates of turnout by sex, based on registration data for the years involved, tell essentially the same story.

It is most suggestive that this machine-caused inversion in participation rates from the modern class-skewed "norm" was itself inverted in 1928. This was a situation of classic cross-pressure. The machine, like most, was based on dependent, largely ethnic voters. Al Smith's Catholicism created a pull on these voters which—especially because of the absence of any major local issues in that year—the Organization's leadership decided to evade as far as possible. The undertow was much more powerful in largely Italian South Philadelphia than in the river wards: boss William Vare's wife, Flora, was defeated in a state senatorial race by an obscure Democrat who rode in on Smith's coattails in what had been an impregnably Republican district. In the river wards, Coolidge had received 90.7 percent of the vote in 1924; Hoover polled only 59.3 percent four years later. The contrast between the turnout rates of the two groups of wards—and especially their movement from 1924 to 1928—is profound. The participation rate in the machine-controlled wards *declined* in 1928 from both 1924 and 1926 levels (1926, characteristically, was the peak year), while it underwent massive increase in the anti-machine wards. This was especially true of turnout among women, which in these latter wards leaped from about one-quarter to well over half of the estimated potential electorate.

The electoral picture presented by the 1920s, then, is a complex one. In certain places, such as parts of Philadelphia, organizational incentives for the production of votes resulted in inversions of the now established (and generally massive) class skew in voting participation. The 1928 election, as has been recognized ever since, forms a landmark at the end of this period and gives important clues to subsequent electoral development. What might have happened to the longer-term party identification of the new voters who poured into the active electorate in 1928 had the economy not collapsed a year later will never be known. But it is clear that a very large proportion of them came to stay—more or less.

Incomplete Remobilization and the (So-called) New Deal Revolution. As we have tried to suggest here, the "system of 1896" was composed of a series of mutually reinforcing parts. It was destroyed by the economic cataclysm which destroyed its ideological and operational base—laissez faire. The period from 1932 to 1940 was marked both by the emergence of the Democrats as the new majority party and by a relatively huge remobilization of the electorate. The two were obviously linked together, though it must not be forgotten that in some areas the remobilization was disproportionately on the Republican side. A few examples of what happened in this period must suffice to give some sense of the general picture.

Two wards in Pittsburgh formed the extremes of the socioeconomic continuum in this period. In the first of these, Ward 14 (Squirrel Hill), 49.6 percent of its 1940 male labor force worked in professional or managerial occupations, and 21.0 percent was found in semiskilled and unskilled manual labor occupations. Ward 6 was largely a Polish slum area adjacent to the Allegheny River and some of the area's many steel mills. In 1940, only 6.6 percent of its male labor force was in professional-managerial occupations, while fully 84.0 percent worked in semiskilled or unskilled manual labor occupations. As was the case in Philadelphia, Pittsburgh had been politically dominated by an entrenched Republican machine since before the turn of the century. Unlike Philadelphia, the Pittsburgh Republican organization was ousted abruptly with the onset of the New Deal and the local elections of 1933. Table 11 presents the data in percentages of estimated potential electorate.

The Democratic mobilization in Ward 6 is particularly striking, but associated with it, to take the 1940 data as a postrealignment benchmark, was an aggregate shift of about 45 percent of the 1920–28 Republican electorate to the Democratic column. Parallel movements are also evident in Ward 14, but at a much lower level. Indeed, if the 1940 Democratic share of the ward's electorate was six as the 1920–24 average, the Republican share was nevertheless 15.7 percent larger than in 1920–24. Save in 1916 and 1932, the participation rate in Ward 14 was significantly larger than in Ward 6. But the most impressive feature of all is the drastic reduction in the size of the "party of nonvoters": by 1940, it had shrunk 33.3 percent from 1924 levels in Ward 14, and 34.0 percent in Ward 6.

A similar profile of mobilization and conversion can be found in San Francisco, though available data do not permit any within-city analysis of socioeconomic areas. In this case, Table 12 reports movement in registration (and partisan division of registration) as a proportion of the potential electorate from 1928 through 1940. If we were to make two most unrealistic assump-

Table 11. Voting and partisanship in two Pittsburgh wards, 1912–40

Year	% Nonvoting	Ward 14 % D	% R	% Other	% Nonvoting	Ward 6 % D	% R	% Other
1912	39.3	16.2	40.9*	3.7	51.0	14.4	29.5*	5.1
1916	35.8	17.7	45.1	1.3	37.3	26.7	34.9	1.2
1920	54.0	7.0	37.4	1.7	64.0	11.5	21.4	3.2
1924	55.8	5.2	32.8	6.3	61.5	1.8	19.1	17.7
1928	44.6	19.8	35.3	1.7	50.9	28.2	20.8	0.7
1932	57.6	18.0	22.8	1.1	58.4	29.7	11.3	0.6
1936	32.5	36.8	29.6	0.4	37.1	50.3	9.4	3.1
1940	23.8	36.4	39.4	0.2	27.5	61.1	11.2	0.1

*1912, combined Republican and Progressive. Progressive, Ward 14, 22.2 percent; Ward 6, 14.4 percent Republican, Ward 14, 18.7 percent; Ward 6, 15.1 percent.

Table 12. Mobilization and conversion in the New Deal period: The case of San Francisco, 1928–1940

Year	Estimated potential electorate	Percentage of potential electorate			Misc., decl. to state
		Nonregistered	Registered D	Registered R	
1928	393,600	35.7	19.1	41.9	3.3
1930	401,933	43.1	8.4	46.2	2.3
1932	410,200	33.3	30.5	34.5	1.7
1934	418,500	25.1	38.9	34.0	2.0
1936	426,800	25.9	47.0	25.5	1.6
1938	435,100	21.7	50.7	25.7	1.9
1940	443,386	12.9	56.2	29.1	1.8

tions—that all conversions and mobilizations favored the Democrats only and that the electorate was wholly stationary during this twelve-year period—we could conclude that the 56 percent Democratic registration of 1940 had the following origins in the 1928–30 period: 14 percent Democratic, 15 percent Republican, 26 percent nonregistered, and 1 percent miscellaneous. Obviously these are in no way the true figures. But in a city where growth was comparatively tightly constrained, they suggest something of what happened. It is probably not wholly off the mark to suggest that about three-fifths of the accretions to the Democratic registration pool came from mobilization of people not registered before and about two-fifths from Republican conversions.

But if remobilizations formed a most central and well-known part of the process of constructing a nationwide Democratic majority in the 1930s, there are important parts of the story which are less frequently emphasized. In the first place, one extremely important component of the "system of 1896" not only survived the New Deal era but was reinforced: the "solid South," composed of eleven ex-Confederate states, all of which retained their antipartisan, antiparticipatory structure until after 1950. Indeed, this massive "extreme case" was reinforced in two ways. In the first place, the pro-Democratic thrust of the New Deal realignment drastically weakened what was left of the Republican opposition in the few states—like Tennessee and North Carolina—where it had been relatively strong in the preceding period. Turnouts in these states, as in the Border states north of them where similar realignment effects occurred, fell subsequently below the levels reached in the 1920s. Secondly, the ascendancy of the Democratic party nationally entailed normal Democratic majorities in both houses of Congress. Granted the salience of seniority and the effects of occasional Republican sweeps in the North (as in 1946 and 1952), the result was that Democratic ascendancy nationally meant Southern ascendancy in Congress. This, as events after 1938 were to demonstrate, involved a critical limitation on the reformist potential of the New Deal. From 1939 until well into the 1960s, the "conservative coalition" saw to that.

This brings us to another part of the post–New Deal story. We earlier saw that the turnout patterns of the late nineteenth century involved not only very high levels of mass mobilization but generally extremely stable levels as well (see Table 9). By contrast, the period between 1940 and 1950 showed major fluctuations, while the post-1970 period has witnessed a precipitous decline from already mediocre levels. Thus, while turnout in the North and West reached 58.6 percent in the congressional election of 1938 and 73.7 percent in the presidential election of 1940, participation skidded to 43.2 percent in 1942 and 48.4 percent in 1946, while by 1948 only 62.2 percent of the Northern-Western electorate bothered to vote for president. Certainly neither war nor woman suffrage can be held accountable for turnout in the latter two elections.[32] One is led to suspect that the replacement of "Dr. New Deal" by "Dr. Win the War" had something rather long-term to do with turnout rates. In this connection, it is perhaps worth stressing that turnouts outside the Southern states have *never since* reached the off-year level of 1938 or the presidential-year level of 1940.

The Contemporary Period: 1952 to Present. In a sense, we return full circle to our initial discussion. The most significant features affecting political participation in the United States during the past quarter-century may be briefly summarized.

(1) In a variety of ways, federal law since 1965 has significantly lowered access thresholds. The final elimination of the network of Southern devices for excluding blacks from the electorate was of course the stated purpose of the Civil Rights Act of 1965 and 1970. As Southern turnout figures in Table 13 indicate, these laws have succeeded in their purpose—helped along by the fact that the demographic shift of blacks to cities and federal pressure have worked to create the conditions for rapid growth of two-party competition in the old Confederacy.

These contemporary turnouts are not precisely breathtaking, but are nevertheless a massive improvement over those of the 1940s and 1950s.

(2) These civil rights acts, coupled with changes in legislation in many states, have significantly lowered access thresholds in non-Southern jurisdictions as well.[33] It is well, in thinking about contemporary non-Southern turnouts, to bear in mind that they are not only lower than in 1939–40, but that the act of voting is much simpler in many jurisdictions now than it was then, or even as late as 1960. This makes the steep decline in voting participation during the past decade all the more remarkable.

The actual gap between the 1960 and 1976 turnout rates is 14.6 percent (put another way, 21.4 percent of the active-electorate base in 1960 was not actively voting in 1976). Moreover, turnout for House elections is declining

Table 13. Mandated remobilization: Southern turnouts, 1948–76

| Year | President | Percentage of potential electorate voting for: | | |
		U.S. house, presidential years	U.S. house, off years	Drop-off
1948	24.6	21.2		
1950			12.4	49.6
1952	38.5	32.2		
1954			16.1	58.2
1956	36.6	29.2		
1958			15.1	58.7
1960	41.2	33.6		
1962			24.0	41.7
1964	45.6	39.3		
1966			29.0	36.4
1968	51.0	41.5		
1970			32.2	34.9
1972	44.9	39.3		
1974			25.1	44.1
1976	47.6	41.8		

Table 14. Surges, declines and decays: Turnouts in the Northern and Western states, 1948–76

| Year | President | Percentage of potential electorate voting for: | | |
		U.S. house, presidential years	U.S. house, off years	Drop-off
1948	62.2	59.8		
1950			52.8	15.1
1952	71.1	68.2		
1954			51.4	27.7
1956	68.4	65.6		
1958			53.8	21.3
1960	71.7	68.2		
1962			54.7	23.7
1964	67.3	64.9		
1966			52.3	22.3
1968	64.1	60.7		
1970			50.7	20.9
1972	59.7	56.8		
1974			41.9	29.8
1976	57.2	53.4		

at a somewhat more rapid rate in presidential years than is presidential turn-out itself. It should be stressed that it is a long time since the 1920s, and much political socialization has supposedly taken place in the meantime. Yet the 57.2 percent rate achieved in 1976 in the North and West is the second lowest in American electoral history, surpassed (negatively) by the 56.3 percent rate in 1920, the first election at which women were enfranchised. Similarly, only once before in the electoral history of the country since Andrew Jackson's inauguration was off-year congressional turnout lower than the 1974 figure of 41.9 percent. This was in 1926, the participatory nadir of the "system of 1896," when only 39.7 percent of the electorate of the Northern and Western states bothered to vote.

Decline in voting participation outside the special mandated case of the South has been nearly universal during the past decade. But the census survey data and aggregate data make it clear that the decline is particularly concentrated in two ways. It has been most extensive both absolutely and relatively among those who already participated least, i.e., the lower half of the social class structure. It has also been most extensive in the most industrially concentrated states and urban regions within states. In other words, the massive class skew in electoral participation which is America's chief peculiarity is becoming more and more accentuated. As our final tabular presentation, we show turnout rates in New York State, stratified among three geographical areas: New York City, the three counties of its suburbs (Nassau, Suffolk, and Westchester), and forty-five nonmetropolitan counties in upstate New York. The period covered in Table 15 is the past century, 1876–1976.

A few aspects of rules-of-the-game in this state should be noted in examining this array. In 1911 the legislature perfected changes, initiated in 1905, in the conduct of elections by which cities about a specified size were to have personal registration and the administrative machinery to enforce it. This lasted essentially unchanged until 1965, when the legislature extended personal registration to the whole state. The law as applied to New York City was deliberately made as onerous as possible: until 1957, voters were required not only to register but to reregister whether or not they had voted in the recent past. By 1948, toward the end of this period of mixed electoral law, 100 percent of the election districts in New York City were covered by (periodic) personal-registration requirements, 48.4 percent were covered by (permanent) registration requirements in the suburban counties, while only 26.1 percent of the election districts in the forty-five upstate nonmetropolitan counties were covered by these (permanent) registration requirements. ·

The turnout rates in the nineteenth century correspond to our earlier discussion, though it may be parenthetically added here that urban jurisdictions appear generally to have had lower turnouts than did rural districts. The subsequent decline is also without surprises. The remobilization from 1928 onward is striking—most of all in New York City, where the turnout rate by 1940 rose to 72.6 percent (the highest since 1908) despite the exceptionally

Table 15. Mobilization patterns in a century of presidential elections: New York, 1876–1976

Year	New York City	New York suburban counties	Upstate nonmetropolitan counties
		Percentage of potential electorate voting in:	
1876	78.9	84.1	93.7
1880	83.2	89.5	93.5
1884	78.2	83.6	92.3
1888	84.3	88.4	95.8
1892	81.5	82.7	87.9
1896	76.9	76.6	88.3
1900	78.8	81.2	89.0
1904	77.4	81.0	87.7
1908	73.8	78.3	86.6
1912	69.9	77.7	78.0
1916	65.7	78.2	81.5
1920	53.7	63.5	62.4
1924	50.4	61.5	66.7
1928	61.8	75.3	80.0
1932	61.2	75.4	75.7
1936	68.9	78.2	79.9
1940	72.6	82.0	80.2
1944	69.3	71.3	71.0
1948	62.1	67.6	68.4
1952	67.3	80.3	78.3
1956	63.3	80.2	77.4
1960	63.4	77.6	79.7
1964	60.9	75.2	73.1
1968	54.1	73.0	67.9
1972	50.9	69.3	66.6
1976	42.1	61.3	63.4

heavy burden which the personal-registration law imposed. The variability in turnout during the "modern" era is also well captured—perhaps especially in the forty-five upstate counties, three-quarters of whose election districts had no personal-registration requirement until the mid-1960s. Thus the standard deviation of their turnout between 1876 and 1900 was 2.9 (3.1 percent of the 91.5 mean), while in the 1940–64 period it was 4.3 (5.7 percent of the 75.4 mean).

But the most impressive feature of Table 15 is the precipitous decline in turnout since 1960. If one may account for part of this decline in the forty-five upstate counties and the suburbs by the introduction of personal-registration requirements, the argument should work the other way for New

York City, where the more burdensome periodic requirement had been replaced by permanent registration prior to the 1960 election. Yet we find a 1960–76 decline of 21.3 percent (33.6 percent of the 1960 turnout base) in the city, compared with declines of 16.3 percent (21.0 percent of the 1960 base) and 16.3 percent (20.5 percent of the 1960 base) in the suburbs and upstate counties respectively.[34] It is important to remember that all of these marginal percentages convert into hundreds of thousands or millions of persons. Taking New York City alone, it is evident that if as large a proportion of its electorate had voted in 1976 as did in 1940, another one and a half million persons would have participated in choosing a president than actually did. As it was, the city's turnout rate in 1976 was easily the lowest since New York's presidential electors were first chosen by popular vote in the 1828 election; it was lower by a considerable margin than in the depths of the 1920s; and it was even lower (for the first time in at least a century) than presidential turnout in the Southern states.[35]

IV. Concluding Observations

As we are always reminded, New York is not the nation. But if it represents an extreme case of electoral disappearance in our time, it is neither wholly unique nor unimportant. One wonders whether the plummeting participation rate may have something to do with the fiscal crisis of 1974–75—when democratically elected government was essentially suspended in important particulars by a visible and narrow consortium of elites. Whether or not it did, there appears very little question that our present low turnouts are quite different in flavor and probably importance from those of the 1920s. In that decade, women had just been enfranchised and their participation rates appear to have been between 20 and 30 percent lower than men's. Moreover, the disappearance of effective general-election competition of any kind in about three-fifths of the country (including, notoriously, the whole of the South) made its contribution. What are we to say of the present situation, five decades later?

The evidence is pretty overwhelming that the current low levels of participation correspond no longer to sudden inflations of the eligible population base, or to the piling up of procedural barriers designed to discourage participation by the lower orders. They correspond instead to a degeneration, now very far advanced, in the collective structures of electoral politics: in short, to the degeneration of the political parties. As we have tried to suggest before, the relevance of these parties to any possible, potentially organizable needs of the lower classes in American society had been problematic at best in postagrarian America. These doubts may have been to some extent temporarily dispelled during the New Deal era. To the extent that they were, it was because electoral politics was part of a larger collective enterprise, and because the divisions between the political parties were as "real" as they ever have been in this century. Since then, however, the needs of state management at home and of imperial management abroad have increasingly crowded out such oppositions and have drastically reduced even the apparent role of the mass of Americans in determining their own destiny. In its heyday, the New Deal alignment provided in the Northern Democratic party a surrogate for a political movement which was relevant to the needs of the"party of nonvoters." How temporary and insubstantial that surrogate was became clear even in the 1942–48 period, to judge from the upwelling of abstention in that period.

It seems to me scarcely coincidental that the most recent growth in the "party of nonvoters" is paralleled chronologically, step by step, with the spectacular growth in indices of political alienation described by Louis Harris and the authors of *The Changing American Voter*.[36] As the disruptive events of the past decade are linked by a now largely media-oriented campaign style, the little man may perhaps be pardoned for thinking, as a majority of little men now appear to think, that while the rhetoric of democracy continues, a very different reality is decisive. To judge from these polls, this is a reality in which the basic decisions are made by a small power elite; elections change little or nothing in what really happens; people in government "don't really care what people like me think"; and the day-to-day encounters with the commanding heights of our political economy (whether nominally private or governmental) are encounters with faceless, unresponsive, and oppressive bureaucracies. "Washington" and what it represents has become so remote from the country and so pervasively unpopular that the present incumbent of the White House got his job, very largely, by running against it: a hopeless task in the long run, since in order to govern he too must become part of that problem. It is also quite characteristic that he won the presidency with the support of 27.2 percent of the potential electorate.

We may conclude this review by giving some estimate of the sheer number of people who are now missing from the world of active electoral politics. How many more people would have voted in 1976 than actually did so, in the context of earlier participation rates? If the criterion is 1888 (or an average European turnout rate of today) the answer is a shade under 40 million. If we accept something more recent like 1940, it is a more modest 12 million nationwide (but 19 million if the South's 1968 rate is used instead of 1940). This reflects part of the cost of maintaining uncontested hegemony into the contemporary era. As we commented earlier, it was highly fashionable in some academic quarters during the 1950s to assume that the heavy nonparticipation rate in American elections bespoke a "politics of happiness." It corresponded, supposedly, to an "end of ideology" which sprang both from immemorial American tradition and more immediately from the apparent elimination of the tendency of capitalist political economies to stagnate. If today such views are much less fashionable, it is in large measure because both society and economy are undergoing a profound and adverse conjunctural change.

This is not the occasion to develop a discussion of that change and its origins. Its symptoms are everywhere, from the New York fiscal crisis to the Snowbelt-Sunbelt controversy, a double-digit food inflation rate in the spring of 1978, and the Jarvis referendum on property tax rates in California. All that can be said here is that there is very strong circumstantial evidence that the declining turnout rates of recent years are *also* symptomatic, and are hence much more politically significant (as a measure of crisis) than were the low turnouts of the 1920s or the 1942–48 period.

There may be those who believe that this represents not a problem but an opportunity, or at least a phenomenon about which too much has been made. Certainly the conservatives who have fought ballot-access devices such as postcard registration in the past few years operate on some such view. After all, the preservation of the existing stratification of power, status, and income in the United States could only be promoted by a politics so organized that there are vast pools of nonparticipant lower-class people at one end of the participation continuum and what Verba and Nie have called "hyperactive Republican" participants on the other.[37]

This, however, may turn out to be too narrow a view of the problem. We may note that it is, of course, antidemocratic and forms part of an obscure but never-ending struggle over the first principles of electoral democracy in the United States—a struggle which has continued for decades after its overt counterparts elsewhere in the West were concluded.[38] But these transcendent ethical issues apart, there are, one should have thought, very practical reasons for concern. There is nothing without cost in politics. Giovanni Sartori, himself no left-wing ideologue, has very recently restated a well-known political truth.[39] Political parties and large-scale mass participation in politics are functional requisites of a modern state. First, they provide a means of channeling demands and socializing adult citizens into the possibilities of politics and the limits of politics in a given system. Second, they provide a means of communication or linkage between rulers and ruled, closing to some extent the great distance which can produce popular alienation from the rulers if it is not reduced. Third, if they are important to rulers as means through which they can penetrate the society for their own purposes, they are also essential feedback mechanisms through which the rulers can gain legitimacy for themselves, their policies, and their rule.

The converse is also the case. It also seems, at the moment, to be our political destiny. Decay of these channels is intimately linked to an explosion of highly intense sectoral demands on political elites. Hyperpluralism coexists with a steep decline in mass participation and in popular support for governing elites. In the process, vital control functions of the state become seriously impaired at a time when the longer-term interests of the political economy— both at home and abroad—require their strengthening. It is hardly surprising that we hear so much these days of a "governability crisis" from certain academics and publicists. But it has come into being not because of the demands of fickle and inconstant masses, as these thinkers appear to believe. It has come into being because of a contradiction, now acute, between the maintenance of uncontested hegemony on one hand and the requirements of the modern state for coherence, stability and legitimacy on the other.

This is tantamount to saying that the contemporary participation crisis in the United States *very probably cannot be resolved within the existing framework of organizable political alternatives—that it is deep-structural in origin, and that a major change in political consciousness would probably be needed to overcome it*. That, it seems to me, is a primary lesson to be drawn from an historical-comparative review of political participation in the United States. But another lesson—not unimportant in its own right—is that electoral law makes a difference. Its current structure (especially the personal registration requirement) is as important as an artifact of uncontested hegemony in the United States as is the degeneration of party structures in the modern period. This legal structure did not descend from the skies. It was made by men, and is now defended by men who see no particular advantage to be gained by making it easier for average Americans to get to the polls. Clearly, then, a thoroughgoing reform of electoral law—eliminating procedural barriers which we now have excellent reason to believe work differentially against the lower classes in American society—is a necessary first step. Such a reform is not likely to be effectuated very soon, or very much before the change in political consciousness of which I have spoken. At the least, the struggle over this question must be brought out of the obscure twilight in which it has been waged and into the light of day. There are abundant signs—including the existence of this conference—that this is beginning to happen. But such reform, essential as it is, would be at most a beginning, the opening of a door. By itself it would resolve little.

What would we find if we opened that door? There is very much that we do not know, and cannot profess to know, in advance of opening it. From the Federalists in the 1790s to the framers of the New York constitution in the 1890s and down to the present day, there have been many conservatives who have feared the end of their liberty and property at the hands of a fickle and dangerous electorate. If the course of American history teaches us anything at all, it is that such fears have been almost wholly illusory. But no one can guarantee that the future will look like the past or even necessarily resemble it very closely. Most of us are aware that somehow we are in the midst of a very profound transition in the basic conditions of American political and social life. It seems pretty clear that the most immediate symptomatic problem is not mass electoral upheaval but hyperpluralism of an exploding universe of intense, particularized, and unconstrained interests. I suspect that the taming of this hyperpluralism, if it is possible at all any longer, can come about short of dictatorship only by mass mobilization linked to the creation of vehicles of collective action—that is, political parties.

But what would be the terms? What would it really take to bring back the missing adults, 40 million or more of them, into active participation in electoral politics? What would they demand as a price for that participation?

What unthinkable things would we have to get used to in order to accommodate their presence in the nation's political life? These, it seems to me, are some of the questions which such a historical review as we have undertaken presents to us. They represent risks which may or may not be very large. The ultimate test is empirical. An assessment of these risks, if such they are, might be undertaken as a calculus of estimation—one's sense of discomfort or threat arising from the change, discounted by assessment of the efficacy of the present functioning of our political system, and weighted decisively at the end by one's relative level of commitment to the egalitarian human principles which underlie any experiment in political democracy.

For my own part, I can do no better than to conclude with the words of a great scholar, the late E. E. Schattschneider, on this subject:

> A greatly expanded popular base of political participation is the essential condition for public support of the government. The price of support is participation. The choice is between participation and propaganda, between democratic and dictatorial ways of *changing consent into support, because consent is no longer enough.*[40]

Postscript

Since this paper was written the 1978 elections have come and gone. They are chiefly memorable, perhaps, for the continuing sag in voter participation. Indeed, 1978 participation in House races appears to have broken all abstention records in the three-quarters of the country which lies outside the South.

To be sure, turnout in 1978 was much higher in the eleven ex-Confederate states than it had been prior to the enactment of civil rights legislation. In years such as 1926 and 1942, Southern turnout rates of 8.5 percent and 7.1 percent respectively had dragged the entire national participation rate down to or near historic lows (32.9 percent in 1926, 33.8 percent in 1942; compared with 35.0 percent in 1978). But even in the post-civil-rights South the 1978 congressional turnout of 26.3 percent was 5.9 percent—or nearly one-fifth—lower than its modern high of 32.2 percent, achieved in 1970.

For many purposes of long-term analysis, it is helpful to exclude the Southern states because of their extremely deviant electoral history. When this is done, we find that turnout rates in the rest of the country have now reached the *all-time* low of 37.9 percent.*

---

*It should be noted in passing that the turnout for all offices was somewhat higher in 1978 (as it is usually) than that for congressional candidates. Both senatorial and gubernatorial elections—and for that matter, some referenda questions—are more visible and thus draw voters who do not cast ballots in House races. Thus the Committee for the Study of the American Electorate estimates an aggregate participation rate of 37.9 percent nationally (down from 39.0 percent in 1974), compared with our estimate of 35.0 percent nationally for the House races. It should also be noted that our estimates are based in part on preliminary data, but it is very unlikely that the final estimates will be more than 0.5 percent out of line from those given here.

There have been thirty-nine off-year congressional elections since 1826, the first date for which comparable estimates can be made. Non-Southern turnout has fallen below 46 percent of the potential electorate on eight occasions during this 152-year period. Table 16 summarizes and comments briefly on these cases.

Perhaps the most striking feature of 1978 in this context is the "ordinariness" of the contexts surrounding this election. All of the others have a clearcut explanation—either particular circumstances which could be expected to have adverse consequences for electoral participation, or structural explanations associated with the culmination of the "system of 1896" during the 1922–30 "normalcy" era. This continued decline in participation is the more noteworthy in that the short-term adverse pressures of Vietnam and Watergate have disappeared and the structure of electoral law has been modified to some extent over recent years to make access to the ballot box somewhat easier.

It should now be obvious that low and declining turnouts such as these are

Table 16. Low turnout elections in non-Southern states, 1826–1978

Year	Non-Southern turnout	Comments
1826	45.4	A "no-party system" prior to Jacksonian democratization of American politics
1918	45.8	World War I; Spanish influenza epidemic; sectional skew and voter demobilization associated with the "system of 1896"
1922	42.7	First election in "normalcy-era" phase of sectionally skewed, demobilized "system of 1896"; first off-year election with full woman suffrage
1926	39.9	Middle "normalcy-era" election; prior to 1978, the all-time low in non-Southern turnout
1930	43.7	Last "normalcy-era" election within the sectionally skewed, demobilized "system of 1896"
1942	41.6	World War II election, with millions of voters in armed forces and little or no provision for absentee service voting
1974	41.5	Post-Watergate; first off-year election following enfranchisement of 18–20 age group
1978	37.9	All-time low reached despite marginal relaxation of procedural barriers during the preceding fifteen years

not *primarily* the artifacts of legal and procedural structure in the conduct of elections in the United States, although these continue to play an important and measurable part. They reflect instead a pervasive and deepening crisis of political integration. Other symptoms of this crisis include the continuing decay of the political parties and the proliferation of intense single-interest lobbies in the political process. As matters now stand, non-Southern turnout in 1978 stands 15.4 percent below the postwar high set in 1962; or, to put the matter even more remarkably, fully 29 percent of 1962's active electoral base has disappeared. The basic internal relationships in voting participation do not appear to have changed significantly from those described in this essay. Neither have the basic trends: the American voter is continuing to disappear.

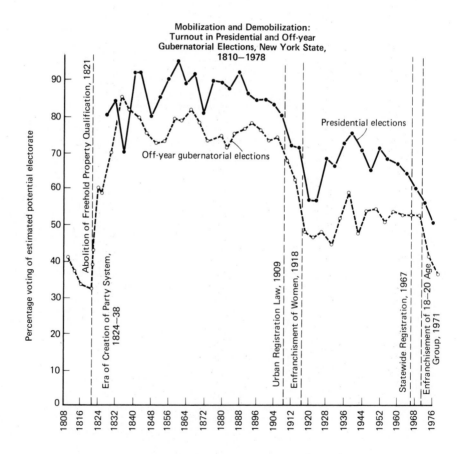

Mobilization and Demobilization:
Turnout in Presidential and Off-year
Gubernatorial Elections, New York State,
1810–1978

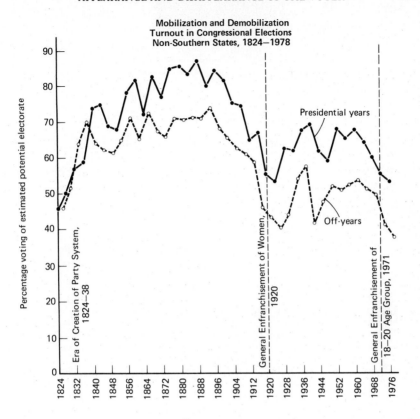

Mobilization and Demobilization
Turnout in Congressional Elections
Non-Southern States, 1824–1978

New York: Turnout for president, governor, and house, 1810–1978

Year	President	Governor	House	Year	President	Governor	House
Freehold property qualification				1900	84.6	84.7	83.3
				1902		73.4	72.4
1810		41.5		1904	83.3	83.3	80.7
1813		37.7		1906		74.2	72.5
1816		33.8		1908	79.7	79.8	79.4
1817		17.3					
1820		32.6		Registration, 1909			
Abolition of property qualification				1910		68.1	67.4
				1912	72.1	70.1	70.2
1822		43.7		1914		62.8	58.9
1824		60.4	58.1	1916	71.6	67.7	64.0
1826		59.4	55.1				
1828	80.2	80.0	56.8	Woman suffrage, 1918			
1830		70.0	71.2				
1832	84.2	84.1	82.7	1918		47.8	45.8
1834		85.7	85.3	1920	56.4	55.7	52.7
1836	70.5	70.4	68.9	1922		46.3	47.2
1838		82.1	81.8	1924	56.3	56.2	53.1
1840	91.9	91.8	91.5	1926		47.6	45.5
1842		79.6	77.3	1928	68.3	67.4	65.3
1844	92.1	92.3	91.5	1930		44.2	44.6
1846		75.4	72.8	1932	66.1	66.2	61.7
1848	79.6	80.3	79.0	1934		51.3	48.8
1850		72.6	70.7	1936	72.6	72.1	68.3
1852	84.7	84.9	81.4	1938		59.2	56.3
1854		73.2	70.1	1940	75.7		72.5
1856	89.9	89.5	89.2	1942		47.8	45.8
1858		79.4	78.2	1944	70.9		67.6
1860	95.5	94.9	94.5	1946		53.9	51.2
1862		78.9	78.8	1948	65.0		63.1
1864	89.3	89.4	87.3	1950		54.2	51.6
1866		82.4	81.7	1952	71.2		69.0
1868	91.7	91.8	90.6	1954		50.5	48.9
1870		78.4	76.9	1956	67.9		65.9
1872	80.5	81.3	81.3	1958		53.6	51.8
1874		73.4	71.3	1960	66.9		64.5
1876	89.6	89.5	87.7	1962		52.4	50.5
1878		74.5 (79)	70.0	1964	64.4		60.8
1880	89.3		87.6	1966		52.6	
1882		71.3	67.7	1968	59.7		53.7
1884	87.5		85.0				
1886		75.5 (85)	67.5	Suffrage for 18–20 age group, 1971			
1888	92.3	92.0	89.0				
1890		76.8 (91)	65.3	1970		52.4	47.4
1892	86.3		83.9	1972	56.6		51.7
1894		78.3	75.3	1974		41.4	38.3
1896	84.3	84.4	79.4	1976	50.7		46.5
1898		76.4	75.3	1978		36.3	32.6

New York: Turnout for president, governor, and house, 1810–1978

Presidential years		Off years					
Year	Turnout	Year	Turnout				
1824	45.1	1826	45.4	1904	74.6	1906	62.1
1828	49.7	1830	51.3	1908	74.0	1910	60.9
1832	56.5	1834	63.5	1912	64.2	1914	58.5
1836	58.4	1838	69.4	1916	66.4	1918	45.8
1840	73.2	1842	63.3	1920	54.5	1922	42.7
1844	74.4	1846	61.8	1924	52.7	1926	39.9
1848	68.0	1850	60.6	1928	61.9	1930	43.7
1852	67.3	1854	64.6	1932	61.7	1934	53.5
1856	77.4	1858	70.9	1936	67.3	1938	56.9
1860	81.0	1862	64.4	1940	68.4	1942	41.6
1864	71.2	1866	71.7	1944	61.4	1946	47.2
1868	82.2	1870	67.0	1948	58.7	1950	51.6
1872	76.0	1874	65.1	1952	67.7	1954	50.4
1876	84.3	1878	70.5	1956	64.9	1958	52.2
1880	84.9	1882	70.0	1960	67.4	1962	53.3
1884	82.8	1886	70.6	1964	64.0	1966	50.7
1888	87.0	1890	70.4	1968	59.8	1970	49.2
1892	79.6	1894	73.4	1972	54.9	1974	41.0
1896	84.0	1898	67.8	1976	53.1	1978	37.9
1900	81.5	1902	65.1				

NOTES

1. Thus the American presidential turnout shows a mean of 58.2 percent for the period 1964–1972, compared with, for example, 75.2 percent for Britain (1964–74), 80.7 percent for France (1967–74), 88.9 percent for West Germany (1965–76), 75.0 percent for Ireland (1965–73), and 74.9 percent for Canada (1965–72)—to take only countries which choose at least a part of their parliaments through single-member constituencies. Switzerland, with its 1963–71 mean turnout of 60.7 percent, constitutes the only significant exception to this generalization. For technical reasons, especially the "hardness" of electoral rolls abroad as denominators in the turnout equation compared with population estimates here, American turnout is comparatively not quite as bad as it looks: to bring it into comparability with the European data, one should add between 3 and 4 percent to the American estimate. On the other hand, the presidential vote represents a maximum: the mean presidential-year turnout for House of Representatives elections is 54.0 percent for this period and 43.3 percent for off years (1966–74).

2. For good discussions of the European cases, see Seymour M. Lipset and Stein Rokkan, eds., *Party Systems and Voter Alignments* (New York: Free Press, 1967); and Stein Rokkan, *Citizens, Elections, Parties* (New York: David McKay, 1970). An early long-range review of American turnout, in which the developmental "movie" runs backwards, is Walter Dean Burnham, "The Changing Shape of the American Political Universe," *American Political Science Review* 59, (1965): 7–28 [Chapter One of the present collection.] The generalization is of course limited by a countercyclical improvement in Southern turnout in the USA from 1952—and especially 1965—onward.

3. The most complete contemporary analysis thus far is Sidney Verba and Norman Nie, *Participation in America: Political Democracy and Social Equality* (New York: Harper & Row, 1972). See also, inter alia, the controversy between the present author and Professors Philip E. Converse and Jerrold G. Rusk: "Theory and Voting Research: Some Reflections on Converse's 'Change in the American Electorate,'" *American Political Science Review* 68 (1974): 1002–23 [Chapter Two of the present collection], comments by Converse and Rusk, pp. 1024–49; and author's rejoinder, pp. 1050–57.

4. Cf. Philip E.Converse's "Change in the American Electorate," ch. 8 in Angus Campbell and Philip E. Converse, eds., *The Human Meaning of Social Change* (Beverly Hills: Sage, 1972), and the present author's discussion in "Theory and Voting Research" [Chapter Two].

5. Converse, op. cit. Similar reliance on rules changes as a sufficient explanation of post-1900 change is noted in Rusk, "Comment," op. cit.; and, though not directly related to the turn-out question, his "The Effect of the Australian Ballot on Split-Ticket Voting: 1876–1908," *American Political Science Review* 64, (1970): 1220–38.

6. The phrase is Heinz Eulau's, but the *locus classicus* is probably Seymour M. Lipset, *Political Man* (New York: Doubleday, 1960), pp. 179–263.

7. The most recent installment of this story is found in *Public Opinion* 1, no. 2 (May–June 1978): 23. Affirmative answers to the following "alienation" questions asked by Harris are as follows for the 1966–7 period:

	% Agreeing with statement				Shift,
Statement	1966	1972	1976	1977	1966–77
The rich get richer, the poor poorer	45	68	76	77	+32
What you think doesn't count much anymore	37	53	62	61	+24
People running the country don't really care what happens to you	26	50	59	60	+34
I feel left out of things around me	9	25	40	35	+26

8. This point is forcefully made, in particular, in a pioneering work which needs more professional attention than it has thus far received: the chapter by William McPhee and Jack Ferguson, "Political Immunization," in William McPhee et al., *Public Opinion and Congressional Elections* (Glencoe, Ill.: Free Press, 1962), pp. 155–79.

9. See in particular the recent article by Douglas A. Hibbs, Jr., "Political Parties and Macroeconomic Policy," *American Political Science Review* 71 (1977): 1467–87, which makes (and demonstrates) the basic point that American macro-economic policy is much more sensitive to inflation (a particular concern among middle- and upper-status voters) and much less sensitive to unemployment (of particular concern to working-class voters) than is the case in European political contexts.

10. The work of Stein Rokkan and his associates, op. cit., is particularly significant in spelling out these linkages by detailed ecological-historical analysis. See also Dieter Nohlen, Bernhard Vogel et al., *Die Wahl der Parlamente*, vol. 1, *Europa* (Berlin: Walter de Gruyter, 1969).

11. Gunnar Myrdal, *Challenge to Affluence* (New York: Pantheon, 1963), especially pp. 92–103. This is an extremely important statement; it may be said that in many ways this essay is a detailed analysis of one part of the problem which Myrdal states so clearly. We may also note his comment (p. 39) that "it is fatal for democracy, and not only demoralizing for the individual members of this under-class, that they are so mute and without initiative and that they are not becoming organized to fight for their interest." Now that the Great Society has come and gone, we can see that this critique of 1963 remains valid. We may also note the

point which Myrdal stresses more than once in this book: that never have the demands of social justice and of national management of the political economy been more closely linked with each other than in the current historical period.

12. There is a well-known general tendency—though with some interesting detailed exceptions—for blacks to participate less than whites at all income levels. Thus in 1976 the racial stratification by income was as follows:

1976 Income level	Whites		Blacks	
	Registered	Voted	Registered	Voted
Under $5,000	57.7	47.2	54.2	41.7
$5,000–9,999	62.6	54.2	58.5	48.8
$10,000–14,999	68.8	61.4	63.3	54.7
$15,000–24,999	76.9	71.2	69.9	64.6
$25,000 and over	83.4	77.8	77.6	73.1
Not reported	60.9	54.5	48.1	39.9
Total	68.7	61.4	59.1	49.5

Note: In 1976, 13.2 percent of whites reported income of less than $5,000 with 10.4 percent not reporting and 10.7 percent falling in the affluent range ($25,000 and over). For blacks, on the other hand, 32.5 percent reported incomes of less than $5,000, 8.0 percent did not report, and only 3.3 percent fell into the top income level ($25,000 and over).
Source: U.S. Bureau of the Census, "Voting and Registration in the Election of November 1976." op cit., p. 66.

13. Norman H. Nie, Sidney Verba, and John R. Petrocik, *The Changing American Voter* (Cambridge: Harvard University Press, 1976), pp. 59–73.
14. Samuel P. Huntington, *Political Order in Changing Societies* (New Haven: Yale University Press, 1968), pp. 122–33; J. R. Pole, *Political Representation in England and the Origins of the American Republic* (Berkeley: University of California Press, 1966), esp. pp. 172–213.
15. Pole, op. cit., appendix II, "Voting Statistics in America," pp. 543–64.
16. Louis Hartz, *The Liberal Tradition in America* (New York: Harcourt, Brace, 1955); cf. also his theory of the "fragment culture" in Louis Hartz, ed., *The Founding of New Societies* (New York: Harcourt, Brace & World, 1964), pp. 3–122.
17. Maurice Duverger, *Political Parties* (New York: Wiley, 1961), pp. 21–22, 217–20.
18. The heyday of the Pendergast machine and its control of votes in Kansas City was reached in the mid-1930s. The 1936 election is particularly striking: a lot more than the typical New Deal mobilization was involved in this case!

Year and office	Percentage of total potential electorate, Kansas City:					% Above state turnout
	Voting	Nonvoting	D	R	Other	
1920 President	66.1	33.9	31.5	33.9	0.7	−1.5
1924 President	65.1	34.9	27.6	34.2	3.3	+1.8
1928 President	74.6	25.4	33.0	41.5	0.1	+5.5
1932 President	81.0	19.0	55.0	25.9	—	+10.1
1934 Senator (Truman)	69.6	30.4	58.0	11.5	0.1	+12.6
1936 President	93.0	7.0	69.0	23.9	—	+15.7
1938 Senator	57.5	42.5	47.6	9.9	—	+5.8
1940 President	70.8	29.2	40.8	30.0	—	−3.6

(Tom Pendergast went to jail in 1939 for income tax evasion, and the machine broke up.) Philadelphia is discussed more extensively in the text.
19. Barrington Moore, Jr., *Social Origins of Dictatorship and Democracy* (Boston: Beacon Press, 1966), pp. 111–55—a most penetrating analysis.
20. See the extraordinarily important article by Eric McKitrick, "Party Politics and the Union and Confederate War Efforts," in William N. Chambers and Walter Dean Burnham, eds., *The American Party Systems* (New York: Oxford University Press, 1967, 1975), pp. 117–51.
21. Rhode Island remained an anomalous case. It retained an essentially freeholder suffrage qualification until 1888, and the last vestiges of the old restrictive pattern were not eliminated until as late as 1928. It is, therefore, the only state in the union to show a longitudinal *upward* trend in voter participation from the mid-nineteenth century to the 1952–70 period.
22. It is to be noted that in this period and, generally, back to the 1840s, newly admitted states on the frontier typically began their electoral participation at very low rates which subsequently rose substantially. In the period under discussion, moreover, the off-year turnout immediately after admission was actually higher than the presidential-year turnout in five sparsely populated Western states; this anomaly very rapidly disappeared.
23. The mean Southern presidential turnouts were: 1912–28, 24.8 percent; 1932–48, 25.0 percent; 1952–68, 42.6 percent.
24. V. O. Key, Jr., *Southern Politics in State and Nation* (New York: Knopf, 1949). See especially part 5, "Restrictions on Voting," pp. 531–663, for a comprehensive discussion of the full range of suffrage-restricting devices then in place in the ex-Confederate states.
25. The basic argument is made, a bit crudely, in Burnham, "Changing Shape of the American Political Universe," op. cit.
26. This is perhaps the chief theme of the so-called "Bible of Progressivism," Herbert Croly's *The Promise of American Life* (New York: Macmillan, 1909).
27. Burnham, "Theory and Voting Research," op. cit.
28. H. L. A. Tingsten, *Political Behavior* (London: King, 1937). A complete review of the comparative data for the period prior to that date is found on pp. 10–78.
29. See for example, Samuel Lubell, *The Future of American Politics* (New York: Harper, 1952), which concentrates its attention on the influx of foreign-born and foreign-stock voters into the Democratic party from 1928 on; and Kristi Anderson, "Generation, Partisan Shift, and Realignment: A Glance Back to the New Deal," in Nie et al., op. cit., pp. 74–95.
30. Tingsten, op. cit., pp. 157–58.
31. A brief comment about the 1910 and 1926 elections may be in order. In the gubernatorial election of 1910 a massive popular revolt developed against the state Republican machine, focussed on the "Keystone" party candidate, since the Democrats were by now little more than a tool of the hegemonic GOP organization. Its candidate, William H. Berry, lost to the Republican Organization's candidate, John K. Tener (a former major-league ballplayer), by 33,487 votes statewide, a margin of 3.3 percent. Though either would have sufficed, both of two factors contributed to Berry's defeat: (1) the Philadelphia machine generated surplus votes in its controlled wards, despite the 1906 registration law; (2) it arranged to field a "captive" Democratic candidate who siphoned off 129,395 largely traditionalist backwoods opposition votes. In 1926 the Philadelphia boss, William S. Vare, ran for the United States Senate, winning the Republican primary (tantamount to election then) by 86,000 votes. Outside the city of Philadelphia, his chief opponent had a commanding lead of nearly 150,000, but a leading Progressive, Gifford Pinchot, won the gubernatorial primary and refused to certify Vare's election. The Senate finally decided in 1929 not to seat him on grounds of corruption. The data reported in Table 10, section C, help to explain why.
32. It may be noted that despite the intensity of the demographic effort involved in the Civil War to which we have alluded, and despite the fact that laws in many states did not permit soldiers in the field to vote, 68.8 percent of the estimated potential electorate voted in the North and West in the 1862 election. In 1866 this turnout reached 75.8 percent, an all-time record for an off-year election and the result, no doubt, of the intense partisan-political

struggle then going on between Andrew Johnson and the Radical Republicans over Reconstruction policy.

33. These include the suspension of literacy tests throughout the United States and the reduction of residence requirements for voting in presidential elections to a uniform thirty days prior to the date of election.

34. The basic methodology employed in the construction of the population-base denominators for turnout estimates is explained in detail in the present author's note "Voter Participation in Presidential Elections by State, 1824–1968," in Bureau of the Census, *Historical Statistics of the United States, Colonial Times to 1970* (Washington: Government Printing Office, 1975); 2: 1067–69. For the most recent period, problems exist not only with census undercounts of the "underclass population" but with the identification of aliens in 1960 and in some jurisdictions in 1970. For 1970 a precise exclusion of those aliens counted can be made for New York City, and a proportionate approximation made for 1960 in constructing the denominator. The larger undercount problem involving "underclass" adults is a serious one. See David M. Heer, ed., *Social Statistics and the City* (Cambridge: Joint Center, 1968). It almost certainly means that the turnouts reported for New York City and other central city areas are substantially *higher* than the true rates of participation; though one wonders how much lower they can get than the 24.7 percent of estimated adult population voting for president (1976) in Shirley Chisholm's Brooklyn district (CD 12).

35. The same can be said of the city's 34.4 percent participation rate in the 1974 gubernatorial election: this was easily the lowest turnout there since the abolition of property qualifications in 1821.

36. See especially Nie et al., op. cit., pp. 277–88, 345–56; and cf. the Harris data reported in note 7.

37. Verba and Nie, op. cit., pp. 224–28.

38. The fullest recent discussion of the philosophical issues involved in this obscure struggle of which I am aware is found in my "A Political Scientist and Voting-Rights Litigation: The Case of the 1966 Texas Registration Statute," in *Washington University Law Quarterly* 1971 (1971): 335–58.

39. Giovanni Sartori, *Parties and Party Systems* (Cambridge: Cambridge University Press, 1976), 39–51.

40. E. E. Schattschneider, *The Semi-Sovereign People* (New York: Holt, Rinehart & Winston, 1960), p. 109.

5

Shifting Patterns
of Congressional Voting
Participation
1981

I

Not too many years ago the systematic study of voter participation in American elections was relatively rare in American political science. To be sure, H. L. A. Tingsten, a Swede, developed a pathbreaking comparative analysis of participation as early as 1937, and scholars and statisticians—notably in Germany—devoted considerable attention to the subject in the Weimar period.[1] And the study of abstentions has always been an integral part of the *géographie électorale* pioneered in France by André Siegfried at an even earlier date.[2] But Tingsten, virtually the only one of these scholars to employ American data, had extremely little of it to work with. To a very large extent this was in the nature of the aggregate data on which most election studies were based in the presurvey period.[3] It was, one suspects, also because the problematics involved were seen widely by most American scholars as civics problems rather than as fundamental issues of political structure. There were some exceptions, to be sure, notably V. O. Key, Jr., and E. E. Schattschneider.[4] How, after all, could one study old-style Southern politics without confronting the ingenious network of rules and norms developed by the region's elites to destroy party competition and reduce the size of the active electorate? As for Schattschneider, it was certainly to his credit that he insisted upon the systemic implications of our enormous pool of nonvoters; but he was content to state the problem without detailed analysis.

In the wake of the behavioral revolution, much of the primary emphasis in the voting studies literature lay elsewhere than on the participation problem. There were important reasons for this. Among these was the need to clarify the motivational bases of voter behavior in a world that became nota-

Prepared for delivery at the 1981 Annual Meeting of the American Political Science Association, New York, September 3–6, 1981. Copyright © by The American Political Science Association.

bly charged with flux after 1964. Thus the immense scholarly effort that went into the still unresolved problematic of "rational" voting and the extent of ideological penetration into the American mass electorate. It was also unfortunately the case that the Michigan studies had a distressingly large response bias, resulting in a vast inflation in the proportion of respondents who claimed to have voted by comparison with the actual turnout.[5] But more than that, it was also true that the period from 1952 through the mid-1960s was an exceptionally quiet time in American political history. One aspect of this is that voter turnout in this period appeared to be both higher and more stable than at any time since before World War I, if not high enough to satisfy either Professor Schattschneider or President Kennedy's 1963 Commission on Registration and Voting Participation. A few studies, like Seymour M. Lipset's *Political Man*, dealt in a useful comparative way with the question of who votes and who doesn't, but they were exceptions in this period.[6]

The situation began to change around 1965. The years since have seen a spate of analyses of American nonvoting. Here too one can see an historically conditioned change at work in the field of scholarly vision. One important element in this was the rediscovery of Southern voting discrimination as an officially recognized fact, after a lapse of seventy years or so, and the adoption of the Voting Rights Act of 1965 and its successor extensions. Another component of change lay in the drift of the political system as a whole into a deepening general crisis and, associated with this, a powerful trend toward abstention. The decay or supression of political parties and the exclusion of large parts of the adult population from any form of active participation in politics have always and everywhere gone hand in hand. By the end of the 1970s voting participation in the non-Southern states had receded to—and in many places below—the all-time lows that had been plumbed in the 1920s. Moreover, this decline materialized in the face of a substantial overall relaxation of procedural barriers to voting outside of the South as well as within it.

The response of the scholarly community to these and other attention-getting changes has been technically proficient, but substantively limited to a peculiar degree. Faced with a large-scale challenge that raises substantive questions of some importance, a great many students of American voting behavior appear to have an unerring instinct for pursuing the small and the procedural.[7] To be sure, as a matter of method or technique this inclination may well be defended on the grounds favored by William of Ockham, namely, parsimony. Such parsimony is certainly to be recommended for all the standard reasons, but only to the extent that one truly explains the *explanandum* thereby. If, as in this case, most of the variance is left unexplained when all is said and done, then one must either attempt to shove the problem under the carpet altogether or resort to less parsimonious if messier explanatory schemes.

The problem here may very often be put in an admittedly controversial way. Much of American political science is characterized by the fact that it lacks the intellectual depth that sociology provides; the sense of other possibilities that comparative analysis at once raises; and the sense of situational transformations that historical analysis contributes. Much of it is, in short, elegantly crafted within its limitations, but atheoretical, ahistorical, and lacking comparative depth. This sort of thing will lead to misspecifying problems or failing altogether to see that there are problems in a given data set. We may take as one case in point a fine study on American turnout by Raymond Wolfinger and Steven Rosenstone, *Who Votes?*[8] This study lays out all the more important variables bearing on the question, as derived from the massive surveys provided to us since 1964 by the Census Bureau. It thus deals with the absolutely central issue of social class differentials in American participation. In this, at least, it goes far beyond studies such as Hadley's *The Empty Polling Booth* that fail to address the class issue at all.[9] What *Who Votes?* says on this subject is clearly correct so far as it goes and within this specific kind of explanatory tradition. The most salient of the class-related variables, of course, is education level. This is perceived as an autonomous "conditioner," and in a wholly nonproblematic way. To quote from the authors' conclusions on the point:

> The implication of these findings is that an explanation for the relationship between "socioeconomic status" and turnout must have at its core a theory about education's role as a facilitator of voter participation. Education, we have argued, does three things. First, it increases cognitive skills, which facilitates learning about politics. . . .
> Second, better educated people are likely to get more gratification from political participation. . . .
> Finally, schooling imparts experience with a variety of bureaucratic relationships: learning requirements, filling out forms, waiting in lines, and meeting deadlines.[10]

One looks in vain for some sense that what the authors are describing might be a *ceteris paribus* situation. That is to say, the institutional mechanisms for social learning among the lower classes—chiefly, of course, the political parties—are so defective that one is left with a kind of apolitical "state of nature" in which formal schooling is the chief thing that matters. And of course it is, to the extent that collective political consciousness and its essential institutional channel, the political party, essentially do not exist. The situation is just accepted as such, and accordingly is not regarded as seriously problematic. Moreover, the modal response to strong circumstantial demonstrations of, say, much higher turnouts in nineteenth-century America is one of mystification or rejection of the evidence. For after all, levels of formal education move toward the vanishing point compared with today's levels as one goes back across the decades, yet turnout goes up and up. Moreover, the verbal evidence that knowledgeable historians regularly cite about the political market of a century ago makes it as nearly certain as anything can be that this was a world in which voters paid attention to and were engaged in political struggles. Something other than schools seems to have been educating them—the parties, perhaps?

When one juxtaposes education-level participation rates in the United States and in other countries in the contemporary period, one notes exactly the same kind of discrepancy. In 1968, for example 75 percent of the Barnes-CISER survey of Italian respondents had had five years or less of formal education, compared with 8 percent in the United States. But with a national turnout rate of 92 percent, the Italian participation rate is virtually flattened out along lines of formal education levels. If about 90 percent of Italians with this low education level voted, only 38 percent of Americans with similar levels did so. The explanation Barnes gives is overwhelmingly simple and powerful: the "educators" and mobilizers in Italy are, essentially, the Catholic Church and the mass parties of the Left.[11] And the whole thrust of Barnes's study is that this kind of organization of an electoral market not only provides the crucial substitute for the all-important American education variable, but permits voters of all education levels to calculate their interests with precision.[12]

If we have laid some emphasis on the Wolfinger-Rosenstone study, there are several important reasons for doing so. As we have said, this is a fine and well-crafted study with the intellectual limits that the authors set for themselves. While much more can be done with the voluminous Census Bureau participation surveys to fill in the details, such further work is not likely to alter the main descriptive findings of *Who Votes?* in any fundamental respect. Accordingly, there would be little point in attempting to replicate their findings here, extending them for the sake of differentiation to congressional elections. These findings include the identification of major class-, race-, and age-related differentials in American voting participation. But they go beyond these familiar phenomena to stress two other points of note. First, the autonomous effect of personal-registration statutes on depressing voting turnout—and doing so in a sociologically selective manner—is quite substantial. In the authors' judgment, it was worth about 9 percent nationally in the 1972 election. One may quibble about details here, but this judgment is broadly correct. The obvious policy implication—especially for the civic-minded, whose worries about our turnouts appear to be largely aesthetically grounded—is that personal-registration laws should be replaced by some automatic enrollment procedure that removes the burden of qualification from the individual voter. Some states have in fact moved part of the way in this direction over the past few years, with comparatively positive effects on their turnout rates.[13]

The second point of major interest to us here is that there is practically no evidence to support the view that these sharp turnout differentials along class, race, and age dimensions are associated with anything of concrete political significance. In other words, when policy or "position" attitudes of voters and nonvoters are compared via the 1972 University of Michigan Survey Research Center/Center for Political Studies survey, differentials tend to be vanishingly small.[14] One may wonder whether the real differences are quite as small as the authors say they are, especially because of the very strong contamination effects that produce much higher rates of supposed turnout among respondents than among the actual population. Thus, for example, a 1977 poll of New Jersey respondents carried out by the Eagleton Institute found some marginally larger differences along a left-right position-issue dimension between the one-third of the electorate "most likely" to vote and the one-third that were "probable nonvoters."[15] Taking the two ends of this five-point continuum, liberals comprised 19 percent of the "most likely" category, but 23 percent of the "party of nonvoters." Conversely, conservatives formed 35 percent of the likely voters, but only 20 percent of the probable nonvoters. Nevertheless, the most compelling aspect of this poll is the *marked conservative skew* among the population *as a whole,* and—in differing degrees, of course—among both likely voters and likely nonvoters. If the two "liberal" position categories were supported by 36 percent of all respondents, 32 percent of likely voters and 42 percent of likely nonvoters, the two "conservative" categories secured 54 percent, 59 percent, and 49 percent respectively. Viewed numerically on a 1–5 scale (left to right), the median value was 4.37 among likely voters, 3.89 among probable abstainers, and 4.16 among the sample as a whole. Basically, then, the argument advanced by Wolfinger and Rosenstone is supported by this subsequent poll, as of course it is by the national attitude surveys they employed in *Who Votes?*

Such a finding is a absolutely first-rate importance to the understanding of American elections. One may read it in either of two ways: as representing a significant theoretical problem, or as an essentially nonproblematic phenomenon. The latter course is chosen by the authors of *Who Votes?* For them it suffices to make the point that if American politics ever becomes engaged with manifest issues that have a strong class thrust, issues of representativeness may arise; but they don't now, because there is little or no detectable policy difference between voters and nonvoters. While true, this argument is also unenlightening. It is our hope here to place this matter at the center of the analysis that will conclude this essay. For now, it is enough to observe that the stance taken in *Who Votes?* naturally enough leads to an implicit judgment that nonvoting in the United States is at most an aesthetic and public-relations problem (it certainly looks bad internationally, for example), and a preoccupation with procedural adjustments for dealing with it.

The essay that follows includes several elements. First, an overview of historical trends in congressional election turnouts since the creation of the American party system in the 1830s is presented. Second, a brief comparative analysis of contemporary participation levels is developed. Third, we examine the patterns of change in American congressional turnout from 1964 through the most recent available data (1976, 1978, or 1980, depending on the specific data set). Finally, an effort is made to present an analysis of the substantive problematics which are central to the whole story.

A few further comments are in order before proceeding. For one thing, it is obvious that congressional turnout patterns are, as it were, subroutines of American turnout as a whole. But the recent availability of relatively reliable data at the congressional district level as well as the presidential off-year sequencing provided by the election calendar permit certain refinements of analysis to be made. Another point to bear in mind is that—not surprisingly—a basic part of the problem to be unraveled is the relationship, both cross-sectional and longitudinal, between turnout and party. As a matter of definition, the splendid post-1964 Census Bureau surveys do not include any information on partisanship at all. And, indeed, most turnout studies on recent years have concentrated on demographic and attitudinal characteristics much more than on the partisan variable. Finally, the obvious should be stressed: this essay makes no pretensions at all either to comprehensiveness or to methodological elegance. It is necessarily selective, and the aggregate material is presented in an indicative way, to make a number of points that should have been obvious to students of the problem long before now.

II. A Historical Overview of Congressional Turnout

The American Civil War was an experience whose long-term effects on American electoral history were second to none. For more than a century after 1854, sectionalism was of supreme importance in American party politics—and especially so far as congressional elections were concerned. This sectionalism reached its peak during the "system of 1896," extending from 1894 through 1930; it has begun fully to vanish only in our own time. Accordingly, it is important in presenting historical data to separate out the electoral history of the eleven states that seceded to form the Confederacy in 1860–61 from that of the rest of the country. For, following the very short-lived Reconstruction episode, Southern politics involved a well-documented search for one-party hegemony and extensive suffrage restrictions. Among other things, this meant that most of the uncontested congressional elections in the United States have been concentrated in this region.

The basic historical data are presented in Table 1, stratified by region. Broadly speaking, the history of non-Southern turnouts falls into several distinct periods, each intimately connected with the nature and scope of party competition. From 1834 through 1852 is the "mature" Jacksonian (or second) party-system era, with competition between Democrats and Whigs. Aggregate participation rates were substantially lower than in the subsequent period, but much higher than they were prior to the mid-1830s or than they

Table 1. Aggregate participation and partisan distribution in U.S. congressional elections (Percentages of potential electorate)

Year	Non-south				South			
	Nonvoting	D	Whig/R*	Other	Nonvoting	D	Whig/R*	Other
1834	42.7	28.6	25.4	3.3	38.7	26.0	27.2	8.0
1836	40.0	29.6	30.0	0.5	42.2	27.6	30.0	0.3
1838	30.2	34.9	34.7	0.2	35.7	33.2	31.1	—
1840	27.1	34.7	38.0	0.2	36.0	32.1	31.8	0.1
1842	37.5	30.9	29.2	2.4	37.4	35.7	26.7	0.3
1844	24.8	36.6	34.9	3.6	31.6	38.5	29.5	0.4
1846	37.0	30.1	29.8	3.1	46.6	27.9	23.3	2.3
1848	31.5	28.4	31.3	8.8	34.4	34.6	30.6	0.4
1850	39.2	30.1	28.2	2.4	40.7	36.8	20.9	1.5
1852	32.4	33.4	30.3	3.9	40.1	35.2	23.1	1.6
1854	35.5	26.7	27.6	10.2	26.7	41.4	—	31.9
1856	22.0	34.6	34.4	9.0	36.1	42.1	—	21.8
1858	29.0	32.5	33.6	4.8	37.9	39.8	—	22.3
1860	19.1	36.0	40.2	4.8	—	—	—	—
1862	25.1	30.5	32.8	1.6	—	—	—	—
1864	29.1	30.8	40.1	0.0	(66.3)	(15.2)	(14.7)	(3.8)
1866	28.3	32.8	38.7	0.2	(49.4)	(11.8)	(38.8)	(—)
1868	17.9	39.2	42.9	0.0	30.7	32.2	35.7	1.4
1870	32.9	31.8	33.8	1.5	33.1	32.0	33.6	0.3
1872	23.9	35.4	40.5	0.2	31.1	32.3	34.9	1.7
1874	34.0	33.8	30.8	1.4	37.1	36.3	26.2	0.4
1876	15.5	41.2	41.9	1.4	26.4	43.9	29.0	0.6
1878	29.4	29.0	32.0	9.6	51.7	31.7	9.5	7.1
1880	15.1	38.6	43.1	3.3	36.3	38.0	18.7	7.0
1882	30.0	33.8	33.0	3.2	48.1	30.6	18.7	2.6
1884	17.2	38.9	41.7	2.2	38.2	39.2	22.2	0.3
1886	29.4	32.6	34.5	3.5	58.0	26.9	13.0	2.1
1888	12.7	40.0	44.5	2.8	38.4	38.8	20.9	1.9
1890	29.6	33.2	32.3	4.8	55.3	31.0	11.1	2.6
1892	20.4	35.3	37.3	7.0	42.2	35.1	9.3	13.4
1894	26.3	25.8	39.3	8.6	52.8	26.1	9.5	11.6
1896	16.0	37.1	45.5	1.4	43.0	39.3	16.8	1.0
1898	32.2	31.1	35.2	1.6	66.4	25.0	8.5	0.1
1900	18.5	35.7	44.0	1.7	58.1	29.3	12.3	0.3
1902	34.1	28.3	34.8	2.8	76.2	18.3	5.3	0.2
1904	25.4	27.9	43.0	3.8	72.1	20.4	7.3	0.2
1906	37.9	25.3	33.2	3.6	81.4	13.6	4.8	0.2
1908	26.0	31.2	39.1	3.7	71.3	21.2	7.1	0.4
1910	38.7	27.2	29.9	4.2	79.4	15.4	4.6	0.6
1912	35.8	25.7	33.0**	5.5	74.6	19.6	4.9***	0.8
1914	41.5	22.9	31.5	4.0	81.4	14.3	3.5	0.7
1916	33.6	27.8	34.2	4.4	70.1	22.7	6.5	0.7
1918	54.2	18.2	25.2	2.4	85.2	12.3	2.4	0.0
			(Woman suffrage, 1920)					
1920	45.6	17.5	33.3	3.6	79.9	13.7	6.2	0.2
1922	57.3	17.7	23.3	1.7	88.2	9.0	2.8	0.0
1924	47.3	19.3	31.0	2.3	82.1	14.0	3.8	0.1
1926	60.1	15.1	23.6	1.1	91.5	6.6	2.0	0.0
1928	38.1	23.6	37.0	1.2	78.7	16.1	5.2	0.0

Table 1-Continued

Year	Non-south				South			
	Nonvoting	D	Whig/R*	Other	Nonvoting	D	Whig/R*	Other
1930	56.3	18.4	24.0	1.3	87.8	9.4	2.7	0.1
1932	38.3	31.5	27.8	2.5	76.4	19.6	3.8	0.2
1934	46.5	27.7	23.2	2.5	86.9	11.1	1.9	0.2
1936	32.7	36.0	28.3	3.0	76.8	19.7	3.4	0.1
1938	43.2	27.1	27.7	2.0	88.7	9.5	1.6	0.2
1940	31.6	33.0	33.7	1.7	76.7	20.4	2.7	0.2
1942	58.0	19.0	22.0	1.0	92.8	6.2	0.9	0.0
1944	38.2	30.1	31.1	0.6	78.1	18.6	3.2	0.1
1946	52.8	20.2	26.3	0.7	89.6	8.3	1.9	0.1
1948	41.3	29.0	28.5	1.2	78.8	17.7	3.4	0.1
1950	48.4	24.2	26.7	0.7	87.6	10.7	1.7	0.1
1952	31.7	30.7	36.9	0.7	67.7	25.5	6.6	0.1
1954	48.7	25.6	25.5	0.3	83.9	13.1	3.0	0.1
1956	33.6	31.6	34.7	0.1	70.7	23.1	6.2	0.1
1958	46.5	29.0	24.3	0.2	84.8	12.5	2.5	0.3
1960	30.7	35.5	33.4	0.4	66.1	26.4	7.1	0.4
1962	45.2	27.4	27.2	0.1	75.8	15.8	8.1	0.3
1964	34.3	36.7	28.8	0.2	60.0	26.9	12.7	0.4
1966	47.9	25.2	26.5	0.4	70.6	19.5	9.5	0.4
1968	38.3	29.2	31.7	0.7	58.0	27.3	14.4	0.3
1970	49.5	26.0	23.8	0.7	67.3	22.1	10.0	0.5
1972	42.3	29.2	27.8	0.7	60.2	24.4	14.9	0.5
1974	58.0	23.9	17.4	0.8	74.4	16.6	9.8	0.3
1976	45.3	30.0	23.8	0.9	57.1	27.2	15.1	0.6
1978	59.6	21.3	18.5	0.6	72.1	16.3	11.3	0.3
1980	46.8	26.2	26.2	0.8	58.3	23.0	18.0	0.6

*This column indicates the Whig vote, 1834–54, and the Republican vote thereafter.
**Republican, 22.1; Progressive, 10.9.
***Republican, 3.8; Progressive, 1.

are today. Competition tended to be close across a very high percentage of districts, and membership turnover was much higher than it was ever to be again.

This period from 1854 through the early 1890s was in all important respects the high-water mark for the American party system organizationally and as a structure of mass mobilization. In view of the extraordinary depth and pervasiveness of the collective conflict agendas of this period—extending from bitter ethnocultural battles at the local level to the epochal struggle over slavery and secession at the national level—the rise of participation to historic highs should not be a matter of surprise. Parties of the convention-structure type had now been fully developed and institutionalized. The norms of the time encouraged, where they did not enjoin, partisanship. Floating or "independent" voters were widely (and probably not inaccurately) perceived as for sale or socially incompetent in some way. If secesssion from major parties developed in this period, this did not take the primary form of the "floating vote," but involved the creation of competing convention-structure third-party organizations. The issues of the time were sharply polarizing or "fixative" in their character, very much as is the case with left and right in countries like France or Sweden today, but of course with quite different foci of polarization. Levels of formal education were very low by present-day standards, but the "educational" (and mobilization) role was evidently amply filled by the churches, the press, and the parties themselves—very much as is the case in the fully mobilized electoral markets of present-day Europe. Neither the mass-consumption society nor its ancillaries, advertising, mass merchandising, and mass apolitical entertainments (e.g., organized professional sports), existed in this period. Communication very probably was far more saturated with mutually reinforcing religious and political themes than it is today. Needless to say, major rules-of-game changes such as direct primaries and personal-registration statutes had not yet come into existence.

In light of all this, it would not be inaccurate to make the general argument that the structures and behavior patterns of American politics in the last third of the nineteenth century had more in common with those of Europe in the present period than the contemporary American electoral market has with either. It is not surprising in this regard that all-time turnout highs were reached for off-year congressional elections outside the South in 1866 and 1894. The 1866 election reflected the acute crises between President Johnson and the congressional Republican party over Reconstruction policy. The 1894 election was fought in an atmosphere of general economic, social, and political crisis on a scale only reached again after 1930. It was, concurrently, a devastating defeat for the Democrats that was to have very long-term consequences, and the high-water mark of Populist protest against industrial capitalism and its political agents. Both reflected the crisis of dynamics of a fully mobilized electoral market saturated with very important collective political agendas and conflicts. Similarly, the presidential-year mobilizing capacity of parties and campaigns reached its apogee in the period extending from 1868 through 1896. Except for the very special situation of 1872, electoral turnouts in the former free states (North and West) regularly approached the 85–90 percent range that is now common in such countries as France, West Germany, and Sweden. By contrast, it is suggestive that the 1894 *off-year* turnout in the non-Southern states—73.7 percent of the estimated potential electorate—stands more than twenty points higher than the *presidential-year* turnout of 1980 in these regions of the country (53.2 percent).

When all these patterns are woven together, there is no real mystery at all about these enormous participation rates, nor is it necessary to introduce tortured explanations to account for them. At that specific historical moment, social structure and political organization alike worked to permit members of an electorate to identify their collective interests and motivated them to act appropriately.[16] As Samuel Popkin and his associates have recently pointed out, the contemporary electoral market in the United States does not;[17] in this, in all probability, lies a very important part of the story.

The general political crisis of the 1890s and the "battle of the standards" in 1896 inaugurated the most extensively repressive and exclusionist political era in our post-Jacksonian history. The processes involved—to be sure, with ups and downs—deepened and intensified down through and a little beyond the collapse of free-market industrial capitalism in 1929. This "system of 1896" has been too frequently and extensively studied to require much further commentary here.[18] Based upon the rapid spread of one-party sectional hegemony not only in the South but in much of the North and West as well, this system was strongly marked by acute signs of party decomposition and vanishing turnout. The real key to the system seems to have been the growth of *depoliticization* at all levels of culture and organization. The press, following the leads developed by Pulitzer and Hearst, dissolved its links with party and rapidly evolved into a cafeteria-style commercial enterprise. Advertising developed as a separate mass-market activity shortly after the turn of the century, and was extended rapidly enough to political campaigns. Mass entertainments explosively developed in scope and importance after 1900. Norms of professionalist expertise and "technocracy" spread to the activities of urban government as these grew increasingly complex. The parties, at the center of American political like in the preceding period, were increasingly attacked in the name of good government and the rescue of democracy from the "ills of democracy."

The proceduralist analysis of systemic problems, much loved by American political scientists then as later, produced decisive changes in the rules of the game: direct primaries, the direct election of U.S. senators, and personal-registration statutes, among others. With the disruption of local party structures, the spread of one-partyism and primary nominations as "tantamount to election," the rules changes and basic shifts in public norms toward the business corporation rather than the party as a core of social enterprise, turnouts naturally could be expected to plummet. So they did, especially with the enfranchisement of women, a notoriously nonparticipant category in the early years after woman suffrange was introduced. About 74 percent of potential voters has cast ballots in the congressional election of 1894, but by 1926 the all-time non-Southern low was reached: 39.9 percent. A comparatively enormous differential developed along sex and class lines—and if we had adequate data we could claim along age lines as well. Data from Philadelphia and Chicago for the 1920s suggest that turnouts among women in these cities ran between 30 and 35 percentage points lower than among men. This might be considered simply to be a matter of acculturation-socialization dynamics, eventually resolved as Andersen and others have pointed out during the massive remobilizations of the New Deal period.[19] On the other hand,

Tingsten's invaluable compilation makes it crystal clear that the sex differential in the first elections after enfranchisement was typically about one-third as large as this in certain other countries.[20] It was, for example, 9 percent in Finland (1908), 12 percent in Norway (1909), about 11 percent in Germany (1919–20), 5 percent in Austria, and 8 percent in New Zealand (1896).

Tingsten's American data for class, as one might expect, are much more sparse for this period than they are for Europe. His reporting of a Delaware, Ohio, local study yields a 1924 participation figure for upper-status citizens of about 85 percent, for middle-status voters of 75 percent, and for working-class voters of 65 percent.[21] This is the more interesting in view of the fact that in this city personal-registration requirements were not imposed until nearly fifty years later. In any event, there seems no question at all that as turnout declined, a very large class gap in voting participation grew as well. There is very little reason to suppose that major *manifest* political interests were any less unrepresented in the 1920s than they were, according to Wolfinger and Rosenstone, in 1972. Rather, when manifest political interests underwent major change under the traumatic blows of the Great Depression, the former nonvoters responded to this by pouring into the active electorate, largely (but by no means exclusively) on the side of the New Deal and the Democratic party. In the 1920s a kind of stable, sectionally reinforced oligarchy had come into being. It was based in large part, in all probability, upon the development of consensus around the beneficence and ascendancy of market-industrial capitalism, which only the shattering collapse of 1929 could disrupt. As a recent study by Louis Galambos suggests, generalized public support for big business had been far more problematic in the 1890s— or indeed down until about World War I—than it was ever to be again, the New Deal period not excepted.[22] The overall issue for participation that this presents in the 1920s, as in the 1970s, is therefore not one of differential *manifest* interests between voters and nonvoters in America, with some important but limited exceptions (Southern blacks in the 1920s, for example). The issue lies in *the nature and range of manifest interests that are capable of being organized in the electoral market*—a very different question indeed, and one to which we shall return in the concluding part of this paper.

The decay of participation after 1896 was rooted in a comprehensive reorganization of the electoral market, aimed at excluding the political from politics, so far as possible. The success of this demobilization and elite strategies designed to accomplish it rested in the last analysis upon uncontested hegemony in the ideological-cultural domain: the Lockean-liberal "monolith" of Louis Hartz's analysis.[23] But within this very broad and unspecific popular attachment to the "American way of life" in general, wide variations exist over time in the extent to which this uncontested hegemony is, as it were, politically concentrated. It was conspicuously so concentrated from World War I to the Great Depression, aided immensely by the effects on Americans of the Russian Revolution and the rise of Communism on the ruins of the old Socialist Second International. This process was also aided even more immensely by the extraordinary dynamism and "progressivism" of American industrial capitalism in the 1920s by any comparative standard—a point that helps to explain the striking influence that America had upon the European left in the 1920s.[24] On the domestic front, little organized opposition remained by the 1920s. Thus low turnout rates of this period broadly reflect not only the specific effects of such things as personal-registration laws and widespread lack of party competition in elections. More generally, they were an artifact of a narrowly uncontested political hegemony, based upon a lack of organizable alternatives to the existing (and apparently successful) politico-economic order.

Basic electoral realignment marked the 1930s, as did massive remobilizations in the electorate outside the one-party South. As Kristi Andersen has recently and usefully pointed out, the realignment and the mobilization were importantly linked.[25] Yet it is essential in reconstructing the past not to read back into it assumptions derived from a more recent period of electoral history. This point needs a little development here, since it is clear from the aggregate evidence that (1) conversions among already participant voters were more significant to the total realignment outcome than is indicated in Andersen's study; and (2), more important for our purposes, practically all accounts of the mobilization sequence that started in 1928 look at the influx on the Democratic side occasioned by Al Smith's candidacy and fail to deal with the remarkable countermobilizations that developed on the Republican side, even in heavily industrial-urban (but relatively non-Catholic) areas.

One major element in Andersen's study is a conjectural reconstruction, based on survey data subsequent to the Great Crash, of partisanship and involvement in the 1920s. This restrospective reconstruction raises major issues concerning the use of counterfactuals in historical explanation. If the analysis offered here is correct, then the organization of alternatives in the 1920s was such as to preclude *as of that time* even a stable *potential* commitment to Democrats among the "party of nonvoters" as a whole. The best way to drive this point home is to evaluate partisan and nonvoting percentages of the whole potential electorate across the period 1920–40. We may note at the outset that, as Table 1 reveals, between 1924 and 1928 Democratic congressional candidates increased their share of the non-Southern potential electorate by 4.3 percent, but Republicans increased theirs by 6.0 percent, reaching indeed the highest percentage of potential electorate voting Republican in any congressional election from the enfranchisement of women (1920) to the present time. This, interestingly, works out to a gross two-party split among the net increases of 42 percent Democratic, 58 percent Republican—not far from the aggregate two-party non-Southern totals for the election as a whole (39 percent Democratic, 61 percent Republican).

But a more microscopic, and necessarily highly selective, analysis of the presidential data from 1920 through 1940 makes this point even more strongly (see Table 2). Naturally, all eight of the metropolitan examples in Table 2 show the powerful trend toward a long-term emptying of the "party of nonvoters" and the rise of relatively huge Democratic percentages of the potential electorate. But the four that were relatively heavily Catholic (Hudson, Cuyahoga, Ramsey, and northern California) behaved very differently in 1928 from the four that were at the time largely non-Catholic (Camden, Summit, St. Louis, and southern California). In the first four cases, clear evidence exists for what might be called the Lubell hypothesis,[26] i.e., a huge 1928 Democratic surge from the former party of nonvoters, LaFollette voters in 1924,

Table 2. Mobilizations, "surges," conversions: Some metropolitan-area examples

County (city)		1920	1924	1928	1932	1936	1940
Camden, N.J.	Nonvoting	38.7	38.8	20.4	26.2	21.5	22.5
(Camden)	D	17.7	14.8	23.7	32.8	54.8	51.0
	R	40.3	40.6	55.6	37.5	22.8	26.1
	Other	3.4	5.8	0.4	3.5	1.0	0.4
Hudson, N.J.	Nonvoting	39.9	38.1	25.8	29.8	21.9	21.5
(Jersey City)	D	22.1	29.1	44.7	50.4	60.7	51.7
	R	35.8	25.8	29.2	18.3	16.9	26.7
	Other	2.3	7.0	0.4	1.5	0.5	0.1
Cuyahoga, Ohio	Nonvoting	48.6	56.2	40.3	45.0	34.2	27.8
(Cleveland)	D	15.7	4.5	27.2	27.5	43.1	45.1
	R	31.1	24.5	31.9	24.7	17.8	27.1
	Other	2.6	20.7	0.6	2.7	4.9	. . .
Summit, Ohio	Nonvoting	52.1	49.9	37.0	43.4	32.6	27.0
(Akron)	D	18.2	10.7	17.9	28.8	46.3	42.8
	R	28.5	32.7	44.6	25.5	19.7	30.3
	Other	1.2	6.7	0.4	2.3	1.4	. . .
Ramsey, Minn. (St.	Nonvoting	51.0	46.2	34.3	40.6	32.1	32.1
Paul)	D	15.1	5.4	33.6	36.4	44.8	38.8
	R	28.7	25.6	31.4	21.2	15.9	28.0
	Other	5.2	22.7	0.6	1.8	7.2	1.0
St. Louis, Minn.	Nonvoting	43.4	30.4	26.8	22.1	20.8	20.2
(St. Louis)	D	17.0	2.8	25.6	37.4	59.0	53.4
	R	32.3	39.9	44.8	32.5	18.9	25.1
	Other	7.3	26.9	2.8	8.1	1.5	1.4
N. California	Nonvoting	49.9	49.3	40.0	37.0	32.9	25.9
Metro (San	D	11.2	3.3	25.0	37.0	46.4	42.8
Francisco)	R	33.5	27.0	34.3	23.6	20.0	30.5
	Other	5.5	20.4	0.7	2.3	0.8	0.9
S. California	Nonvoting	55.4	48.3	39.3	34.2	32.8	25.3
Metro (Los	D	10.5	4.0	17.5	37.2	44.4	42.9
Angeles)	R	30.1	32.4	42.5	26.3	21.9	30.9
	Other	4.0	15.3	0.6	2.8	0.9	0.9

Table 3. Mobilizations, conversions, and realignment: The case of California and San Francisco, 1922–40 (Percentage of potential electorate)

Year	Registered	Not registered	Party Registration		
			D	R	Other*
A. California					
1922	67.8	32.2	14.1	42.9	10.8
1924	71.4	28.6	15.6	46.4	9.4
1926	70.0	30.0	15.0	47.6	7.5
1928	75.9	24.1	19.4	50.4	6.1
1930	68.2	31.8	13.9	49.8	4.6
1932	81.6	18.4	32.8	44.2	4.6
1934	83.3	16.7	41.3	38.0	4.1
1936	81.4	18.6	47.1	31.1	3.2
1938	85.4	14.6	50.7	30.6	4.1
1940	90.9	9.1	54.3	32.7	3.9
B. San Francisco					
1922	62.1	37.9	11.2	44.5	6.4
1924	63.9	36.1	11.9	46.1	5.9
1926	60.6	39.4	10.6	45.5	4.6
1928	65.1	34.1	19.4	42.4	3.3
1930	55.7	44.3	8.2	45.3	2.2
1932	65.8	34.2	30.1	34.1	1.6
1934	74.5	25.5	38.7	33.8	2.0
1936	74.2	25.8	47.1	25.6	1.6
1938	79.0	21.0	51.1	26.0	1.9
1940	88.4	11.6	57.0	29.5	1.9

*Includes those who declined to state a party affiliation.

and former Democrats. But in the other four, the most conspicuous feature was a very large mobilization surge toward Hoover rather than Smith. Needless to say, evidence of aggregate conversions can be found in most of these jurisdictions as well.

The combined effect of conversion and mobilization is perhaps best perceived in such data files as California and San Francisco registrations from 1922 through 1940, shown in Table 3. These constitute, after all, the closest aggregate approximation we have to a "party identification" measure.

In the aggregate, mobilizations from former nonregistrants and conversions from Republican registrations share about equally in the Democratic increment after 1930. This is true both for the state as a whole, whose potential electorate grew very rapidly in this period, and in San Francisco, where it did not. In California as a whole, moreover, the timing of the two increments is of some interest: in 1932–34, on a net basis, rather more than half of the pro-Democratic change appears to have come from previous nonregistrant groups, less than one-third from former Republicans, and about one-sixth from registrant groups that had either supported other parties or declined to state a party affiliation. In the 1936–40 phase, however, about one-quarter of the additional Democratic increment seems to have come from former nonregistrants, while close to three-quarters came from former Republican registrants. In San Francisco, on the other hand, the overall pro-Democratic shift was somewhat steeper than in the state as a whole, and the two contributing groups—nonregistrants and former Republicans—contributed fairly equally to the Democratic increase in both periods.

The bottom line in this particular part of our discussion is that, at least in this state and this city, conversions seem to have been of about as much importance as mobilizations. A more refined analysis will doubtless show that in many areas of the North and West (notably in suburbs and certain rural areas), remobilizations by 1940 had come to benefit the Republicans as much as, or more than, the Democrats—which, if true, would generate still further problems for any view that there was an overwhelming pool of latent Democrats in the party of nonvoters prior to 1932.

But our basic concern here is with this 1928 Hoover surge in non-Catholic industrial areas. How to account for it? We could, of course, assimilate it into V. O. Key's judgment that 1928 was a mixture of permanent realignment (in strongly Catholic urban and mining areas) and short-term deviation—notably in the South, where five states went Republican for the first time since reconstruction.[27] At least by implication, this rationale is isomorphic to Andersen's discussion. But viewed from a politico-historical perspective, one important counterfact immediately presents itself: what would the shape of this universe have looked like after 1928 had the Great Depression never happened? The 1928 election was not fought over issues of capitalism or the role of the state in managing the economic system. It was centered upon profound ethnocultural conflicts within and between the two major parties, especially the Democrats. The deepening consensus about economics and the relation of the state to private enterprise that marked the entire history of the "system of 1896" was not interrupted in 1928. Accordingly, we would expect that there were few or no major representational differences between the "party of voters" and the "party of nonvoters" during this period, very much in the same way as there were very few such differences in the 1970s.

This would involve the proposition that there might well be exceptionally stimulating election situations that could produce major remobilizations during the lifetime of this "system." But this would also mean that if there were (as in 1928), the mobilizations would correspond rather closely to the expressions of opinion among the already active electorate, though with concentrations along the primary axis of cleavage (in 1928, Protestant-Catholic for the most part). As a review of the strange history of the *Literary Digest* poll reveals, an autonomous class-related variable in partisanship did not apparently exist in American elections prior to 1934. The conclusion would seem to be that no such thing as a huge implicit or potential Democratic majority existed among the "party of nonvoters" *until the Great Depression and the New Deal created one.*

The subsequent history of non-Southern turnout is divided into four rather short periods. The first, 1932–40, is the era of New Deal mobilization (and again, it should be noted, Republican countermobilizations in 1938 and 1940). In 1938, turnout reached a height never subsequently achieved for an off-year election (56.8 percent), and more than 16 percentage points above voter participation in the 1978 congressional election. The second, even shorter period extends from 1942 through about 1950, and is marked by declines that are particularly marked in the "good" Republican years of 1942 and 1946. Very little effort has gone into serious study of this peculiar era. Clearly the war had something to do with low turnout (especially in 1942, for procedural and other reasons); it seems likely as well that the development of institutionalization pertinent to the post-1929 dispensation was quite incomplete in a large number of states.

From 1952 through the mid-1960s, participation rates rose to highs comparable to those achieved in the final mobilization elections of the 1932–40 period. These highs, of course, remain very far below the normal turnout rates of the later nineteenth century, but almost certainly reflect a significant closing of the class gap in participation that had opened so widely after 1900. This era of generally stable and—for the time and context—rather high participation rates corresponds to the quiet, "allegiant-participant" electoral politics that marked the Eisenhower era.

Beginning in the mid-1960s and continuing down to the most recent elections, turnout has continually and severely declined. By 1978 non-Southern congressional turnout, at 40.4 percent of the estimated potential electorate, had fallen to very near the all-time low achieved in 1926. This represented a disappearance of more than one-quarter of the electorate that had participated in the 1962 election. Similarly, the 46.9 percent abstention rate of 1980 is second in this history only to the 47.3 percent high that was achieved in 1924. Even if we correct by about 2 percent for the abysmal turnout rates among the 18–20 age cohort universally enfranchised in 1971, these participation rates are comparable only with the troughs of the 1920s and such special cases as the 1942 election. The difference this time is that one cannot account for it by reference to a huge participation differential between the sexes: by 1980 this difference had entirely disappeared. Still more striking is that some aspects of federal voting-rights legislation since 1965 and the actions of at least a dozen states have significantly lowered the procedural thresholds of access to the ballot box. It was generally much easier to cast a vote in 1980 than it was in 1960, and yet the 1980 congressional turnout outside the South stood 16.1 percent (or, with the age-correction estimate, aobut 14 percent) below 1960 levels.

There can be little question that this turnout decline is closely linked to the enormous political upheavals that have gripped the United States since the beginning of the country's overt participation in the Vietnam War in 1965. A full unraveling of the causal links remains an important unfinished item on the agenda of scholarly research. Following a short comparative overview of American turnout in the most recent period, we shall explore post-1964 participation dynamics in somewhat greater detail. It is enough to say at this point that there is at the least a strong chronological connection between this rapid disappearance of the American voter and such well-known developments as decline in public trust of government, the decomposition of the parties (especially the Democratic party), the emergence of the "new" domestic public policy after 1963, and the development of "disappearing marginals" and incumbent insulation in congressional elections.[28]

A Note about Southern Turnout. The profile of Southern congressional turnout since Reconstruction is on the whole much more simply described. From the readmission of the Southern states to the union in 1868–69 until the civil rights revolution a century later, the objective of most organized Southern white opinion was the political destruction of the black electorate and its original political vehicle, the Republican party. During the general economic and social crisis of the 1890s, Southern elites in addition became fully aware of the dangers to themselves of permitting *any* partisan competition in general elections. Then, and for long afterwards, the social and economic structure of the South had much more in common with those of underdeveloped countries south of the border than with those of the rest of the United States. Populism was a reflection of the conflicts possible in such a setting as, locally, was the phenomenon of Longism in Louisiana. The history and consequences of Southern disfranchisements have been thoroughly documented, and need not detain us here.[29] Essentially, congressional elections in the South can be divided into four periods of very unequal length: Reconstruction, 1868–76; partial suppression, corruption, and "home rule," 1878–1900; full-fledged disfranchisement and the near-total end of party competition, 1902–50; and the stepwise remobilization, return of a very differently grounded Republican party, and end of the "solid South" from 1952 to the present.

The high-water mark of voter participation in the postbellum South was reached at the end of Reconstruction, 1876, when 73.6 percent of the estimated Southern electorate participated (frequently under less than free circumstances). Disfranchisements were consolidated between 1896 and 1904, with regional turnout plummeting to 25.4 percent of the potential electorate in the latter year, and 18.6 percent in 1906. Characteristically, the region's all-time records for abstention were achieved in 1926 and 1942, when more than nine-tenths of Southerners failed to vote.

The most recent period, on the other hand, has been marked by significant if limited remobilizations, operating countercyclically after 1964 against the declines in the rest of the country. Viewed at the aggregate-global level, this remobilization has been almost entirely a question of the rise of Republicanism to levels not seen since the pre-Populist era a century ago. The countercyclicality owes its existence to the effects of the voting-rights acts and the Republican gains in a region that has been decisively transformed in its socioeconomic structure over the past generation. For the first time in history, for example, Mississippi had a higher congressional turnout in 1980 than did New York (47.7 percent to 43.4 percent). But while Southern turnout has grown considerably from the basement figures common before 1952, it remains very far below that of the pre-Populist era, not to mention that of Reconstruction. There is good reason to suppose that it will expand little more from now on unless or until effective partisan channels of collective political conflict are built. More probably, it will completely converge with non-Southern participation rates both in aggregate terms and in terms of the social characteristics that surveys illuminate.

America versus Europe: Contemporary Participation Levels

In 1980, the nationwide turnout for U.S. House elections was 49.8 percent of the estimated potential citizen electorate of voting age, compared with a presidential turnout rate of 55.1 percent. If the forty-six uncontested seats are excluded from consideration, the congressional turnout is a little better: 52.0 percent. This notoriously contrasts with the situation in most other democratic countries: 90.7 percent in Sweden (1979), 90.4 percent in West Germany (1976), 90.4 percent in Italy (1979), 85.9 percent in France (presidential 1981), and 75.9 percent in Britain (1979). Moreover, while Congress is not a parliament, its significance is rather greater than that of a local government body. Yet the national turnout rate for the House of Representatives in 1974 was 36.8 percent in 1974 and 37.5 percent in 1978—considerably lower than the usual turnouts for such local bodies on the other side of the Atlantic.

It requires no elaborate demonstration to show that there are fundamental economic class, education level, and age differences in American voting participation, and that these differences are minimal to nonexistent in most European settings. For American age-related data we have the preliminary census survey for 1980, and the more detailed reports for 1976 for other indicators. The European data is taken from official or quasi-official publications.[30] Turning first to age: the most noteworthy differences of course involve the youngest age cohorts (see Table 4). In the United States, the youngest age groups (18–20, 21–24) had 1980 turnout rates of 35.7 percent and 43.1 percent respectively, between 40 and 50 percent below the levels reached in Sweden and West Germany. Similarly, if the youngest voters (18–20) in Sweden and Germany participated at rates 7 percent and 10 percent

Table 4. Turnouts in the United States, West Germany, and Sweden by age group (Percentage of potential electorate)

	United States			West Germany		Sweden	
Age	1964	1980	Shift	Age	1976	Age	1979
18–20	39.2	35.7	−3.5	18–20	84.1	18–20	87
21–24	51.3	43.1	−8.2	21–24	82.9	21–30	90
25–34	64.7	53.2	−11.5	25–34	87.8	31–40	93
35–44	72.8	64.4	−8.4	35–44	92.0	41–50	93
45–54	76.1	67.5	−8.6	45–49	93.6	51–60	94
55–64	75.6	71.3	−4.3	50–59	93.8	61–70	90
65–74	71.4	69.3	−2.1	60 and over	91.1	71 and over	86
75 and over	56.7	57.6	+0.9				
Total*	69.3	59.2	−10.1	Total*	90.4	Total*	91
21 and over only	69.4	61.2	−8.2	21 and over only	90.8	21 and over only	91
Actual turnout**	63.5	55.1	−8.4		90.4		90.7
Gap, survey/actual	5.9	6.1			Nil		Nil
Gap, most to least voting	36.9	35.6			10.9		8

*As reported by the relevant survey.
**Estimates based on election returns.
Sources: United States—Bureau of the Census, Current Population Reports, Population Characteristics, Series P-20. *Voting and Registration in the Election of* . . . (for 1980; this is the advance report, Series P-20, No. 359 [1981]).
West Germany—Statistisches Bundesamt, *Statistisches Jahrbuch 1980 für die BRD* (Stuttgart, 1980), p. 85.
Sweden—*Statistisk årsbok for Sverige, 1980* (Stockholm, 1980), p. 420

below those of the most participant middle-aged strata, in the American case the gap between the youngest cohort and the 55–64 age group was nearly 37 percent. If one feature of "political socialization" is the incorporation of the youngest age groups into the world of active politics, then by this criterion the American political system fails as dramatically as these European systems succeed. No matter how much may be allowed for the supposedly autonomous effects of American election procedures or high mobility, the discrepancy remains stark. Without much doubt, it is related in some way to the strong tendency of young and new voters in the United States to be particularly uninterested in political parties and party identifications.[31]

The same considerations apply *pari passu* to education levels or socioeconomic position. When aggregate turnout levels are in the high 80s or low 90s, there can be little or no room for major class differentials in voting participation. Indeed, this point was noted years ago by Seymour Martin Lipset.[32] It implies, among other things, that in the high participation elections of the late nineteenth century such differentials were vanishingly small in the United States too. Conversely, as we move historically toward the ascendancy of industrial capitalism and the first major period of party decomposition and voter demobilization, the class gap clearly opened wide. Narrowed again (outside the South) during and after the New Deal realignment, it has once again increased markedly since the early 1960s as turnout has slumped again toward its historic lows. For the sake of the record, we may examine this point by a comparison between the United States in 1964 and 1976 and Sweden in 1979, shown in Table 5.

The overall level of formal education in Sweden is markedly lower than in the United States, though of course higher than in countries like Italy. Adults with only a primary-school education constitute two-fifths of the 1979 Swedish sample, but only one-sixth of the American example. Turnout in Sweden is invariant along educational lines, but steeply skewed in the American case where, as modern survey analysts from the authors of *The American Voter* to the authors of *Who Votes?* agree, it forms a variable of first-rate explanatory importance.

The ultimate explanation for such discrepancies is, of course, systemic rather than procedural. They are rooted in profound differences in the density of lower-class and democratic organizations in general, and in the interest in or capacity for mobilization by political parties in particular. As suggested earlier, there are many ways in which education relevant to political consciousness and choice can take place in society. When organized political alternatives exist and are echeloned throughout the social structure, formal education may well become practically irrelevant to the extent of mobiliza-

Table 5. Mobilization and education: a tale of two countries (Percentage of potential electorate)

Education level	United States			Sweden 1979	U.S.-Sweden differential	
	1964	1976	Shift		1964	1976
1	59.0	44.1	−14.9	95	36	51
2	65.4	47.2	−18.2	95	30	48
3	76.1	59.4	−16.7	93	17	34
4	82.1	68.1	−14.0	94	12	26
5	86.5	78.6	−17.9	97	10	18
6	89.2	81.9	−7.3	97	8	15
Total	70.8	59.2	−11.6	92	21	33
Gap between top and bottom	30.2	37.8	+7.6	4		

Note: It is notoriously difficult to obtain an exact correspondence between American education-level data and their European counterparts. The American level stages are 1, 0–8 years of education; 2, 9–11 years; 3, 12 years (completed high school); 4, 13–15 years; 5, 16 years (completed college); 6, 17 years and over. The Swedish category 1 is that of "folk school," with six or seven grade levels; category 6 is university level. It is worth noting in this regard that this basic-education "folk school" was the level attained by 39.4 percent of Swedish respondents, compared with 29.3 percent of 1964 American residents. Very strong partisan differentials exist along this education dimension in both countries. Percentages voting for the nonsocialist parties in Sweden were, by level: 1, 38 percent; 2, 39 percent; 3, 51 percent; 4, 52 percent; 5, 66 percent; 6, 66 percent; total, 48 percent. In the Unites States, using the CPS 1976 data with collapsed categories, the percentage Republican by education level was for grade school, 33 percent; for high school, 46 percent; for college educated, 57 percent; total, 49 percent.

It is also to be noted that the Swedish sample overreports the actual turnout somewhat for this variable only. On the other hand, the U.S. Census Bureau survey, while far more accurate than others, had an overreporting of 7 percent in 1964 and 4 percent in 1976, which largely cancels out the Swedish overreporting.

Table 6. A tale of three countries: Correlates of recent legislative elections in major metropolitan areas

		N	$\bar{M}_x$	$\bar{M}_y$	$Y_c =$	$r =$	$r^2 =$
1980:	USA, three largest conurbations (New York, Chicago, southern California)						
	X = % $15,000;						
	Y = % NV*	66	30.4	54.4	78.951 − 0.808X	−.752	.565
	Y = % Dem	66	30.4	23.3	26.247 − 0.096X	−.136	.019
	Y = % Rep	66	30.4	21.4	−6.409 + 0.916X	+.727	.528
	X = % NV; Y = % Dem	66	54.4	23.3	24.933 − 0.029X	−.045	.002
	Y = % Rep	66	54.4	21.4	74.688 − 0.979X	−.834	.696
1979:	Britain, London metro						
	X = % Nonmanual;						
	Y = % NV	40	44.8	28.3	36.794 − 0.189X	−.322	.104
	Y = % Lab	40	44.8	27.6	52.247 − 0.551X	−.838	.702
	Y = Cons	40	44.8	33.4	7.611 + 0.575X	+.793	.629
	Y = % Lib	40	44.8	9.3	−0.053 + 0.209X	+.453	.205
	X = NV; Y = % Lab	40	28.3	27.6	18.460 + 0.321X	+.286	.082
	Y = % Cons	40	28.3	33.4	59.836 − 0.934X	−.755	.570
	Y = % Lib	40	28.3	9.3	21.667 − 0.437X	−.556	.310
1978:	France, Paris metro						
	X = % NV Y = % Left	83	20.8	36.4	49.021 − 0.607X	−.200	.040
	Y = % Right	83	20.8	41.4	51.251 − 0.474X	−.178	.032

Note: For the United States, political data are district-by-district percentages in *contested* congressional elections in districts comprising the greater New York, greater Chicago, and southern California conurbations. "Class" data is percentage of population (1970) in families, etc., with incomes of $15,000 or more. It should be noted that exclusion of nine largely black or Hispanic districts yields little change in r^2, and none in direction of relationships. In order, r^2s for the five categories above with these nine districts excluded .456, .014, .413, .000, and .697, respectively. For Britain, only the forty of ninety-two constituencies could be included where boundaries were essentially unchanged after the 1971 reapportionment, and for which 1966 census data as to the percentage of nonmanual workers by district could be included. See David Butler and Michael Pinto-Duschinsky, *The British General Election of 1970* (London: Macmillan, 1971), pp. 360–73. For France, official returns from *L'Année statistique 1978* (Paris, 1978), pp. 487–560; elections on the second round were used except in those districts where a candidate achieved election on the first.

*Nonvoting.

186

tion and rational interest choice in a given electorate. Such was almost certainly the case in the most socioeconomically advanced parts of the United States a century ago; such is obviously the case in Europe today. Turnout is, therefore, a comparative social indicator of prime importance for the evaluation of systemwide problematics. In the American case, its generally low level constitutes a critical limit on the mobilizing capacity of the major parties in general, and of the Democratic party in particular. Fluctuations in it, as we shall see, are a sensitive reflection of changes in the manifest content of electoral politics.

We now turn to a more detailed comparative exploration for a preliminary overview that must be regarded as indicative at best—but suggestive, it is hoped, of the mileage that might be gained from a more thorough study. There are three sets of observations in Table 6. First, there are 1980 congressional turnouts and related phenomena for the three largest conurbations in the United States (New York, Chicago, and southern California), since very sharp differentials in class-related patterns show up at the congressional-district level. Secondly, turnout and related phenomena are presented for the forty constituencies with available data that were located in the greater London area in 1979. Third, turnout and its relationship to left-right voting in the 1978 election are explored for the eighty-three seats that make up the greater Paris area. For the latter, unfortunately, class-related data were not readily available; but the pattern is quite striking enough for inclusion here on this preliminary basis.

There are certain evident relational dissimilarities between the American data of 1980 and that for London in 1979. The percentage middle-class (nonmanual) by constituency was powerfully related negatively to the percentage Labour of the potential electorate, and almost equally positively to the percentage Conservative. The relationship with the "party of nonvoters" is also visible, but it is low, explaining on a simple correlation basis only 10 percent of the observed variance. In the American conurbations, the income measure employed shows an importantly different pattern of relationship. In the first place, it is much more strongly and negatively related to the "party of nonvoters" than in the London case (an r^2 of .565 or, with black and Hispanic districts excluded, .456 compared with .104 in London). Second, the relationship between this measure of middle-classness and the Democratic percentage of the potential electorate across these districts is also negative, but vanishingly small compared with the British example (r^2 of .019 in all districts, or .014 in the largely white non-Hispanic districts, compared with the r^2 of .702 for Labour voting). Thirdly, on the other hand, the relationships are extremely strong and positive between middle-classness and Republican voting, through less so than with 1979 Conservative voting. Regression performed on the relationship between the size of abstention and the major parties sharpens the contrast on the American side (r^2 of essentially zero between nonvoting and percentage Democratic, but of nearly .7 between nonvoting and percentage Republican). On the other hand, the relationship between percentage nonvoting and percentage Labour in the British case is somewhat larger and positive: as abstentions rise, so does Labour voting— up to a point, of course!

In the Paris case, we only have nonvoting and the left-right dichotomy to deal with. But this, as a continental European political setting, is a case in which we have every reason to suppose that the political data constitute a straightforward (if not perfectly accurate) census of underlying social characteristics. The obvious invariance of abstention levels with district propensity to vote left or right is revealed in the near approximation of the regressions to each other (with abstention as the "independent" variable), as well as the fact that the r's are negative for both left and right and the r^2's are tiny (.040 and .032 respectively). Thus in the American case there are strong aggregate relationships between social class and nonvoting on one side and Republican voting on the other, but virtually none between class and Democratic voting. Similarly, the relationship between nonvoting and Democratic voting is essentially zero, but is very strongly negative when Republican voting is substituted. In the London example, there is a weak negative relationship between class and nonvoting but very strong relationships between class and party voting (negative with Labour, positive with Conservative), and a weak (positive) relationship involving Conservative voting. In Paris, the relationships between nonvoting and political preference are isomorphic to each other, negative, and extremely small.

While one ought not to generalize excessively from cases of such limited range and methodological sophistication as this, these results broadly conform to what could be expected from comparative survey and other evidence. To put it in the most deliberately provocative way, the whole range of data suggests that if there is class struggle in American politics, it is almost entirely one-sided. Or, to put the matter another way, there is certainly a right in American electoral politics, with the Republican party its dominant organizational vehicle. But there is no left party, sociologically speaking. If it is true that there are important and persistent modal differences between the parties on public policy roll calls in Congress, it is also true that the Democratic party is not remotely to be confused with a left party in organizational structure, or in terms of serious motivation to mobilize the "party of nonvoters." Accordingly, the party of nonvoters as a whole corresponds comparatively to the place in the entire potential position array that a genuinely left party would occupy if it had been historically possible to organize one in this country. And the larger this abstentionism grows, the greater the class skew and the more likely it is that, sooner or later, conservatives who occupy only a minority position in the whole population can prevail at the top of the political vacuum that is thus created. It is worth observing, for example, that Ronald Reagan's famous "mandate" of 1980 rested upon the support of just 28.0 percent of the potential electorate. This compares with the 28.3 percent that Wendell Willkie received in 1940 while being heavily defeated by Franklin D. Roosevelt. For that matter, it contrasts with the 40.2 percent of the French potential electorate who supported Giscard d'Estaing in his *losing* bid for reelection in the 1981 presidential election.

IV. Recent Turnout Change in the United States

The cumulative thrust of the foregoing argument is that, essentially, the "party of nonvoters" fills up approximately to the extent that the Democratic party loses coherence and drifts away from the leftward end of our narrow political spectrum. Conversely, it shrinks to the extent that this party gains coherence and moves somewhat left of center. The period since the mid-1960s has been chiefly marked by the disruption of the old Democratic coalition. This disruption has been manifest in one way or another in every election from 1968 through 1980: by the rise of George Wallace in 1968, the nomination and crushing defeat of George McGovern in 1972, the 1976 nomination and election of the most conservative and inept Democrat in at least a half century, and the shattering presidential and senatorial defeats of 1980. This process of degeneration has been conspicuously associated of late with a rightward shift in the party's elites and a manifest bankruptcy of ideas on the part of its associated intellectuals. The latter is properly the subject of another essay. It is enough to argue here that if the "party decomposition" much discussed in recent years really turns out in the end to be *Democratic* party decomposition, this should be circumstantially linked to appropriately concentrated changes in voter participation since the mid-1960s. And so indeed it is.

Let us begin by a straightforward tabulation of the Census Bureau's survey of turnout rates at the beginning and end of this period, stratified by occupational class. (The end, in this case, will be 1976 for presidential years, with 1980 data still absent, and 1978 for off-year elections. The data are for males only.)

As is very clear from Table 7 the turnout declines in presidential and off-year elections across this period become larger the further down the class structure one travels. There is very evidently a major break in this regard between all middle-class occupations taken as a whole and all working-class occupations taken as a whole. The same general pattern also appears when the drop-offs between a presidential election and the immediately succeeding off-year election are examined. Cross-sectionally, drop-off grows, if not wholly monotonically, as one moves down the class structure. Longitudinally, growth in the amplitude of drop-off from the 1964–66 pair to the 1976–78 pair follows the same course. As turnout decays in either presidential elections or in off-year elections, the gap in participation between the top and the bottom of the class structure also grows. Put another way, by 1976 the very broadly middle-class component of the entire adult male employed

Table 7. Occupational class, turnouts, and changes in turnout, 1964–78

Occupational class	Presidential-year turnouts 1964	1976	Shifts Absolute	Relative	Off-year turnouts 1966	1978	Shifts Absolute	Relative	Drop-off rates 1964–66	1976–78
1. Professional-managerial	84.0	76.4	−7.6	−9.0	69.5	60.9	−8.6	−12.4	17.3	20.3
2. Clerical-sales	81.2	69.9	−11.3	−13.9	66.3	54.7	−11.6	−17.5	18.3	21.7
Middle class subtotal	83.1	74.6	−8.5	−10.2	68.5	59.1	−9.4	−13.7	17.6	20.8
3. Craftsmen, etc	72.0	56.0	−16.0	−22.2	56.7	40.7	−16.0	−28.2	21.3	27.3
4. Operatives (semiskilled)	63.2	47.4	−15.9	−25.0	48.3	32.3	−16.0	−33.1	23.6	31.9
5. Laborers (unskilled)	57.3	43.7	−13.6	−23.7	41.1	28.1	−13.0	−31.6	28.3	35.7
Working class (manual) subtotal	66.6	50.9	−15.7	−23.6	51.2	35.5	−15.7	−30.7	23.1	30.3
6. Service workers	72.8	57.4	−15.4	−21.2	57.9	44.3	−13.6	−23.5	20.5	22.8
7. Farm owners, managers	79.5	77.2	−2.3	−2.9	71.0	64.0	−7.0	−9.9	10.7	17.1
8. Farm laborers	37.4	34.8	2.6	−7.0	32.7	24.5	−8.2	−25.1	12.6	23.6
Recombined										
Propertied (1,7)	83.3	76.5	−6.8	−8.2	69.7	61.2	−8.5	−12.2	16.3	20.0
Subaltern middle class (2)	81.2	69.9	−11.3	−13.9	66.3	54.7	−11.6	−17.5	18.3	21.7
Subtotal	82.7	74.7	−8.0	−9.7	68.8	59.4	−9.4	−15.7	16.8	20.5
Upper working class (3, 6)	72.2	56.4	−15.8	−21.9	57.0	41.7	−15.3	−26.8	21.1	26.1
Lower working class (4,5,8)	60.0	45.7	−14.3	−23.8	45.9	30.8	−15.1	−32.9	23.5	32.0
Gap, propertied/lower	23.3	30.8			23.8	30.4				
All respondents	73.5	61.9	−11.6	−15.8	59.2	46.4	−12.8	−21.6	19.5	23.0

Note: Data compiled from U.S. Census surveys of registration and voting participation, 1964–78 (series P-20, Current Population Reports); 1964 detailed stratification based on estimates derived from 1966 and 1968 surveys. The "drop-off" measure was first discussed in Walter Dean Burnham, "The Changing Shape of the American Political Universe," *American Political Science Review* 59 (1965), 7–28: [Chapter One in the present collection]. It measures the relative magnitude of voting participation between a presidential election and the immediately succeeding off-year election; e.g., for professional-managerial, 1976–78, the absolute decline is 15.5 percent, but the dropoff is 20.3 percent (15.5/76.4).

labor force was about 47 percent, while it constituted 56.1 percent of those who voted, but only 31.6 percent of the "party of nonvoters." It may well be said that such a skew does not matter so long as manifest economic class issues are not of primary importance in our politics. This was certainly true in the "normalcy era" before 1929, as it was equally true of the 1972 situation that Wolfinger and Rosenstone analyze. But it is quite obvious that zero-sum economics is central to the political situation in the eary 1980s. In such a case (as the Depression elections most vividly demonstrate), representation-mobilization phenomena based upon a class struggle extending from the Republican upper and upper-middle classes to the rest of the population may well put in an appearance.

Viewed geographically, the Census Bureau's data, now provided in the Congressional District Data Book series and including estimated-population denominators annually in the pages of the Federal Register, give us the basis for inquiry from the mid- to late 1960s through 1980. Where has change been concentrated globally from 1968 through 1980 (using the latter year's district boundaries)? The answer is provided very broadly in Tables 8A and 8B. The 1968–80 decline is broadly based, but with important "extreme cases." Of the fifty districts where turnout actually increased in this period, more than half were in the former Confederacy, and nearly four-fifths in the South and border states taken together. Of the seventy-six seats where the decline was 20 percent or more from the 1968 base, nearly half were concentrated in the central-city core areas of the nation's three largest conurbations, and the large majority of the others were found in smaller but equally proletarian settings. Rather similar considerations apply to the extremes of absolute participation in the 1980 congressional election. The two districts with the lowest participation in the country in contested congressional races were the 12th New York (Bedford-Stuyvesant) at 18.8 percent of estimated potential electorate, and the 22d New York (Grand Concourse-Bronx), with a participation rate of 21.8 percent. Both are represented by liberal Democrats, one black and one white. The two turnout leaders were both wealthy suburban districts: the 10th Illinois (73.6 percent), just outside Chicago, and the 3d Minnesota (71.0 percent), suburban Minneapolis. Both are represented by Republicans. Truly, the overall profile of 1980 participation in American congressional elections is a case of the rich getting richer, the poor getting poorer!

Before turning to our final in-depth data discussion of turnout change and its relationships in the Chicago metropolitan area, let us once more place these 1980 congressional turnouts in their comparative setting (Table 9). Only in some such way can the true magnitude of the American problem be grasped to some extent. Further detailed comment seems unnecessary.

Our last data exploration is as tentative as the earlier ones, but it may cast some light on aggregate change patterns in a single metropolitan area, Chicago. We begin by noting a very strong differential rate of decline from 1964

Table 8A. United States: Turnout shifts, 1968–80 (Percentage of total districts by shift category)

Region	−30 and over	−20 to −29	−10 to −19	−0.1 to −9	+0.1 to +9	+10 and over	Total N
Northeast	6.7	36.0	50.6	5.6	1.1	—	89
Midwest	3.8	8.5	46.2	37.7	3.8	—	106
West	5.9	16.2	27.9	19.1	1.5	1.5	68
North and West	5.3	20.5	45.2	24.7	3.8	0.4	263
Border	—	15.6	12.5	34.4	28.1	9.4	32
Non-South	4.7	20.0	41.7	25.8	6.4	1.4	295
South	1.9	3.8	11.5	30.8	23.1	28.8	52
Total USA	4.3	17.6	37.2	26.5	8.9	5.5	347
Three largest conurbations	19.4	35.5	33.9	11.3	—	—	62

Table 8B. United States: Congressional turnout in 1980 (Percentage of total districts by turnout category)

Region	0–29	30–39	40–49	50–59	60–69	70 and over	Total N
Northeast	6.3	6.3	27.1	52.1	8.3	—	96
Midwest	—	3.7	11.2	53.3	29.9	1.9	107
West	1.3	13.2	26.3	47.4	11.8	—	76
North and West	2.5	7.2	20.8	51.3	17.6	0.7	279
Border	—	9.4	34.4	46.9	9.4	—	32
Non-South	2.3	7.4	22.2	50.8	16.7	0.6	311
South	3.8	23.1	50.0	23.1	—	—	78
Total USA	2.6	10.5	27.8	45.2	13.4	0.5	389
Three largest conurbations	10.6	21.2	30.3	25.8	10.6	1.5	66

Note: Contested races only.

to 1980 in mean congressional turnouts as between the region's central-city districts and its suburban districts, a phenomenon that finds many parallels elsewhere. The first thing to note in Table 10 is that the collapse of voting participation in presidential years is much heavier in Chicago proper than in its suburbs. The turnout gap between the two, which was 3.6 percent in 1964, increased to 19.7 percent by 1980. The city, of course, was preponderantly Democratic throughout, as the suburban districts were heavily Republican (the Chicago suburbs were among the very few areas outside the South to give majorities even for Barry Goldwater in 1964). The decline in Chicago in both presidential and off years was particularly concentrated on the Republican side—not surprising on both racial-ethnic or economic-interest lines. Conversely, in presidential years at any rate the suburban pattern is one of rather striking continuity, with Democrats winning typically a little more than one-fifth of the potential electorate and Republicans not far short of

Table 9. Turnout rates in three countries by district and metropolitan status

A. Turnout rates across all districts: N of districts

Country and metro/ nonmetro status	0–29	30–39	40–49	50–59	60–69	70–79	80–89	90–100	Total
USA, 1980									
Conurbations	7	14	20	17	7	1	—	—	66
All others	3	27	88	159	45	1	—	—	323
Britain, 1979									
Greater London	—	—	—	8	23	56	5	—	92
All others	—	—	—	1	30	315	78	—	424
France, 1978									
Greater Paris	—	—	—	—	—	34	49	—	83
All others (metropole only)	—	—	—	—	4	41	337	9	391

B. Turnout rates across all districts: % of districts

	0–29	30–39	40–49	50–59	60–69	70–79	80–89	90–100	Total
USA, 1980									
Conurbations	10.6	21.2	30.3	25.8	10.6	1.5	—	—	100
All others	0.9	8.4	27.2	49.2	13.9	0.3	—	—	100
Britain, 1979									
Greater London	—	—	—	8.7	25.0	60.9	5.4	—	100
All others	—	—	—	0.7	7.1	74.3	18.4	—	100
France, 1978									
Greater Paris	—	—	—	—	—	41.0	59.0	—	100
All others	—	—	—	—	1.0	10.5	86.2	2.3	100

Note: To this should be added the point that in the United States in 1980, four conurbation seats were unopposed by the second major party, and forty-two seats in the rest of the country. There were no uncontested parliamentary seats either in Great Britain in 1979 or in France in 1978.

Table 10. Turnout and aggregate partisanship in the Chicago, metropolitan area (Percentage of potential electorate)

A. Presidential years

	Central-city districts			Suburban districts		
	Nonvoting	D	R	Nonvoting	D	R
1964	30.4	46.9	22.7	26.8	28.8	44.4
1968	36.7	40.6	22.7	31.9	20.5	47.6
1972	45.0	37.1	17.9	30.8	23.3	44.9
1976	48.9	38.9	12.1	37.8	22.1	40.1
1980	53.9	36.8	9.3	34.2	20.3	45.5

B. Off years

	Nonvoting	D	R	Nonvoting	D	R
1966	43.8	33.8	22.4	40.7	16.4	42.9
1970	50.2	36.7	13.1	48.6	19.0	32.4
1974	67.6	26.3	6.1	60.9	16.8	22.3
1978	64.2	27.4	8.4	55.9	15.0	29.1

one-half. It would appear unlikely that the decay of the Daley machine in Chicago accounts very convincingly for the decline. Daley did not die until 1977, by which time the city's turnout collapse had largely run its course.

Not surprisingly as these gaps in participation opened up and as turnout declined in presidential years, the relationships between class indicators and nonvoting tended to increase. Taking all the thirteen Illinois districts in the Chicago area as a whole, the (negative) relationship between percentage white-collar (or middle-class) and abstention rates yielded an r^2 of .107 in 1964, .288 in 1968, .615 in 1972, .473 in 1976, and .567 in 1980. Similarly, an interesting evolution can be seen in the increasingly one-sided relationship between the percentage nonvoting and percentages Democratic and Republican of the potential electorate (Table 11). This series of "snapshots" reveals little change in the relationship between Democratic voting and nonvoting, but *pari passu* with the very substantial urban declines in participation—an increasingly strong negative relationship between abstentions and Republican shares of the potential electorate. This is but a cumbersome way of conveying the point that, in 1980, the variation across district in the Democratic percentage of the potential electorate was not very large compared with the situation with Republican candidates. What this implies is that, in the poorer urban districts of both races, the Democrats may or may not be "left" in some meaningful sense, but the Republicans as a "right" party do not constitute a very frequented alternative at the level of House elections. With aggregate two-party competition declining and in the absence of any organized party to the left of the Democrats, the "party of nonvoters" in Chicago has swelled from less than one-third to nearly one-half of the city's total electorate. In the suburbs, on the other hand, turnout holds up much more robustly in presidential years at any rate, and the relationship between Democratic and Republican shares of the potential electorate has tended to remain quite stably defined in congressional elections there.

Taken all in all, then, the evidence is quite substantial that the long-term decline in voting participation has been particularly concentrated geographically in central-city and sociologically analogous districts (both minority and white, it may be added) outside the South. These are marked by the usual

Table 11. Regressions between nonvoting and partisanship, in the Chicago metropolitan area, 1964–80

Year	Nonvoting and democrat			Nonvoting and republican		
	$Y_c =$	$r =$	$r^2 =$	$Y_c =$	$r =$	$r^2 =$
1964	39.051 ± 0.078X	+.048	.002	60.869 − 1.0751X	−.550	.302
1968	17.601 + 0.478X	+.321	.103	82.330 − 1.476X	−.723	.523
1972	10.018 + 0.545X	+.562	.315	89.937 − 1.544X	−.888	.789
1976	14.165 + 0.409X	+.365	.133	85.364 − 1.401X	−.799	.638
1980	13.120 − 0.374X	+.467	.218	86.891 − 1.374X	−.888	.789

complex of social problems; they had already tended to have relatively low participation rates at the beginning of the contemporary decline period. Similarly, the decline has been smallest outside the South in the most affluent districts, which originally tended to have high participation rates. (In the Chicago metropolitan case, the r^2 between the percentage white-collar among employed males in 1970 and the net 1968–80 turnout shift was .536). Viewed in terms of the Census Bureau's surveys, it is equally clear that this decline has been concentrated—both in presidential and in off-year elections—among people of lower education and lower social class, people who would vote Democratic if it were still worth their while to do so. By almost any yardstick, then, we can say that in the very long run American electoral politics has moved de facto toward an oligarchy dominated by middle-class and propertied population elements to a degree wholly unknown either to contemporary Europe or to the nineteenth-century United States. The year 1980, indeed, comes closer to such a condition than any election in recent history. This fact may be not randomly related to its candidates, issues, and outcome. For that matter—though no evidence can be adduced here to support the proposition—it may well be that the recent quest by congressional incumbents to move from political-partisan to ombudsmen strategies, documented by Mayhew and Fiorina among others, has made its own contribution to discouraging participation among the lower orders in American society.[33]

V

But what does it all matter? If the dominant tradition of scholarship in this field is so doggedly insensitive to the substantive problematics, might there not be reasons for this? In particular, it is worthwhile to recall again that Wolfinger and Rosenstone are very probably correct in their view that there are few detectable policy or manifest-interest differences between those who vote and those who don't, and to remind ourselves that the nonvoters in the 1977 New Jersey sample are a remarkably conservative group overall, differing only in small degree from the voters on many class-related dimensions. Such findings reflect the current realities, so much so that, even if Ronald Reagan won his famous "mandate" with only 28.0 percent of the potential electorate actually voting for him, a broader claim for such a mandate can be made than simple analysis of voting participation data might suggest, especially to enthusiasts of the Left. If the Democratic party is amorphous, relatively disorganized, unable and unwilling to develop seriously the resources needed to reach this apparent potential constituency, there are important reasons for that too. For this party, no less than the resurgent Republicans, remains an accurate barometer of the actual structure of public opinion in the here and now as practicing and very practical professionals in politics comprehend it.

This leads us to certain speculative propositions about the American turn-out problem. These entail a close approach to a chicken-and-egg problem on one hand, and—even more dubious, no doubt—to the whole tangled web of "false" and other kinds of consciousness on the other. But this can't be helped.

Very broadly, American politics is based upon a mass electorate that is seriously baffled by the complexity of our constitutional arrangements, has extremely low levels of information, and has not been educated by any social instrumentality—certainly not by the schools!—to an understanding of *politics*. Associated with this are other characteristics that serve to morselize social experience, and to focus attention very concretely on the immediate milieu in which the individual attempts to live out his or her life. In this regard, analysts as different as C. Wright Mills, Robert E. Lane, and the authors of *The American Voter* are all describing the same phenomenon.[34] Other electorates, of course, have quite different characteristics, shaped by the history and political organizations peculiar to their societies. This is one reason among others for denying that there is necessarily any such thing as *the* nature of belief systems in mass publics. Granted the objective reality that modern industrial-capitalist society is everywhere dominated by power concentrations and more or less extreme and meaningful social inequality, the question arises as to how, if at all, this reality forms a part of the *organized* electoral market.

In the United States the pattern of politics has been rooted in the uncontested hegemony of a single idea set, a single overarching "liberal tradition." This is the kind of political culture that bourgeoisies around the world would give much to have implanted in their own countries. In the specific context of class society as mediated by and through this individualist-proprietarian liberal tradition, reinforced heavily by an extraordinary comparative pervasiveness and intensity of religion, what emerges politically is an enormous case of *ceteris paribus*. If all other things were equal in modern class society, if there were no historically grounded, hence legitimate, alternatives to this bourgeois hegemony in political culture, then only the propertied Right would be effectively organized to the extent that it needed to be to achieve its ends through the state. In such a context the lower orders would be socially and politically disorganized. In the electoral market, the alternatives presented to them would be so defined that they would tend to abstain. The so-called left party in this universe would be deeply shot through with producer-class elites and followings, and would tend toward incoherence and, especially, organizational-mobilizing ineffectiveness in the long run. There would of course be other, not so pleasant, by-products, such as an extremely high comparative level of crime and other forms of "anomic interest articulation."

If these arguments are correct, it would follow that the huge party of non-voters, like the poor, ye will always have with you in America so long as this

Lockean-liberal hegemony remains uncontested. It would be speculative in the extreme to attempt here to address the question of whether, or to what extent, its *universality* in the culture might be overthrown. Nothing less than the rediscovery of politics would be necessary to such overthrow. Yet the entire weight of our social, no less than our political, history in this century has been in the direction not of rediscovering politics but of suppressing it.[35] But even short of that, it has sometimes happened in this modern history that the lower orders have been decisively remobilized, and that the Democratic party has had a rather more leftish position in the spectrum than it occupied in the 1970s. Unfortunately, it required nothing less than the economic holocaust of 1929–33 to provide the catalyst on the last occasion when this occurred.

Perhaps in the 1980s less cosmic traumas could produce similar effects; perhaps not. It is not difficult to construct economic and political scenarios that would involve huge oppositional "surges" out of the nonvoting pool and into the active electorate, especially with the Right now in power. On the other hand, there is equally no a priori reason to doubt that the newly dynamic Republicans of the 1980s could coexist in power for quite a while with a huge and growing vacuum at the bottom of the political system. As E. E. Schattschneider put the matter, referring to an earlier such period, "There is very little in the history of the Republican party from 1896 to 1932 to support the conclusion that Americans cannot create the kinds of parties they want."[36]

Put in the most oversimplified terms: so long as what Americans *normally* want is Lockean liberalism, but in the context of an advanced industrial-capitalist class society, then enormously large abstentions from the electoral market concentrated among the lower orders of this society will almost certainly continue indefinitely. They will also probably continue to grow somewhat, assuming of course that no runaway catastrophic processes overwhelm the economy. If large numbers of Americans come to want something different, the effects on the shape and characteristics of this electoral market will be of very large magnitude, including both an evacuation of the nonvoting pool and the development of modes of political organization appropriate to the social needs of these new constituents.[37] If not, then American politics will become even more dominated by issues of interest to the haves than it is now. The "party of nonvoters" will continue, in its enormous size and social concentration, to be a profound but entirely silent critique of this state of affairs.

But, by the same token, political scientists would probably find incentives to shift their scholarly imaginations and energies toward more profitable lines of inquiry. After a time what would be left to discuss, even on a procedural-methodological level? Still, nature appears to abhor a vacuum in politics as elsewhere.The choice appears to be between a continuing, gradual euthanasia of the American electorate, or the revitalization of democratic forces in this country. As in the past, the size, shape, and composition of the "party of nonvoters" will quite sensitively reflect which path is being taken.

Addendum

The current intensity of scholarly interest in the American turnout problem is reflected in a special issue of *American Politics Quarterly* (volume 9, no. 2, April 1981) that is entirely devoted to this subject. Regrettably, the text of this essay was developed during the late spring of 1981, and did not therefore incorporate or refer to the discussions of turnout in this issue. However, a review of the six contributions involved, all exploring various aspects of the problem with satisfactory methodological rigor, appears to require no major changes in the arguments developed in the preceding pages.

Richard Boyd's article "Decline of U.S. Voter Turnout: Structural Explanationa" (pp. 133–60) "looks optimistically on levels of turnout in the United States" (p. 153) for the following reasons: (a) changes in the age composition of the electorate, concentrating relative weight among the youngest and oldest (therefore, the most abstentionist) cohorts; (b) a participatory political culture, which has led concretely to a huge proliferation in the sheer number of elections (a by-product of this, presumably, is that there are far more voters who vote in *some* election than in any *given* election, including the presumably high-stimulus presidential elections); (c) the presence, detected via the Center for Political Studies Voter Validation study and a sample drawn from Middlefield, Connecticut (1970 population 4,132), of a large core electorate that votes with some frequency. One may note as to (a) that the comparative dimension is of course not covered; as to (b), that the argument appears to imply the presence of substantial numbers of voters who vote in local elections, primaries, etc., but also abstain in presidential elections, a proposition that might need more support than is provided in (c), especially in light of the fact that abstentions and the tendency of abstention to increase since 1960 are particularly concentrated in metropolitan cores rather than in small towns. Of course, all things are relative.

John R. Petrocik's essay "Electoral Oscillation" (pp. 161–80) adds light to the "overreporting" problem in surveys first addressed by Clausen, and also—undoubtedly correctly—reiterates the Wolfinger-Rosenstone argument that "a policymaker contemplating a change in the election laws should not be led to believe that nonvoting necessarily distorts the voice of the people in elections" (p. 177). The article by Carol A. Cassel and David B. Hill, "Explanations of Turnout Decline: A Multivariate Test" (pp. 181–95), attempts directly to attack the recognized fact among practitioners of the art that the recent turnout decline is a "puzzle." The authors approach the usual variables through multiple regression and probit analysis, only to find at the end that

> the analysis leaves the puzzle largely unsolved. Our findings suggest that changing political attitudes, like declining partisanship and concern over

the election outcome have contributed somewhat to lower turnout. At the same time, however, the electorate's increased education level should have boosted turnout. Since these effects counteract each other, our explanation of the decline in turnout is minimal [p. 192].

One hazards the supposition that the puzzle remains unsolved because the underlying change phenomena reflect a complex interaction between very broad systemic change and individual political cognitions. Unfortunately, no one knows better than I how very difficult it is to formulate macro-systemic indicators and build the bridges between them and micro-level attributes in a way that is plausible to the research community in this field. On the other hand, we ought not forever to be limited in our inquiries by Abraham Kaplan's "law of the instrument."

The article by Herbert F. Weisberg and Bernard Grofman, "Candidate Evaluations and Turnout" (pp. 197–219), arrives at the very interesting finding based on the 1976 CPS study that the most important cause of candidate-based abstention in 1976 was satisfaction. Citizens who like *both* major candidates are less likely to vote than others; paralleling this, alienation had a more limited effect (few citizens disliked both candidates in 1976) and indifference had practically none at all. However, the global range of explanation using candidate cognitions as the independent variable is vanishingly small: "Overall, the evidence indicates very limited effects of the candidate choice on abstention in 1976. The bulk of the abstention simply cannot be explained by the candidate set. . . . Abstention is predominately exogenous to citizen evaluations of the candidates" (pp. 216–17). So the puzzle very much remains a puzzle, at least on this dimension.

M. Margaret Conway contributed an essay, "Political Participation in Midterm Congressional Elections" (pp. 221–44), that elegantly portrays the basic sociodemographic and attitudinal bases of voting participation—again based on the CPS surveys of 1970, 1974, and 1978. While there are no overwhelming novelties in her findings, the essay adds welcome precision to the current discussion. As Professor Conway observes, "Voting participation in congressional elections can be seen as a process which develops out of the pattern of life experienced of the individual, with those whose social position (as indicated by age, education and income) results in their having a potentially greater stake in the outcome of the election being more likely to vote" (p. 240). Quite so; as we have suggested, this is just what one would expect to encounter in a *ceteris paribus* political universe.

The concluding short essay by Raymond E. Wolfinger et al., "Presidential and Congressional Voters Compared" (pp. 245–56), utilizes the Census Bureau Surveys studied at length in *Who Votes?* to do the comparative job indicated in the title. The specific question is, with drop-offs of from one-fifth to one-quarter of the previous presidential year's turnout levels, whether major compositional differences exist between presidential-year and off-year electorates. The major differential appears to be that of age (the presidential-year turnout gap between young and middle-aged voters grows even larger in off-year election situations). Otherwise, "the most prominent feature of a midterm election is its lower turnout. Once this has been taken into account, conclusions about the demographic correlates of turnout in presidential elections are also germane to midterm years" (p. 255). In short, off-year electorates are very much like presidential-year electorates, only a little more so.

A review of these essays as a whole brings us back to the discussion with which our own study began. The command of rigorous technique is admirable. The aesthetic objections to low turnout rates are either clearly stated (as in Petrocik) or are in fact reversed by a cheery optimism that asserts that turnout rates are much higher than they look (Boyd). Substantive problematics are conspicuous by their absence. And the "puzzle" of declining turnout remains overwhelmingly a puzzle after the most sophisticated methodologies have been brought to bear on sophisticated (but hardly fault-free) survey data. One continues to look in vain for any sign of insight that might be provided by historical or comparative analysis, or by social theory that is grounded in something other than individual-plus-group social psychology. The "puzzle," I suspect, is very probably insoluble within the limitations this tradition imposes. Contributions of this sort have enormously enriched our knowledge of details, some of which are of no small importance. But it would appear that they leave the authors as baffled about the ultimate causalities involved as they leave the reader.

NOTES

1. Herbert L. A. Tingsten, *Political Behavior* (London: King, 1937).
2. André Siegfried, *Tableau politique de la France sous la IIIe République* (Paris, 1913). See also F. Goguel, *Géographie des élections françaises de 1870 à 1951* (Paris, 1951); and Claude Leleu, *Géographie des élections françaises depuis 1936* (Paris, 1971).
3. There are occasional important exceptions, for example the detailed information on registration and voting by sex in Chicago, 1916 and 1920, reported in the *Chicago Daily News Almanac* for 1917 and 1921. But these are rare. So far as turnout is concerned, of course, European state-enrollment systems necessarily provide reasonably accurate and explicit statements of the potential electorate at all levels of reporting. In the United States, such electorates must be reconstructed from census data.
4. See V. O. Key, Jr., *Southern Politics in State and Nation* (New York: Knopf, 1949), pp. 491–663; and E. E. Schattschneider, *The Semi-Sovereign People* (New York: Holt, Rinehart & Winston, 1960), pp. 97–113.
5. In 1964, for example, the actual national turnout was 63.5 percent of estimated citizen voting-age population, but the University of Michigan Center for Political Studies rate was 77 percent. This discrepancy has produced at least one research article: see Aage Clausen, "Response Validity: Vote Report," *Public Opinion Quarterly* 32 (Winter 1968–69): 588–606.
6. Seymour M. Lipset, *Political Man* (New York: Doubleday, 1960).
7. A case in point is the controversy over reconstructing the 19th-century American electorate (including its turnout rates) between Philip E. Converse and Jerrold G. Rusk on one side and myself on the other. See Walter Dean Burnham, "Theory and Voting Research: Some Reflections on Converse's 'Change in the American Electorate,'" *American Political Science*

Review 68 (1974): 1002–23, and the subsequent exchange among the three participants, pp. 1024–57. ["Theory and Voting Research" appears as Chapter Two in the present collection.]

8. Raymond Wolfinger and Steven J. Rosenstone, *Who Votes?* (New Haven: Yale University Press, 1980).

9. Arthur T. Hadley, *The Empty Polling Booth* (Englewood Cliffs, N. J.: Prentice-Hall, 1978). But Hadley may well have uncovered something of first-rate importance: that the chief attitudinal discriminator between a certain rather substantial group of voters and their sociological counterparts among nonvoters is the belief on the part of the latter that chance rather than purposive activity determines what happens to them in life. As Hadley observes, "America's present problem is not a violently split society where everyone is voting. It is an apathetic, cross-pressured society with strong feelings of political impotence, where more and more people find their lives out of control, believe in luck, and refrain from voting" (p. 113).

10. Wolfinger and Rosenstone, *Who Votes?* pp. 36–37.

11. Samuel H. Barnes, *Representation in Italy* (Chicago: University of Chicago Press, 1977), pp. 61–64.

12. Which of course is very largely the explanation for the good heuristic fit of Anthony Downs's spatial model to the Italian case, ibid., pp. 97–115, and its lack of fit to the American, at least during the Eisenhower era. See Donald E. Stokes, "Spatial Models of Party Competition," in Angus Campbell, Philip E. Converse, Warren E. Miller, and Donald E. Stokes, *Elections and the Political Order* (New York: Wiley, 1966), pp. 161–79.

13. This has recently taken the form in about a dozen states of such modifications of the personal-registration requirement as same-day registration, driver's license registration, registration by mail, or combinations of these. These states are located primarily in the upper Midwest and the West, and have had a history of reform and early democratization of election procedures and requirements. However, viewed across the twenty years 1960–80, the basic finding is that these relaxations of personal-registration hurdles have worked to slow down the rate of post-1960 decline compared with other non-Southern jurisdictions. Thus, four states that have adopted one or another variant of these reforms are Michigan, Minnesota, Oregon, and Wisconsin. Their 1960–80 turnout regression slopes are −2.86, −1.61, −2.23, and −1.55 respectively (corresponding to a 1980 turnout level that was 13.6 percent, 8.3 percent, 10.1 percent and 7.8 percent respectively, below the 1960 level). This compares with North Dakota, the only state in the Union that still lacks personal-registration requirements altogether, with a slope of −2.73 and an absolute 1970–80 decline of 15.2 percent. On the other hand, this decline in the non-Southern states as a whole yields a slope of −3.36 (a loss of 16.1 percent between 1960 and 1980). And in unreformed New York—one of the most rapidly declining states in the country—the 1960–80 turnout slope was −4.25, reflecting an absolute loss across the period of 20.4 percent. So there is some reason to suppose that these changes have cushioned the fall in the states where they have been adopted. But—as the North Dakota "control" case makes particularly clear—the 1960–80 turnout decline outside the South is quite substantial irrespective of rules of the game or changes in them; it is, of course, mostly behavioral rather than procedural.

14. Wolfinger and Rosenstone, *Who Votes?*, pp. 102–14.

15. Reported in New Jersey Department of State, *Electoral Participation in New Jersey* (Trenton, 1979), pp. 21–22.

16. This was nineteenth-century America, to be sure; the interests in question were not class interests as such. See, for example, Lee Benson, *The Concept of Jacksonian Democracy: New York as a Test Case* (Princeton: Princeton University Press, 1961), especially pp. 270–328.

17. Samuel Popkin et al., "Comment: What Have You Done for Me Lately? Toward an Investment Theory of Voting," *American Political Science Review* 70 (1976): 779–805 The overall perspectives of the present essay closely parallel the analysis offered in this "comment."

18. See, for example, Bruce A. Campbell and Richard J. Trilling, eds., *Realignment in American Politics: Toward a Theory* (Austin: University of Texas Press, 1980), pp. 176–201, 263–87,

and the comprehensive bibliography contained in this work; and my "The System of 1896: An Analysis," in Paul Kleppner et al., *The Evolution of American Electoral Systems* (Westport, Conn.: Greenwood Press, 1981).

19. Kristi Andersen, *The Creation of a Democratic Majority, 1928–1936* (Chicago: University of Chicago Press, 1979).

20. Tingsten, *Political Behavior,* pp. 10–78.

21. Ibid., pp. 156–57.

22. Louis Galambos, *The Public Image of Big Business in America, 1880–1940* (Baltimore: Johns Hopkins University Press, 1975).

23. Louis Hartz, *The Liberal Tradition in America* (New York: Harcourt, Brace, 1955). See also Donald J. Devine, *The Political Culture of the United States* (Chicago: Rand McNally, 1972).

24. See, for example, John Willett, *Art and Politics in the Weimar Period* (New York: Pantheon, 1980), pp. 98–104; see also the section "Americanism and Fordism: From Antonio Gramsci's prison Notebooks," in Q. Hoare and G. N. Smith, eds., *Selections from the Prison Notebooks of Antonio Gramsci* (New York: International, 1971), pp. 279–318.

25. Andersen, *Creation of a Democratic Majority,* especially pp. 53–72.

26. Samuel Lubell, *The Future of American Politics* (New York: Harper, 1952), pp. 28–80.

27. V. O. Key, Jr., *Politics, Parties and Pressure Groups,* 4th ed. (New York: Crowell, 1958), pp. 203–5.

28. Linkages of this sort are (properly) stressed by Hadley, *Empty Polling Booth.* A long-term decline in turnout of the sort that has happened in both presidential and off years since 1960 could perhaps be explained ad hoc, in response to the specific issues and candieates of specific elections (e.g., reactions to Watergate and the turnout slump of 1974), but this makes for important theoretical difficulties. It seems probable, instead, that we are dealing here with the effects of a general, if diffuse, system-wide crisis to which each variable component has made its own contribution. See, for example, Norman H. Nie, Sidney Verba, and John R. Petrocik, *The Changing American Voter* (Cambridge: Harvard University Press, 1976; 2d ed., 1979), pp. 345–56.

29. The most exhaustive treatment of the subject is J. Morgan Kousser, *The Shaping of Southern Politics: Suffrage Restriction and the Establishment of the One-Party South, 1880–1910* (New Haven: Yale University Press, 1974). The story it tells is a testament to political man's ingenuity in seeking out legal means to get what he wants.

30. See sources cited in Table 4.

31. Nie, Verba, and Petrocik, *Changing American Voter,* pp. 47–73.

32. Lipset, *Political Man,* pp. 181–219.

33. David R. Mayhew, *Congress: The Electoral Connection* (New Haven: Yale University Press, 1974); Morris Fiorina, *Congress: The Keystone of the Washington Establishment* (New Haven: Yale University Press, 1977). See also Walter Dean Burnham, "Insulation and Responsiveness in Congressional Elections," *Political Science Quarterly* 90 (1975): 411–35. [Chapter 6 of this Volume].

34. C. Wright Mills, *The Power Elite* (New York: Oxford University Press, 1956), especially "The Mass Society" and "The Conservative Mood," pp. 298–342; Robert E. Lane, *Political Ideology: Why the American Common Man Believes What He Does* (New York: Free Press, 1962), especially pp. 413–59; Angus Campbell, Philip E. Converse, Warren E. Miller, and Donald E. Stokes, *The American Voter* (New York: Wiley, 1960), especially pp. 146–215, 521–58.

35. The literature on this point, especially by historians, has become very substantial in recent decades. See, for example, Robert Wiebe, *The Search for Order, 1877–1920* (New York: Hill & Wang, 1969); and Samuel P. Hays, "Political Parties and the Community-Society Continuum," in William Nisbet Chambers and Walter Dean Burnham, eds., *The American Party Systems,* 2d ed. (New York: Oxford University Press, 1975), pp. 152–81. For an almost uniquely rich case study of some important consequences, see also Robert A. Caro, *The Power Broker: Robert Moses and the Fall of New York* (New York: Random House, 1974).

36. E. E. Schattschneider, "United States: The Functional Approach to Party Government," in Sigmund Neumann, ed., *Modern Political Parties* (Chicago: University of Chicago Press, 1956), pp. 194–215, at p. 201.

37. The organizational forms that seem relevant to such clienteles are spelled out with classic clarity and detail in Maurice Duverger, *Political Parties* (New York: Wiley, 1959). Needless to say, it is genuinely difficult to imagine any changes in consciousness or behavior in the foreseeable American future that could provide the social basis for the creation of such organizations.

III

Premonitions of Upheaval: The 1970s

6

Insulation and Responsiveness
In Congressional Elections
1975

The 1974 congressional election is part of a crisis sequence in the current history of the American political system. The tremendous upheaval which has been going on in our electoral politics since 1964 has had a number of remarkable by-products. Chief among these has been the progressive weakening of the hold which party loyalties have had upon the voters in channeling their voting decisions. This growing dissolution of party-in-the-electorate entails a serious erosion of political parties as basic institutional components of the political system. Closely associated with this, a profoundly important electoral reinforcement of the constitutional separateness of our national policymaking institutions has occurred. The 1974 election has made its own contribution to these trends, and it seems clear that the policymaking vacuum which dramatically emerged in the spring of 1973, when the Watergate cover story fell apart, will continue unabated at least until 1977. Yet this election has some features which do not easily fit a simple model of party decay, and it has had consequences of importance, both substantively and analytically. There is justification, therefore, for giving it a close study.

I. Recent Changes in Congressional Election Patterns

Recent analyses of American elections have made discoveries which are of direct concern to this study. We now know that the portrait of the American electorate which was laid down by the Michigan survey-research group in the early 1960s must be very significantly modified when voting behavior and attitudes in the late 1960s and early 1970s are evaluated.[1] It is now quite clear

From *Political Science Quarterly* 90, no. 3 (Fall 1975): 411–35. Reprinted with permission.

that the pressure of political events during the past decade has produced levels of attitudinal constraint and issue-oriented voting behavior which were largely unsuspected in an earlier period.[2] Not surprisingly, much of this literature has been concerned with attempting to assess the prospects for a critical party realignment as a resolution of the current crisis sequence.[3] Whatever such prospects may be, the dislocation of voting coalitions which has developed during this crisis sequence has had major effects on the structure of congressional-election outcomes, and on the relationship between congressional and other elections. The 1974 election requires detailed analysis in this context.

Two scholars, David Mayhew[4] and Edward Tufte,[5] have addressed themselves to certain basic and very recent changes in the aggregate profile of congressional election outcomes. Since about 1956–60, incumbent members of Congress of both political parties have become increasingly invulnerable to defeat in their districts, more or less regardless of the fate of their party's candidates for other offices in the same election. Incumbents have become quite effectively insulated from the electoral effects, for example, of adverse presidential landslides. As a result, a once notable phenomenon, the so-called coattail effect, has virtually been eliminated.[6]

A profound change has occurred in the shape of congressional district outcomes as the influence of party in shaping voting decisions has decayed. At one time (say, around 1900), the percentage distribution of the vote between the two parties was very similar, or unimodal, tending to coincide with a usually competitive national partisan percentage. These distributions also tended to converge, whether one looks at outcomes involving incumbents running for reelection, contests without incumbents, or the presidential vote at the congressional district level. In particular, there was little or no difference between the concentration of cases in the competitive range for congressional races with or without incumbents. Somewhat later (most notably, perhaps, in 1924, 1952, and 1956) these distributions began to diverge, but all of them remained unimodal. At some point after 1956, however, the pattern of outcomes involving incumbents running for reelection began to change dramatically. Since then the concentration of cases somewhere within the competitive range has been "hollowed out," and replaced by two modes or peaks. One of these is occupied by Democratic incumbents and the other by Republican incumbents, and both are normally located quite far from the competitive center. As a result, the proportion of all incumbents running for reelection who lose their contests has tended to decline sharply, while the proportion of incumbents running to all contests has tended to increase. But while this has been going on, the distribution of outcomes for both presidential elections at the district level and for congressional contests without incumbents has remained heavily concentrated in a single mode which tends to coincide with the national partisan percentages for the two offices.[7] It seems clear that electorally decisive minorities of voters have increasingly been voting for incumbent representatives as incumbents during the past two decades.

These changes in distributional patterns, so obviously favorable to incumbents of both parties, are of large significance. In evaluating their causes, both Mayhew and Tufte have stressed the importance of efforts by elites to control and enhance the predictability of congressional election outcomes. Incumbent members of Congress, with rapidly growing resources of staff, free publicity through the frank and other resources at their disposal, do what they can to entrench themselves personally in their districts. These efforts become increasingly successful the more available such resources become and the more skillfully and singlemindedly the legislator uses them. Additionally, as Tufte points out, many if not most state legislatures prefer the "collusive gerrymander"—the protection of incumbents of both parties—as a conflict-minimizing device when the time comes for them to redraw congressional district boundaries.[8]

There is no question that these explanations are valid as far as they go. But it is very likely that the phenomena with which they deal are more significantly rooted in major changes which have developed in the behavior of voters.[9] The newer literature has clearly identified many of these changes. It is now pretty certain, for example, that the shift toward the professionalization of House careers—including the emergence of the seniority norm—occurred not later than the 1894–1910 period. One assumes an enduring interest after 1910 in a lifetime career on the part of incumbents; yet we need to wait half a century to see the decisive distributional breakthrough which Mayhew and Tufte identify. We need to wait, in short, until the current crisis of party decomposition intersects with a major increase in the availability of resources for incumbents seeking to be reelected. Of course, we do not know nearly enough about the history of reapportionment to be sure as to when "collusive gerrymanders" became common. But we do know that there are a number of states—among them New Hampshire and West Virginia—where no such device was ever needed because of one-party control of the state legislature, but where the shift toward landslides favoring congressional incumbents can easily be traced anyway.

In the early 1960s Warren E. Miller and Donald E. Stokes pointed out that few voters know very much about congressional candidates. What little they know normally tends to favor the incumbent, who has some visibility in the district.[10] This creates a kind of "breakage effect" favoring incumbents, roughly similar to the steady percentage advantage which the owners of an honest gambling house have over the players as a whole. One is probably right in assuming that this breakage effect is now considerably greater than it was when Miller and Stokes first described it. But if so, it would be reasonable to assume that the erosion of party-in-the-electorate—an alternative cue-giver to voting decisions—occurs at a time when the incumbent's available publicity resources are increasing. Of course, in such circumstances the "collusive gerrymander," if used, will tend to be increasingly successful. Added together, the effects of party decomposition in the most recent period would also necessarily erode the quantitative basis for any hypothesis that "nationalization of political effects" was increasing across time.[11]

Instead, we can summarize what we could expect to happen under these circumstances:

The number of incumbents seeking and winning reelection will tend to approximate an actuarial maximum.

Heterogeneity will massively increase across both time (as measured by partisan swing) and space (as measured by electoral outcomes).

The aggregate relationship between congressional and all other elections will systematically decrease, which entails the proposition that the pattern of electoral coalitions will become increasingly discrete by office, with one predominant in a presidential election, another in a congressional election, and so on. This will finally reach the point where the separateness of these coalitions becomes decisive in both electoral and policy terms, above all providing an immense behavioral reinforcement of the separation of powers which the Constitution prescribes.

The structure of American electoral politics came quite close to approximating these conditions in the 1966–72 period. It becomes of some importance to analyze if, and to what extent, the 1974 election has maintained these trends.

A Profile of the 1974 Election

The 1974 election was fought under notoriously abnormal circumstances. It was part of a drama of high constitutional politics ending in Richard Nixon's resignation in disgrace. As it happened, the congressional election can really be said to have begun early in 1974, when six seats in the House which Republicans had won in 1972 became vacant.[12] Most of these were supposedly safe bastions for the GOP, but five of them elected Democrats between February and June. Moreover, these by-elections were subjected to the most intense scrutiny, by politicians and analysts alike, of any since Franklin Roosevelt's attempted purge of conservative Democrats in the 1938 primaries. They thus made their own contribution to the politics of presidential impeachment and resignation. Among other things, they projected a nationwide Democratic swing of between 8 and 10 percent in November, and produced the first Democratic victory in President Ford's old Grand Rapids district since 1910.

It was anticipated both before Nixon's resignation and after his pardon by Ford that the Republican party would suffer major losses in 1974. Throughout the year, the Gallup poll projections indicated that the Democrats would win at least 60 percent of the national congressional vote. In the event, they did somewhat less well than this. But with 59.2 percent of the two-party vote, they still won the largest share of the two-party and total vote in their entire history, and the second largest congressional popular-vote landslide of all time. The Republicans lost forty-eight seats net over 1972, five of them during the by-election sequence in the spring of 1974.

The situation for Republicans was manifestly poor throughout the year, and it was widely argued that this had contributed to a heavy retirement rate among Republican incumbents. Viewed in terms of *two* transitions, 1970–72 and 1972–74, there is clearly something to this argument. The total attrition rate for 1970 incumbents after the 1974 election was 17.6 percent for the Democrats but 31.1 percent for the Republicans. But the attrition resulting from retirement or other preelection causes was 12.5 percent in 1972 and 12.3 percent in 1974 for Democrats, and 15.0 and 17.1 percent respectively, in 1972 and 1974, for Republicans—certainly a small absolute difference.

There was a relatively intense participation of Republican incumbents in the 1974 congressional election. As Table 1 indicates, the bulk of the GOP attrition in 1974 is accounted for by the electoral defeat of an abnormally large number of these incumbents. In November 1974 a total of fifty-five seats changed party hands.[13] Of these, forty were lost by incumbents of both parties, while only fifteen involved contests between nonincumbents.[14] It is obvious, then, that this election did not conform in this vitally important respect to the incumbent-insulation model which we have discussed above. Nevertheless, as we shall see, the outcome of this election much more closely resembles the *recent* large-swing results of 1964 and 1966 than those of earlier "landslide" congressional elections.

Among other things, the distribution of outcomes and vote swings involving Republican incumbents was anything but what a nationalized-effects model might predict. The 1972–74 shift—a swing of 6.0 percent to the Democrats nationwide—was made up of extremely complex elements at the district level. Two things emerge with crystal clarity at the outset. First, the

Table 1. Incumbents and nonincumbents in House elections, 1972 and 1974

Year	Incumbent Democrats			Incumbent Republicans			Nonincumbents			Total House	
	Won	Lost	Total	Won	Lost	Total	D	R	Total	D	R
1972	216	7	223	149	4	153	24	35	59	243	192
1974	210	3	213	124	36	160	45	17	62	291	144

Note: Both years' figures are based on treating all winners of preceding by-elections as nonincumbents. If the winners of 1973–74 by-elections are regarded as incumbents, the figures change marginally:

Year	Incumbent Democrats			Incumbent Republicans			Nonincumbents			Total House	
1974	216	4	220	127	36	163	39	13	52	291	144

Democratic seat gain was not won by anything remotely approximating a uniform swing, for example one which might have tipped over most Republican incumbents who had won relatively close contests in 1972. Second, incumbent Republican losses in 1974 were not randomly distributed across the wide policy spectrum which exists within the congressional Republican party. Using a simple policy-score index based on votes during the 1971–74 period, Table 2 provides a stratification of winning and losing Republican incumbents along this policy dimension.

Very clearly, Republican losses were concentrated at the conservative end of the party's policy spectrum. As the journalists' accounts of the election pointed out, the conservative Republican Steering Committee lost thirty of its seventy members, including its chairman, Lamar Baker of Tennessee (−96 on our index). In this respect too, it would appear that the 1974 election presents similarities to the 1964 situation: in both cases there was a visible relationship between the policy liberalism of Republican incumbents and their relative success at the polls.

This suggests the existence of policy voting among some 1974 voters.

Table 2. Party and policy: Composite policy scores of reelected and defeated Republican incumbents, 1974

Score	Reelected	Defeated	% Defeated
+40 and over	5	0	0
+30 to +39	2	0	0
+20 to +29	3	0	0
+10 to +19	4	0	0
+0 to +9	2	0	0
−0 to −9	5	1	16.7
−10 to −19	1	0	0
−20 to −29	5	1	16.7
−30 to −39	8	1	11.1
−40 to −49	8	1	11.1
−50 to −59	13	4	23.5
−60 to −69	18	6	25.0
−70 to −79	22	5	18.5
−80 to −89	15	10	40.0
−90 to −100	16	7	26.1
− 49 and above	43	4	8.5
−50 to −79	53	15	22.1
−30 to −100	31	17	35.4
Total	127	36	22.1

Note: This index is based on a composite of ratings by Americans for Democratic Action (liberal) and Americans for Constitutional Action (conservative). It varies over a range of 200, from −100 (most conservative) to +100 (most liberal).

Unfortunately, the analysis of aggregate data alone is inadequate to permit solid inference on this point. But it does reveal—quite graphically—the complexities. As in the past, one of the chief elements of complexity is the persistent deviation of the South as a region from national norms. It is not necessary for the purposes of this analysis to do more than point out that this region is asymmetrically conservative. This conservatism is manifested not only in presidential elections such as 1968 and 1972, but in the policy scores of its representatives as a whole, and of its Republican representatives in particular.[15] Thus any study of the relationship between the policy stances of members of Congress and what happened to them electorally between 1972 and 1974 must take account of this Southern regional deviation.

Another complicating factor can be derived almost deductively from the incumbent-insulation model discussed here. It can be expected that, all other things being equal, first-term representatives running for reelection will do markedly better in any trend situation than older incumbents of the same party. The rationale here is that there is an upward trajectory today in an incumbent's career at the polls. The first election (as a nonincumbent) would be relatively close, but margins thereafter would increase until they reached a point somewhere near the noncompetitive mode established for all incumbents of the party. Moreover, we might expect that a very large fraction of this upward shift from competition to electoral insulation would occur between the representative's first and second election campaigns. In the specific situation before us, this implies that Republican first-termers in 1974 should show little or no pro-Democratic swing as a group, and Democratic newcomers should enjoy a much larger swing than their older incumbent party colleagues.

Table 3 demonstrates that the pattern of the data is consistent with these

Table 3. New incumbents versus old incumbents: Partisan percentages and swing, 1972–74

	Mean percentage Democratic 1972	Mean percentage Democratic 1974	1972–74 swing toward Democratic party
	(N)		
Democratic	NI (13) 60.8	73.1	+12.4
districts	OI (114) 65.8	70.7	+ 5.1
t	(−1.69)	(+0.87)	(+2.95)*
Republican	NI (42) 43.4	43.2	− 0.2
districts	OI (96) 33.6	44.2	+10.6
t	(+7.88)**	(−0.63)	(−7.44)**

Note: Mean percentage swing among all incumbents, +6.6; standard deviation, 9.0. Total number excludes all incumbents without major-party opposition in 1972 or in 1974, and all incumbents from California.
*t significant, d.f. = 120, at .005 level.
**t significant, d.f. = 120, at .0005 level.

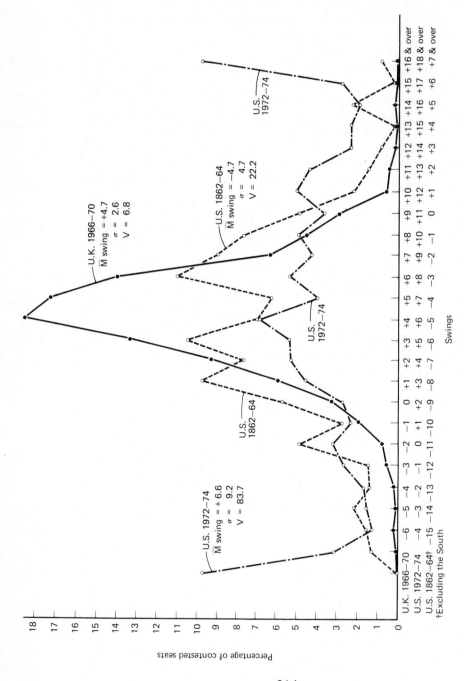

Figure 1. Comparative heterogeneity of partisan swing; United Kingdom, 1966–70; and United States, 1862–64 and 1972–74*

214

expectations. In fact, the first-generation or "new incumbent" Republicans as a group actually *improved* their position very slightly in the face of an exceptionally large national shift toward the Democrats.[16] This generalized pattern captures a trend which is reflected in extreme form by the showings of three 1972 freshmen in the 1974 election: Joel Pritchard, a liberal Republican from Washington (a shift from 49.4 to 29.4 percent Democratic); Trent Lott of Mississippi, a Nixon supporter on the Judiciary Committee and a conservative Republican (a shift from 44.4 to 16.9 percent Democratic); and Gerry Studds, a liberal Democrat from Massachusetts (a shift from 50.2 to 74.8 percent Democratic). Whatever else may have been going on in 1974, the electoral careers of at least some new legislators were following the pattern which the incumbent-protection model presupposes.

A congressional election is supposedly a national event, an event whose outcome has been thought to depend at the margins upon the presence or absence of compelling short-term forces of national scope.[17] One would have thought that such national forces were particularly salient in 1974, and indeed they were to some extent influential, as we have pointed out. Yet one obvious inference to be drawn from the party-decomposition model of contemporary American electoral politics is the expectation that diversity of outcomes, influenced by a host of factors peculiar to each individual contest, will become increasingly important as a party-in-the-electorate fades. An overview of the 1972–74 swing as a whole does nothing to challenge this proposition. As Figure 1 makes graphically clear, the 1972–74 swing outside the South was vastly more heterogeneous than the 1966–70 swing in Great Britain, and considerably more so than the 1862–64 swing in the United States.[18] There is a degree of diversity in this swing which, at other times and places, has been curbed by the central importance of party in shaping electoral outcomes. It follows, of course, that any effort to explain the determinants of the 1972–74 vote swing would require a model of truly formidable complexity.[19]

The distribution of partisan percentages in 1974 is very similar in shape to the pattern described by Mayhew and Tufte; by now, this should not be surprising.[20] For races involving incumbents, there are two peaks or modes— one for Republicans, one for Democrats—while there is a sharp single peak of cases in the set of nonincumbent contests. This peak centers around the national percentage. There is a major positional difference between 1974 and elections in the immediate past, a consequence of the fact that this was a landslide election. By contrast with 1972, the Republican mode is shifted toward a highly competitive range, while the Democratic mode moves into the 70.0–74.9 percent range, with the trough between them centering around the national percentage.[21] Despite this shift, the general shape of the distribution is unchanged from that of 1972. Granted the fact that nationwide congressional election outcomes of 59–41 have been extremely rare, the overall danger to Republican incumbents arising from the one-sided competi-

Table 4. A study in contrast: Distribution of percentages Democratic by incumbency and party, 1866 and 1974

	Percentage of seats by category							
	1866					1974		
	Incumbents			Non-in-cum-bents	Incumbents			Non-in-cum-bents
Percentage Democratic	R	D	All		R	D	All	
0.0–19.9	5.4	0	4.4	0	0.8	0	0.3	0
20.0–29.9	11.7	0	9.6	9.2	4.7	0	2.1	0
30.0–39.9	24.5	0	20.2	12.3	26.4	0	12.1	0
40.0–44.9	18.1	0	14.9	16.9	20.6	0.7	10.3	8.7
45.0–49.9	34.0	15.0	30.7	27.7	25.6	2.0	12.8	15.2
50.0–54.9	5.3	45.0	12.3	12.3	14.7	4.6	9.3	17.4
55.0–59.9	1.1	15.0	3.5	7.7	3.1	4.6	3.9	28.3
60.0–69.9	0	10.0	1.8	6.1	3.1	25.0	14.9	13.0
70.0–79.9	0	5.0	0.9	6.2	0	35.5	19.2	8.7
80.0–100.0	0	10.0	1.8	1.5	0	27.6	14.9	8.7
N	94	20	114	65	129	152	281	46

tiveness of 1974 will probably turn out to be more apparent than real in future elections. On the other hand, there is no convincing evidence that the basic shape of these distributions is likely to change very much in the near future.

It is once again worth emphasizing, by way of contrast, how sharply this complex pattern of outcomes differs from those of the past, when party labels had more operational meaning in our electoral politics than they have now. The contrast chosen here is as between the distributions of outcomes in 1866 and in 1974, excluding the Southern states in both cases.[22]

What Table 4 stresses most fully is the severe decline in competitiveness and the association of this decline with incumbency. Taking the 40.0–59.9 percent range as broadly competitive, we can actually see marginal improvement in the nonincumbent sets: 64.6 percent of such contests fell within this range in 1866 and 69.6 percent fell within it in 1974. For contests with incumbents running for reelection, however, there is a decline from 61.4 percent in the former year to 36.3 percent in the latter. The effect of this change is magnified by the professionalization of House careers over a century: while nonincumbents contested 36.3 percent of the nonsouthern seats in 1866, they contested only 14.1 percent of them in 1974.

III. 1974 in Long Term Perspective

If we are more fully to appreciate the place of the 1974 election—and its immediate aftermath within the House—in the context of contemporary American politics, we need to see it in its broader historical setting. The time is more than ripe for a detailed quantitative history of congressional elections and of the internal processes of the House of Representatives. This, however, must await another occasion; what is offered here is a summary statement of trends over the past century and their relationship to the outcome of the 1974 election.

(1) The competitiveness of congressional elections has been declining persistently ever since the turn of the century, even while the national vote as a whole has remained competitive between the major parties. The South has, of course, been a deviant case. In the aftermath of the region's massively successful movement toward disfranchisement seventy-five years ago, the proportion of seats in Southern and Border states which were won by 60 percent or less of the total vote declined abruptly from 64.0 percent in 1896 to 20.8 percent in 1902.[23] It has remained below one-third of the total in the former slave states ever since. Accordingly, this decline has been particularly marked as a secular trend in the North and West. Between 1874 and 1892 the proportion of contests which were won with 60 percent or less of the total vote in these regions was always more than four-fifths of the total, and was sometimes over nine-tenths.[24]. After the critical realignment of the 1890s a decline set in until, by 1926, competitively won congressional seats reached a low of 35.2 percent of the total.[25] This decline was abruptly and massively reversed with the critical realignment of 1932. But once again, decline in competitiveness reasserted itself: while three-quarters of the seats in the North and West were won by broadly competitive margins in 1932, only half were in 1964.

The bimodal pattern of outcomes favoring incumbents which we have discussed above has been superimposed on this longer-term trend. As a result, the decline in competitiveness since 1964 has not only continued but accelerated. By 1972, only 32.5 percent of the seats in the North and West were won with less than 60 percent of the district vote, the lowest figure since the beginning of the series in 1824. Of course, as the massive Democratic swing ate into the Republican mode in 1974, such seats increased to 43.5 percent of the total; but this remained markedly below the 1964 figure of 51.5 percent and vastly below the 66.1 percent registered in the 1948 election. The improvement in the proportion of competitive seats in 1974 was thus tightly constrained by historical standards. Indeed, this proportion has been *lower* than in 1974 on just six occasions in our history. These characteristically occurred in 1926, 1928, and 1966–72: the former two at the trough of the semioligarchic fourth-party system, and the latter four in the years of party decomposition immediately preceding 1974.

(2) As Professor Tufte has demonstrated, the emergence of a bimodal pattern of outcomes favoring incumbents is naturally associated with a sharp drop in the responsiveness of party representation in the House to shifts in the popular vote. The reader can visualize this for himself. If the distribution of outcomes has a cluster or peak centering around the 50 percent mark, small shifts in aggregate vote percentages will produce considerably larger

shifts in seats toward the party enjoying a vote swing in its favor. The steeper the peak is, the larger will be this seat turnover for any given size of vote swing. If, on the other hand, the national percentage approximates 50 but there are *two* clusters or peaks far away from this point and a shortage of cases near the competitive center, a relatively wide shift in the vote will produce only very limited gains for the advantaged party.[26]

In the late nineteenth century the sensitivity of turnover to changes in the major-party vote was extremely high; very often a shift of 1 percent in a party's vote won it an additional 4 percent or more of seats in the House. Since about 1896 these swing ratios have undergone a generalized but undulating downward trend which closely parallels our other indicators. At the same time, it is clear that as late as 1948 a large swing in the vote could produce a turnover in seats which was higher than the 2.8:1 swing ratio suggested by the so-called cube law, or the rough 2.5:1 ratio suggested by Professor Tufte. The ratio in that year was 3.1:1, for the Democrats gained 7.2 percent of the two-party vote outside the South over 1946, and fully 22.5 percent of the seats.[27] By 1966, on the other hand, the Republicans could gain only 11.9 percent of the non-Southern seats with a vote shift of 7.3 percent in their favor; the swing ratio was only 1.6:1. As one might expect, 1974— with its swing *away from* competitiveness nationally—stands somewhere between the two, but much closer to 1966. The 6.1 percent pro-Democratic swing outside the South produced a seat swing of 12.8 percent, and thus a swing ratio of 2.1:1. It is worth noting that in fifteen of the past sixty congressional elections, a vote swing of 5 percent or more has occurred; of these fifteen, only one—1966, of course—showed a *lower* ratio than 1974. Despite the special factors which led to the defeat of so many Republican incumbents in 1974, this damping down of responsiveness held their losses down. In all probability, considerably more than a dozen Republican incumbents survived the 1974 tide who would have lost under pre-1960 conditions.[28]

(3) It is fascinating to observe that aggregate policy differences between the two parties in the House have followed a pattern of decline in the past century which is very similar to those described above. Moreover, this decline has also speeded up very considerably since the early 1960s; in the 91st Congress (1969–70) the global differences between the parties on roll-call votes reached the lowest point since the creation of the Republican party in the 1850s. The policy measure used here is, of course, a very gross one, an Index of Party Dissimilarity taken across all nonunanimous roll calls in each Congress.[29] Accordingly, it cannot take account of differences in the substantive political importance or weight of any given issue domain. It is nevertheless a useful measure of a very important, and a very long, secular trend.[30] Table 5 summarizes two aspects of change: a general downward trend, but punctuated by upsurges in party polarization or distance. These upsurges chiefly occur in the immediate policy aftermaths of the critical realignments of 1896 and 1932.

Table 5. The case of the disappearing party: Index of party dissimilarity in the House, 1873–1973

Period	Congresses	Mean index of party dissimilarity*	(Standard deviation)
1873–81	43–47	.583	(.019)
1883–93**	48–50, 52–53	.438	(.042)
1895–1911	54–62	.625	(.087)
1913–31	63–72	.424	(.059)
1933–39	73–76	.529	(.038)
1941–51	77–82	.397	(.020)
1953–63	83–88	.357	(.030)
1965–71	89–92	.250	(.049)

*The higher the mean score, the more dissimilar the voting of the two parties in roll-call votes.

**The index of .713 for the 51st Congress (1889–91) is abnormally high; the explanation for this is to be found in the large number of procedural votes taken during Speaker Reed's struggle to establish centralized House rules over bitter Democratic opposition. If it is included, the mean for the period becomes .484 and the standard deviation becomes .109.

In this series too, there has been a speedup over the past decade in a longer-term trend toward the erosion of party in the political process. To say the least, this forms an arresting context for the 1975 effort by House Democratic liberals to restore the importance of party in that body's decisional structure.

(4) All of the long-term trends discussed so far are associated in time with another which promises to be of very great importance to the workings of the American policy process as a whole in the near future. Put at its simplest, there has been greater and greater divergence in presidential and congressional voting coalitions since the turn of the century. One way of measuring this phenomenon is to test the state-by-state relationships between the Democratic percentages of the vote for the two offices in presidential years, when both offices are voted for at the same time. Table 6 summarizes this twentieth-century trend, partitioning the country into Southern and non-Southern regions. As is obvious, the trend is broadly downward, but with many countervailing movements in the non-Southern states. The lag in reintegrating the party system after the New Deal realignment in these states is conspicuously evident. More to the immediate point, this trend toward divergence of electoral coalitions, like so many others, has markedly accelerated in the past fifteen years.

It is particularly worthy of note that the proportion of the variance in the non-Southern congressional vote which can be explained by the presidential vote has declined from over one-half in 1956 to a little more than one-tenth in 1972. Viewing the national system as a whole, however, the impact of the South's secession from the presidential Democratic coalition since the 1950s is of at least equal political importance.

Much has been written about split-ticket voting and the newer indepen-

Table 6. Party decomposition, 1900–72

Year	N	United States Correlation between % Democratic of Congressional and Presidential vote (r)	(r²)	(Sy · x)	N	Non-Southern states Correlation between % Democratic of Congressional and Presidential vote (r)	(r²)	(Sy · x)
1900	45	+.978	(.956)	(3.33)	34	+.984	(.968)	(1.30)
1904	45	+.973	(.946)	(4.77)	34	+.874	(.763)	(3.92)
1908	46	+.928	(.861)	(7.33)	35	+.870	(.757)	(3.64)
1912	48	+.961	(.923)	(5.78)	37	+.824	(.679)	(4.76)
1916	48	+.931	(.867)	(7.03)	37	+.666	(.440)	(6.94)
1920	48	+.950	(9.03)	(7.29)	37	+.857	(.735)	(6.38)
1924	48	+.930	(.865)	(8.70)	37	+.748	(.574)	(9.16)
1928	48	+8.10	(.656)	(12.36)	37	+.490	(.240)	(8.37)
1932	48	+.907	(.822)	(7.20)	37	+.555	(.308)	(6.96)
1936	48	+.908	(.825)	(7.24)	37	+.634	(.402)	(7.37)
1940	48	+.920	(.846)	(8.19)	37	+.692	(.478)	(8.47)
1944	48	+.941	(.885)	(6.69)	37	+.819	(.671)	(4.90)
1948	48	+.941	(.885)	(6.32)	37	+.812	(.659)	(5.56)
1952	48	+.861	(.742)	(9.96)	37	+.871	(.758)	(3.96)
1956	48	+.717	(.514)	(11.77)	37	+.755	(.570)	(4.28)
1960	50	+.368	(.136)	(14.83)	39	+.651	(.424)	(5.36)
1964	50	−.265	(.070)	(10.37)	39	+.579	(.335)	(5.62)
1968	50	−.274	(.075)	(12.27)	39	+.470	(.221)	(7.14)
1972	50	−.071	(.005)	(11.54)	39	+.335	(.112)	(9.24)

Note: In the South, 1948, the figure is based on combined presidential percentages for Democratic and Dixiecratic candidates.

dent element in the electorate. But this partisan decay cannot fail to have massive effects on the political system at the center as well. As Republicans have come to win the White House more often than Democrats in the past two decades, so their base of support in Congress has become increasingly eroded. The dissolution of the partisan nexus that used to like the two branches of government involves an enormous and very recent reinforcement of the separation of powers, and of the mutual conflicts between persons in office, which the drafters of the Constitution sought to prescribe two centuries ago. The result of the 1974 election sharpens the basic problems of governing which this situation presupposes. Despite the large Democratic victory, this pattern of coalitional dissociation also means that the outcome of the 1974 congressional election tells us precisely nothing about which party's candidate will win the presidency in 1976. It would seem more likely than not that this dissociation at the base will be linked at the center with an ongoing concern of presidents of either party: how to get and maintain an adequate base of support in Congress.

(5) In one crucial respect, 1974 reflects not merely similarity to trends of the immediate and more remote past, but their extension. For some years past the rate of participation in American elections has been declining in both presidential and off years. This decline accelerated dramatically in 1974. National turnout, at 37 percent for House contests, fell very nearly to the all-time lows registered in the 1920s and in the wartime election of 1942. Even this low national turnout does not tell the whole story. Recent Southern elections, stimulated by the Civil Rights Acts of 1965 and 1970, have seen increases in off-year congressional turnout to levels not reached since the early twentieth century. When the South is excluded, the 41.0 percent participation rate in 1974 becomes the second lowest since the creation of the party system in the Jacksonian era. For many states, such as Indiana, the 1974 turnout was the lowest of all time.

As the introduction of woman suffrage cannot account for the steep drop in participation in 1918, so the enfranchisement of eighteen- to twenty-year-olds in 1971 cannot begin to account wholly for the 1970–74 decline. Similarly, despite some Republican claims to the contrary, there is no very good reason to suppose that the abysmal participation rate in 1974 was the result of one-sided abstentions of GOP supporters in the wake of Watergate. On the contrary, it is much more likely that the recent decline in voting participation is part of a current syndrome of party decomposition and widespread, generalized public dissatisfaction with politics as such.[31] To the extent that this is so, it would follow that Republican prospects, if any, for recovering the bulk of their congressional losses must lie elsewhere. The year 1974 resembles the immediate past closely enough to permit the prediction that many of the seventy-five Democratic newcomers will have begun successfully to entrench themselves by 1976.

This decline in voting participation is a massive political fact. In the 1870–94 period, outside the South, off-year participation averaged 68.1 percent; as late as 1914 it was 59.6 percent. More recently, it has been poor but stable, averaging 51.3 percent between 1950 and 1970. Whether the drop to 41 percent in 1974 is temporary only the future will show. But insofar as it is associated with major increases in levels of public alienation, this colossal abstention rate intrudes an element of uncertainty into any predictions. It also, of course, raises serious questions about "American democracy" as a whole. It would be reasonable to suppose, on past form, that the system can remain both stable and electorally demobilized. But it would have taken the participation of 11 million more voters than actually came to the polls in 1974 to return participation even to its mediocre 1970 level. One wonders what conditions might stimulate them to enter, or return to, the active electorate.[32]

IV. The Present and the Near Future

So it is that a complex system behaves like a system. As party erodes both in the electorate and in the House, divergence in the electoral coalitions of president and Congress increasingly reinforces the separation of powers; incumbents tend increasingly to be protected or insulated, except to some extent in abnormal cases like 1974; the responsiveness of the representational system to electoral change declines; and abstention from the polls becomes the largest mass movement of our time. To this one may add that the events of the past decade have produced a new institutional balance between executive and legislature, and a larger autonomous importance for Congress than it has had in many decades.

In this context, the immediate aftermath of this election almost startles the observer. The greatest reorganization of power within the House has occurred since the great anti-Speaker revolt of 1910. The importance of central party mechanisms in the majority's decisional processes has been strikingly reasserted. It is just possible that the first steps may have been taken to develop a collective will which alone can permit a serious competition for influence over policy with the executive. Very clearly, the new Democratic freshmen have taken a central part in all this. From the viewpoint developed here, this is the more remarkable, since the proportion of freshmen is not much different than in 1965, and is very much smaller than it was in 1949. To be sure, the totals of newcomers become more impressive when one adds reelected 1972 freshmen to the 1974 newcomers. These constitute 129 members of the House (29.7 percent of the whole), a respectable rate of turnover by contemporary standards. Even so, one would have thought that an anti-seniority generational interest has always existed among newcomers to Congress. Why the change in 1975?

Of course, we do not yet possess a complete voting profile on the 1974 newcomers, but the probabilities are that most of them entered the House on the "left" end of the Democratic policy spectrum. It is fair to assume that these newcomers in the main reflect a new politics, a new age generation, and, in some measure, a public dissatisfaction with the older political styles of the Johnson and Nixon years. But another factor is the long-term regional transformation in party coalitions in Congress, and the cumulative effects of this change are very likely of crucial importance for the events of early 1975. The revolt of 1910 against Speaker Cannon had been preceded by both a long-term solidification of the seniority norm within the House and by the replacement of conventions by direct primaries for congressional nominations outside of it. The current change appears to reflect not only the increased importance of the "new politics" in the House, but also a recent and rapid decline in the South's relative weight in the congressional Democratic party.

After the 1974 election, Southern representatives constituted only 27.8 percent of the total House Democratic membership, and 18.8 percent of all House Republicans. This is the lowest proportion of Southerners to all Democrats since 1870, and the highest proportion of Southerners to all Republicans since 1872. Moreover, there has been a persistent pro-Democratic coun-

Table 7. Shifting party bases in the House by region, 1940–74

Region	Percentage Democratic seats			Y_c		(r^2)	$(Sy \cdot x)$
	1948	1964	1974	a	b		
Northeast	46.7	66.1	67.6	33.99 + 1.55X		(.543)	(7.84)
Midwest	36.4	49.6	54.1	20.32 + 1.34X		(.443)	(8.26)
West	49.0	68.1	65.8	46.62 + 0.64X		(.106)	(10.30)
Border	88.1	80.6	82.9	73.50 + 0.31X		(.033)	(9.23)
South	98.1	84.9	75.0	105.55 − 1.74X		(.841)	(4.16)
U.S.A.	60.6	67.8	66.9	52.76 + 0.47X		(.167)	(5.84)

	Percentage Democratic and Republican seats						
	A. Democrats						
North and West	46.8	59.7	62.2	38.03 + 1.24X		(.709)	(4.36)
South	39.2	30.5	27.8	48.25 − 1.04X		(.619)	(4.49)
	B. Republicans						
North and West	95.9	83.6	77.1	98.48 − 1.04X		(.783)	(3.02)
South	1.2	11.4	18.8	−3.78 + 1.10X		(.863)	(2.40)

Note: Biennial series, based on percentage Democratic of the two-party total of seats. Regions: Northeast—New England, Middle Atlantic, Delaware; Midwest—East North Central, West North Central except Missouri; West—Mountain, Pacific; Border—Kentucky, Maryland, Missouri, Oklahoma, West Virginia; South—eleven ex-Confederate states.

tervailing trend elsewhere over the past generation, especially in the Northeast and the Midwest. These movements parallel, at a much slower pace, well-known changes in the regional bases of party support in recent presidential elections. Assuming that these trends continue, at some point between 1980 and 1985 the South will be represented in both congressional parties in proportion to its share of the whole House for the first time since the 1840s,[33] though this may be only the prelude to an eventual one-party Republicanism outside the region's metropolitan areas (see Table 7).

In connection with these trends, it is important to point out that Southern Democrats are today the most heterogeneous regional grouping in the House, so far as policy scores are concerned. On the other hand, the growth of the Southern Republican contingent has involved the entry of people who are overwhelmingly and cohesively on the far right of the congressional GOP. There are two apparent reasons for this. First, the growth of urbanism in the South has brought with it an influx of moderate and liberal Democratic representatives in such areas, including the three black representatives first elected in 1972 or 1974. Second, ultraconservative Democrats have been increasingly replaced by ultraconservative Republicans. There are some interesting recent examples of this. In 1972 Trent Lott and Thad Cochran won the 4th and 5th Mississippi districts, once held for many years by John Bell Williams and William Colmer, when their incumbents chose to retire.

The 6th Louisiana district had been held since 1966 by John Rarick, a Democrat whom the authors of *The Almanac of American Politics* flatly describe as "the most rabidly right-wing member of Congress."[34] Rarick lost the 1974 Democratic primary to a more moderate Democrat, who in turn lost in November to W. Henson Moore III, a Republican, by an apparent majority of fourteen votes. Moore then resigned, a new election was held, and he won easily.

Perhaps an even more interesting case is that of Congressman John Jarman, who from 1950 to 1975 represented the 5th Oklahoma district (Oklahoma City) as a Democrat. Jarman is a conservative, and appeared to fit the district: it gave Nixon 76.0 percent of its presidential vote in 1972. But the congressman encountered more than usual difficulty that year, winning only 60.4 percent of the vote. Still worse was the result in 1974, when he won reelection with only 51.7 percent, a swing of −8.7 percent in a Democratic year. Following the caucus votes which removed Representatives Hebert, Patman, and Poage from the chairmanships of the Committees on Armed Services, Banking and Currency, and Agriculture, Jarman announced in January 1975 that he was joining the Republican party. He gave as his reason the view that the Democratic party in the House was being captured by the same elements that had nominated George McGovern in 1972. There is no doubt that, with a −76 score on our index, Jarman will feel more at home in the ranks of congressional Republicans. At the same time, he may well improve his chances of reelection in this conservative district.

Such cases may prove to be as isolated as that of Strom Thurmond, who joined what he called the "Goldwater Republican" party in 1964. But there has been an interesting trickle of defections in recent years. These have also included Ogden Reid of New York and, most recently, Donald Riegle of Michigan—both liberal Republicans who crossed the aisle to join the Democrats. It may well be argued that the chief motivating force which has bound conservative Southerners to the congressional Democratic party recently has been the disproportionate power over the legislative process which was given to them by seniority and associated norms of committee activity. The congressional power game is more complex now. Seniority has by no means been wholly abandoned as a selection device. But it prevails only so long as the policy stance and personality of the chairman involved do not deviate too widely from the majority's preferences. The implications of this for the motivations of conservative Southerners to remain Democrats are obvious.

Leaving aside newcomers, there are forty House Democrats—thirty-seven Southerners and three others—who have scores of −40 or lower on our index. Were they all to defect to the Republicans, the Democratic gains of 1974 would be virtually wiped out and a new balance of 255 Democrats to 185 Republicans would be created. But there would be other consequences as well. In the first place, the liberal policy majority in the Democratic caucus would become overwhelmingly large. Secondly, the minority of liberal

Republicans would become even more hopelessly isolated within the congressional GOP than it is now; and this could well prompt other Riegles to join the Democrats. The end product of any such large-scale movement would be, presumably, the creation of a "more responsible two-party system" in Congress and the realization of the dreams of many programmatic reformers. On past form, one would have to suppose that such a shift will tend to remain relatively gradual in the very near term rather than abrupt or massive, but there is very little reason to suppose that the coalitional trends underlying it will be reversed. A necessary paradox of this movement, so far as programmatic liberals are concerned, is that it constitutes the only significant chance the Republicans have to capture the House of Representatives at all within the foreseeable future.[35]

The entire context of the 1974 election, no less than the striking mixture of diverse movements in its result, suggests that it marks a turning point in the current crisis sequence which has been affecting American electoral politics as a whole. Moreover, it is associated with a swing in the institutional balance between president and Congress toward the latter, the first such swing in many decades. For many quite deterministic reasons, this will not mean the emergence of anything approximating "congressional government" even in the short term. But for some time to come, the autonomous policy importance of the legislative branch will probably be enhanced to a degree scarcely imagined by political analysis of even a few years ago. It would seem very probable, therefore, that scholarly attention in the field of electoral politics will also shift toward Congress. To be sure, much important work has been done on congressional elections, both by the Michigan survey research center group and by others.[36]. But there is little doubt that the lion's share of analysis has hitherto been devoted to presidential electorates and elections. As it becomes clearer that electoral coalitions are increasingly diverging on office-specific lines, and that Congress will retain considerable importance in its own right, it can be hoped and expected that more scholarly effort will be made in this area. A great many lacunae still exist in our knowledge of the past development and present structure of American congressional elections.

NOTES

1. See, for example, Gerald Pomper, "From Confusion to Clarity: Issues and American Voters, 1956–1968," *American Political Science Review* 66 (June 1972): 415–28, as well as the contributions of John Kessel, Richard Boyd, and Richard A. Brody and Benjamin I. Page to that symposium issue: Arthur H. Miller, Warren E. Miller, Alden S. Raine, and Thad A. Brown, "A Majority Party in Dissarray: Social and Political Conflict in the 1972 Election" (Paper presented at 1973 meeting of the American Political Science Association, New Orleans, September 1973); and Norman H. Nie and Kristi Andersen, "Mass Belief Systems Revisited," *Journal of Politics* 36 (August 1974).

2. Except, perhaps, by the late V. O. Key, Jr. See his *The Responsible Electorate* (Cambridge, Mass., 1966).

3. One recent analysis of this literature is contained in Walter Dean Burnham, "American Pol-

itics in the 1970s: Beyond Party?" appearing as the final chapter of William Nisbet Chambers and Walter Dean Burnham eds., *The American Party Systems*, 2d ed. (New York, 1975).

4. See David Mayhew, "Congressional Elections: The Case of the Disappearing Marginals" (Paper presented at 1973 meeting of the New England Political Science Association, Boston, April 1973). A more general discussion of his thesis that congressional behavior is shaped by the rational desire to be reelected is found in his *Congress: The Electoral Connection* (New Haven, 1974), but the book contains surprisingly little information on actual electoral patterns.

5. Edward Tufte, "The Relationship between Votes and Seats in Two-Party Systems," *American Political Science Review 68* (1973): 540–54. See also his *Data Analysis for Politics and Policy* (Englewood Cliffs, N.J., 1974), pp. 91–101.

6. For an effort to analyze this phenomenon, see Malcolm Moos, *Politics, Presidents and Coattails* (Baltimore, 1952). In retrospect, it seems probable that the "coattail effect" was part of an intermediate stage in the historical decomposition of American parties as electoral mechanisms. Prior to about 1900, near identity of results at most or all levels of election tended to be the rule, but since the early 1960s the dissociation of electoral coalitions for various offices has become increasingly complete.

7. See the charts in Mayhew, "Congressional Elections," and in Burnham, "American Politics in the 1970s."

8. Tufte, *Data Analysis for Politics and Policy*, p. 99.

9. See the communication of Walter Dean Burnham to the editor, *American Political Science Review 68* (1974): 207–11.

10. Warren E. Miller and Donald E. Stokes, "Party Government and the Salience of Congress," in Angus Campbell, Philip E. Converse, Warren E. Miller, and Donald E. Stokes, *Elections and the Political Order* (New York, 1966), ch. 11.

11. Donald E. Stokes, "Political Parties and the Nationalization of Electoral Forces," in Chambers and Burnham, *American Party Systems*, ch. 7. It must be emphasized that the incumbent-insulation pattern fully emerges only *after* the latest decade of Stokes's analysis (1952–60).

12. In order of their occurrence: Pennsylvania 12th, Michigan 5th and 8th, Ohio 1st, and California 5th and 19th, with only the last retained by the GOP and the Ohio first recaptured by it in November. The mean percentage Democratic of the total vote in the five districts won by the Democrats had been 37.3 percent for the House in 1972 and 39.0 percent for president.

13. All comparisons between 1972 and 1974 are affected by the judicially ordered redistricting of California's forty-three seats in the interim. One entirely new seat was created and was won by a Democrat (the 24th). Of the forty-two others, four had no incumbents; two of these were held by Democrats and two were won by Democrats from the GOP. Of the thirty-eight seats with incumbents running for reelection, twenty-one were defended by Democrats, all of whom won, and seventeen by Republicans, of whom two lost (the 17th and the 35th). In a number of cases, specified where they occur, it has been necessary to exclude California from data presentations.

14. On the basis of incumbency as of November 1974, we find:

Category	Incumbents		Nonincumbents		Total	
	N	%	N	%	N	%
Unchanged R	127	33.2	11	21.2	138	31.7
From D to R	4	1.0	2	3.8	6	1.4
From R to D	36	9.4	13	25.0	49	11.3
Unchanged D	216	56.4	25	48.1	241	55.4
New seat (D)	—	—	1	1.9	1	0.2
Total	383	100.0	52	100.0	435	100.0

15. Outside the South and border states, the mean score for Democratic incumbents was +59, and for Republicans was −48. The mean score in the South was −32 for Democrats and −82 for Republicans—the latter, it should be noted, with a much smaller standard deviation than for any other party/regional grouping in the House.

16. Nevertheless, 11 of the 45 Republicans first elected in 1972 (24.4 percent) lost their seats, compared with 25 (21.2 percent) of the 118 older Republicans.

17. This is the rationale behind Stokes's argument in "Political Parties and the Nationalization of Electoral Forces."

18. The variance in the former case was 11.7 times as large as in the latter. Similarly—and excluding the South for both elections—the variance in the 1972–74 swing was 3.6 times as great as was the variance of the 1862–64 swing.

19. Thus while it is true that Republican *losses* in 1974 were concentrated in suburban districts (twenty-eight out of forty-eight losses in comparable districts, or 58.3 percent, while suburban districts constituted only 42.8 percent of all Republican seats), the *swing* data are much less compelling. The mean 1972–74 swing was +8.3 percent for seventy-one predominantly suburban Republican seats, +5.7 percent for twenty-nine predominantly central-city Republican seats, and +7.9 percent for sixty-six nonmetropolitan Republican seats; the dispersions around these measures of central tendency are very large in all three cases.

20. The basic bimodal shape of 1974 outcomes involving incumbents is substantially the same as that graphed for 1972 in Burnham, "American Politics in the 1970s," and quite different from the 1952 distribution presented there.

21. I.e., in the 55.0–59.9 percent range, precisely the area of *maximum* concentration of outcomes for contests with no incumbents running.

22. Apart from the need for comparability with Civil War elections, the exclusion of the South emphasizes the universality of this trend in areas of the United States which did not undergo the massively deviant political evolution found in the ex-Confederate states.

23. The data for this discussion are from a much larger file provided by courtesy of the Inter-University Consortium for Political Research.

24. The 1874–92 mean was 85.7 percent of all seats in the North and West; the standard deviation was 3.9.

25. There was a temporary reversal of trend in 1912 and 1914, a result of Pregressive and Socialist candidacies in these House elections. But the downward trend resumed dramatically in 1918, an election whose characteristics reveal it to be one of the most important in this century.

26. An unusually good example is the 1968–70 relationship. The popular-vote swing to the Democrats was 4.0 percent (51.1–55.1 percent of the two-party vote), but they gained only an additional 2.7 percent of the seats for a swing ratio of 0.68. The Democrats gained twelve seats; had the distribution been unimodal and the system otherwise unbiased, they could have expected to gain thirty-nine according to the so-called cube law.

27. Calculations of swing are of course affected by many possible sources of bias. One of the most important is the concentration of uncontested (usually Democratic) seats in the eleven ex-Confederate states. Removal of this region from analysis provides a rough but fairly effective correction for this factor, particularly when dealing with long time series.

28. If one uses large-swing election pairs as a criterion—those where the swing was 5.0 percent or more either way—this gain would have ranged from five (using the 1918–20 swing ratio) to thirty-one (using the 1936–38 swing ratio) more seats than actually won in 1974, with a mean of seventeen.

29. Data courtesy of Inter-University Consortium for Political Research.
    The original data are reported in terms of the Index of Party Likeness, a standard roll-call measure and corresponding to the formula

$$1 - \frac{(D_{yes} - R_{yes}) - (D_{no} - R_{no})}{2}$$

where, e.g., D_{yes} is the proportion of Democrats voting yes on an issue. This measure varies from 1.0 (identical) to 0 (total dissimilarity). The Index of Party Dissimilarity simply removes the term 1 −, so that it ranges from 0 (identical) to 1.0 (total dissimilarity).

30. A regression of this index on time from the 43d Congress (1873–75) to the 92d (1971–73) gives an equation of $Y = .624 - .062X$, an r of $-.698$ and an r^2 of .487.

31. The evidence for this growing disillusionment is now quite impressive. For one example, see the first survey ever commissioned by the Congress, Louis Harris's *Confidence and Concern: Citizens View American Government*, Committee on Government Operations. U.S. Senate, 93d Congress, 1st sess. (Washington: Government Printing Office, 1973).

32. Of course, if the mean 1838–94 off-year turnout rate had been realized in the 1974 congressional election, the added increment would have been on the order of 38 million.

33. In the 1972–80 period, 24.8 percent of the whole House.

34. Michael Barone et al., *The Almanac of American Politics 1974* (Boston, 1974), p. 398.

35. A shift of forty conservative Democrats into the Republican party in the House retrojected into the recent past, would have given control of that body to the GOP in four of the last eight congressional elections (1966, 1968, 1970, and 1972), assuming no countervailing defections from the extreme left of the congressional Republican party.

36. See, for example, the articles on congressional elections and electorates in Campbell et al., *Elections and the Political Order:* Stokes, "Political Parties and the Nationalization of Electoral Forces"; Milton Cummings, *Congressmen and the Electorate* (New York, 1966); and, perhaps especially, William McPhee and Bernard Berelson, eds., *Public Opinion and Congressional Elections* (New York, 1962).

7

The 1976 Election:
Has the Crisis Been Adjourned?
1978

Every American election is a discrete event. In due course this event is joy-fully and exhaustively chronicled by journalists and political scientists. The 1976 election is no exception. Think of the personal dramas of this election: the meteoric rise of Jimmy Carter from obscurity to the presidency, the cliff-hanging battle between Gerald Ford and Ronald Reagan for the Republican nomination, the closeness of the contest between Ford and Carter. These dra-mas have evoked an outpouring of books and analyses of remarkable volume in the short time since the election was decided.

But if 1976 is a discrete event, like all other elections it is also part of a flow, a moment in our political evolution. Analysis of an election as a moment in historic time, linked indissolubly with other moments in past and future, must differ from event analysis. Above all, the election must be fitted into the context of social and economic change if it is to be properly under-stood. The need for this is particularly compelling at the present, when many of the traditional, implicitly accepted orderings of electoral politics have dis-integrated. When the contextual landscape is murky—and it has never been murkier than it is today—the digging must go even deeper than usual. That is the object of this essay. It is an exercise in studying the "deep background" of the 1976 election.

Since the early 1960s the United States has passed through major crises of an intensity and variety unknown in recent times. Their names form a famil-iar litany: the Kennedy assassination, the escalation of the Vietnam War and its eventual—and highly predictable—loss, the domestic upheavals of 1967–68, the Nixon presidency and its abrupt, tragicomic end, the oil crisis, and

From *American Politics and Public Policy,* ed. Walter Dean Burnham and Martha Wagner Weinberg (Cambridge: MIT Press, 1978).

the worst economic slump capitalism has known since before World War II. Quite a large number of things in American life, things in which government is directly involved, have gone very wrong in a very short period of time. It is natural, therefore, for the observer to suspect that the crisis as a whole is greater than the sum of these parts—that more than bad luck or the characterological deficiencies of individual leaders must be blamed for this disarray.

A conjunctural crisis of politics occurs when political settlements, and the operational ideologies that justify and sustain them, collapse. This collapse, in turn, occurs when conditions in the society and the economy have made the political settlement irrelevant to the point where large parts of the population come to regard it as oppressive. With desperate brevity one might summarize this deep-background context of the 1976 election as follows. The political settlement which grew out of the New Deal and World War II had two essential elements. Domestically, politics came to be articulated as "interest-group liberalism," as an activity involving the interactions of the acknowledged and legitimate leaders of major peak groups with each other and with government. Party leaders played a pragmatic, "brokerage" role. Enduring differences on economic-allocation issues divided most Democrats from most Republicans, but these differences were narrow and grew narrower over time. At the same time, the interventionist role of the federal government was—and until the early 1960s remained—very limited. It was limited most of all in those domains of potential public policy where "social" rather than economic regulation or intervention was involved. In a very real sense the Lockean individual who stood at the base of the country's hegemonic political ideology was "saved" from the implications of depression, war, and empire by being transformed into the Lockean group.[1]

In the international arena, the political settlement that unfolded after 1945 was imperial or "globalist," and militantly anti-Communist. The United States, confronted with an international challenge to its well-being, saw the new needs of the existing order of things interact with the liberal activism of our leaders to produce, eventually, its nemesis: the Vietnam War. On the way, it necessarily created new structures of power in the "military-industrial complex" and—among many other things—linked the peak trade-union leadership to the warfare state not only on grounds of ideology but, increasingly, on grounds of direct economic interest as well.

The electoral history of the past dozen years and more has been in large part the chronicle of this political settlement's disintegration. The old rules by which politicans and political scientists tend to live break down in such circumstances, and all sorts of peculiar things happen. For underneath the collapse of any political settlement is a refusal of assent to the "old politics," the accepted ways of doing things, a refusal that spreads, like a contagion, from one group of citizens to another. Moreover, the refusal of one group to perform the role stereotypically assigned to it in society will, rapidly enough, engender bitter opposition from members of other groups who find their values threatened.

The big accelerator of this process in the 1960s was, of course, the war. But the Wallace voters of 1968 were not in the main motivated by opposition to the fighting in Vietnam, but by opposition to the tremendous increase of federal intervention into social domains formerly the preserve of local governments or private arrangements. In short, for many who left the old Democratic party coalition by that door—and for others who were tempted to do so but until 1972 did not—the problem was that blacks were not "keeping their place," and, what was worse, they were being supported by Northern middle-class liberals and by the federal government itself.

The wavelike escalation of ghetto riots and student demonstrations moved a very large part of interest articulation altogether outside the well-known (and well-blocked) institutional channels. The very legitimacy of these channels was challenged, with the crescendo of protest centered on the 1968 Democratic convention. The pressure to restabilize that party by making it more representative was overwhelming. Out of the 1968 fiasco came the network of changes in Democratic convention rules that shifted the focus of selection away from the back rooms and toward the primaries and that— very significantly—relied upon quasi-proportional representation formulas for ingesting the proper number of scheduled minorities into the delegations.[2]

In retrospect, it appears to have been inevitable that one of the chief institutional victims of the collapse of the postwar political settlement was the party system. The Goldwater triumph in the 1964 Republican convention had demonstrated how very tenuous the welfarist consensus at the top had been in the United States, and how very powerful the appeals of laissez faire ideology continued to be. But it also led to a party debacle in the short term, and it appears to have been the starting point in the process of erosion of GOP strength among voter groups, including affluent students, which had hitherto supported the party. Was the GOP a "usable" opposition any longer? Despite Nixon's victories of 1968 and 1972, this question has recurred, formulated in various ways, ever since. It remains unresolved to the present day. What is not in dispute is that there is now an extreme imbalance in size between the two major components of the party system, whether measured in terms of individual party identification or in terms of such "grass-roots" support measures as the percentage of state legislature seats won by each party.

Quite naturally, then, much recent analysis of the problem of party disintegration has focused on the internal dynamics of the majority party. It is clear that the war and social-issue cleavages ruptured this party in 1968 and 1972. The so-called pragmatists of the party's prowar wing had as their chief opponents not the Republican opposition but the newer middle-class liberal Democrats for whom the Cold War fomulas of the old political settlement were not merely irrelevant but positively abhorrent in the context of the Vietnam conflict as they understood it. The candidate of the "pragmatists" won the 1968 nomination, but four years later the continuing pressure of war and social-issue cleavages, coupled with revised party rules which more accurately represented the current balance of forces within the party, led to the victory of their opponents. The upshot, among other things, was the demonstration of a political truth often buried in myth. "Pragmatists" and broker politicians act as they do under normal circumstances because the interests they defend are best protected by this style of behavior. When circumstances change and these interests are threatened, they can and do become as ideologically rigid, as polarized, as the veriest purist conducting a political jihad.[3]

But these internecine convulsions exposed even deeper issues. No two-party system can credibly exist for very long if politics becomes acutely polarized along more than a single dimension. As a result of the 1964, 1968, and 1972 elections, large but constantly shifting minorities of voters (and sectoral interests) found themselves losers in elections of unusual general-issue importance.[4] If the Republican party had become unprecedentedly shrunken and unattractive to the electorate as a whole, the Democratic party—by the very fact that it was in executive power from 1961 to 1969 and by virtue of its "incompatible" nominations of 1968 and 1972—had become identified as a central part of the problem. When to all of this was added the disastrous Nixon presidency, hard on the heels of an almost equally disastrous Johnson presidency, it was scarcely surprising that by the mid-1970s the strength of partisan identification had plummeted; that surveys repeatedly demonstrated a massive public loss of confidence in the working of our political institutions, especially the parties; and that writers like me were talking of "American Politics in the 1970s: Beyond Party?" or asserting flatly that The Party's Over.[5]

With this sort of crisis-ridden background, the 1976 election was early the focus of anxious attention by pundits and politicians. On the right loomed the specter of Ronald Reagan, a leader of some charisma who was exceptionally well suited to the image requirements of the electronic media. As for the Democrats, it was very hard to see how the party's antagonistic fragments could ever be welded together behind a nominee who could create enough intraparty consensus to win. At the very least, this fragmentation, coupled with the rules, appeared virtually to ensure a multiballot convention for the first time since the rise of television in 1952, and thus, paradoxically, to give unanticipated scope to power brokers at the convention.

As is usual in disordered and unstructured conditions, the expectations of politicians and pundits were confounded by the event. This was less noticeable on the Republican side. Had Ronald Reagan and his supporters conceded less to President Ford and worked harder, they might have ousted him. As it was, Ford's incumbency and the influence of the more cosmopolitan business interests within the party were just enough to secure his nomination.

Even this, however, could not hide the continuing rightward shift of the GOP. The choice at Kansas City was between conservative and far right: the once prominent liberal wing of the party, though still showing its prowess by winning senatorial elections in industrial states, is now virtually powerless in the councils of the party as a whole.

The real surprise was the remarkable return to consensus in the Democratic nomination process. Jimmy Carter, a former one-term governor of Georgia and a man almost wholly unknown as late as early 1975 to political elites, much less to voters, won a first-ballot nomination. This was such a break with long-term traditions and shorter-term expectations that an adequate explanation of it is essential to understanding what is happening to American electoral politics. The Carter campaign tactics were brilliantly designed to fit the disordered conditions I have discussed. From a strategic point of view it resembled nothing quite so much as a "revitalization movement" aimed at reintegrating the disintegrated, with a minimum of disturbance to the cultural or material values of hegemonic interests. As a Southerner, Carter first had to purchase credibility by demonstrating that he could defeat George Wallace in his home region. Carter did so, but the leverage he won thereby went far beyond this. Because of George Wallace's very existence—and the ruinous popular reaction against Democratic presidential nominees in the South and elsewhere which was linked to him—any successful non-Wallace Southern Democrat would appeal broadly to other Democrats looking for a national winner.

Additionally, Jimmy Carter campaigned as an "outsider," as one not in any way part of the Washington "establishment," and hence as one who could in no way be considered (like all his rivals but Wallace) part of the problem. His was a campaign fought on personalist grounds, stressing his religious beliefs, the purity of his moral commitment, and the need for comprehensive but nonpartisan reform in the way government did its business. From very early on it was clear to liberals and conservatives that the former had not much to hope from, and the latter rather little to fear from, a Carter administration. All of this, it should be stressed, fitted very well the pervasively disgusted mood of the electorate in 1976. It was an electorate heavily populated with voters who were alienated but who at the same time sought credible reassurance that their continuing belief in the old-time symbols of the American political religion was not vain.[6]

Of course, success is often its own justification in politics. But it is hard to escape the judgment, in retrospect, that this centrist, vaguely revitalizationist campaign succeeded within the Democratic party as well as it did because the older structures of opposition were already crumbling away. This crumbling was reflected in the weakness of all of Carter's major initial opponents. In fact, the only men who were to give Carter much trouble after the Massachusetts primary in March were Senator Church of Idaho and, especially, Governor Brown of California. Brown, after all, was a still newer face—but

one preaching the same gospel of a new, retrenched, and self-denying liberalism whose popularity fitted the exhausted and bewildered mood of the
country at large. He remains Carter's most formidable opponent within the
Democratic party. His 1976 career, like Carter's, epitomizes the new politics
of the 1970s: change the faces with ever-increasing speed, but keep the dream
and the system that spawns the dream.

The chief significance of the 1976 Democratic convention was that it was
captured by a campaign identical in many of its essentials to campaigning in
a Southern Democratic gubernatorial primary. There was irony in all this,
the latest of countless ironies in the twentieth-century history of our electoral
politics. The reformers of 1972 were defeated by changing circumstances and
by superior technique in utilizing the reforms which they themselves had
created. Carter's success in gaining the nomination meant the sudden
adjournment of the politics of confrontation within the Democratic party.
Both parties to that confrontation had exhausted and to a considerable
degree discredited themselves. Carter's surge represented what E. E.
Schattschneider long ago called the most devastating of all political strategies: the ability, when the time is ripe, to change the subject.

With the Vietnam War long over, and in the wake of the worst economic
slump since before World War II, it could be expected that the setting of the
1976 election would be utterly different from that of any previous election
of recent years. Even the most cursory review of the survey data reveals the
attitudinal effects of this. Among the respondents in the postelection voting
study at the Michigan Center for Political Studies, the preoccupation with
economic malfunction was enormous. Among those who identified a "single
most important problem" facing the country, 31 percent selected unemployment, 26 percent chose inflation, and another 13 percent gave a variety of
economics-related replies—in all, a total of 70 percent. By contrast, public-
order problems were selected as the "single most important" by 8 percent (5
percent choosing crime/violence specifically); the energy crisis by 4 percent;
social-welfare issues by 4 percent; all foreign-affairs items by 3 percent; all
defense-related items by 2 percent; and race-relations problems by 1 percent.[7]
Foreign-policy and defense issues were probably less salient in 1976 than at
any time since 1948 or earlier. The salience of "social-issue" problems had
largely evaporated. The salience of economic problems was much higher in
1976 than it had been at any point in the preceding quarter-century.

One would expect that these stimuli would work to restore the New Deal
coalition, including the South—largely because of Carter's regional identification. To a limited extent this coalition materialized, especially by contrast
with the losing McGovern coalition of 1972. Thus the Gallup surveys reveal
that the heaviest pro-Democratic swings between 1972 and 1976 were, in
order: Southerners (+26); self-identified Democrats (+22); people who
describe themselves as moderates in ideology (+22); people in blue-collar
occupations (+20); and males (+20). Conversely, the voters among whom

Carter's appeal was weakest relative to McGovern's were: people aged 18 to 21 (a swing of −5); blacks (−4); people living in cities of 500,000 or over (+2); Republicans (+5); people describing themselves as liberals in ideology (+5); people making $20,000 a year or more (+7); and Westerners (+8). Similarly, a review of the internal polarizations in voting group categories shows a maximum *decline* from 1972 to 1976 among age, community size, region, and ideology, and a very sharp *increase* of polarization in the category of party identification and, especially, income and occupation.[8]

All of this sounds at first blush as though "happy days were here again" in 1976. But the striking feature of this election, as of its immediate predecessors, was its pied results at different levels of the election. At the grass roots—state legislative outcomes, or the elections for the House of Representatives—the Democrats secured more than two-thirds of the seats. The Republicans were reduced to depths plumbed before only in the 1932–36 period and in the special case of 1964. They quite failed to bounce back from the Watergate-induced rout of 1974. But the presidential election was extremely close. Carter won only 51.1 percent of the two-party vote. By one criterion of closeness (the minimum number of votes theoretically needed to change the result in the electoral college), 1976 turns out to be the third closest presidential contest in the last century and a half, exceeded in that regard only by the elections of 1876 and 1884. Nor was there any doubt that Ford was an economic conservative, and that the Republican party was significantly more disliked by the electorate than the Democratic party, largely because the GOP in the minds of many voters was associated with big business.[9] Instead of a relatively comfortable victory—let alone a victory by the landslide projected by the polls of midsummer—Carter barely squeaked through, gathering just enough fragments of the New Deal coalition to edge past the incumbent. Why was this so?

First of all, Gerald Ford was the incumbent. In this century only three sitting presidents have been defeated for election (in 1912, when the majority party was split in two, in 1932 in the face of economic disaster, and in 1976). But Ford's pardon of Nixon clearly rankled many voters in 1976, and it is quite possible that this act in the end cost him the election.[10] Moreover, Ford was a nonelected incumbent who had never, even as vice-presidential candidate, faced the voters in a national election. Yet when the Michigan interviewers asked voters what they liked about the candidates, Ford's two chief assets loomed very large in the responses: his personal qualities (as a "nice guy," trustworthy) and, especially, his experience. These two categories together constituted 69 percent of all favorable references to Ford, compared with 34 percent for Carter. Carter's positive rating on personal characteristics was slightly higher than Ford's overall (with heavy emphasis on sincerity and morality), but a much higher proportion of respondents than in Ford's case were favorable to him because they perceived him as friendly to the little man and as the candidate of his party. But in any case an incumbent's stock in trade is his experience, and incumbents are accordingly hard to defeat as a rule.

Second, both candidates were widely perceived as less than satisfactory: 80 percent of all references to Carter's leadership qualities were negative, as were 76 percent of similar references to Ford's leadership qualities. While Carter appears to have "won" the debates overall, his lead was neither large nor "deep."

Third, it is worth stressing again that presidential elections in our time are overwhelmingly plebiscitary events in which voters have been induced over the years to choose on increasingly personalistic grounds. The whole of Carter's campaign was heavily oriented—as in any Southern Democratic primary—toward stressing this personalistic side of campaigning until about five weeks before election day. But "pure" personalism is not yet completely dominant in voter choice. As the summer went on, Carter's overwhelming lead in the polls shrank to the vanishing point; as the campaign went on, Carter's strategy shifted toward group-oriented policy appeals to the bedrock constituencies of the national Democratic coalition. It was at that point, and not before, that his slide stopped—for personalism is a two-edged sword. Ford as a person was at least as attractive overall as Carter was, to judge from the evidence, and being an incumbent of stable and well-known views, he profited from the ambiguity that surrounded Carter's campaign.

Fourth, while the election turned on economic issues, they appear to have cut in different ways than they once did. There is good reason to suppose that the evolution of the American political economy over the past generation has produced a kind of "two nations" phenomenon. It has been true for a long time that the relative concern over inflation and unemployment varies significantly by social class. People in the lower half of the income-occupation structure tend to be preoccupied with unemployment problems, while concern with inflation issues becomes relatively greater the higher the respondent's position is in the class structure.[11] But this propensity was probably heavily reinforced by the extreme unevenness of the impact of the 1974–75 slump, both geographically and occupationally. Recovery from the slump's effects also appears to be moving at very different rates of speed within the occupational structure. Moreover, while factory workers have seen little or no rise in real income after taxes for the past decade, businessmen, professionals, and people in the higher reaches of the federal service have clearly done much better than this. One survey from the NBC poll of voters on election day makes this point quite graphically. Respondents were asked whether their family's finances were generally in better, the same, or worse shape than they were a year earlier. The results, stratified by income, are reported in Table 1.[12]

Differentiation of voting in terms of these responses appears to have been quite sharp. Of those respondents reporting improvement (22 percent of the total), 30 percent voted for Carter. This compares with a Carter percentage

Table 1. "Two Nations?" Responses to NBC poll, by income level, 1976 (Percentages)

Perceived change in family finances since 1975	Family income				
	Less than $5,000	$5,000– 9,999	$10,000– 14,999	$15,000– 24,999	$25,000 and over
Better	9	13	26	27	14
Same	56	50	51	50	46
Worse	35	37	30	27	17
Percentage of respondents	9	20	26	27	14

of 51 among those reporting no change (48 percent of the total), and 77 among those who feel worse off (28 percent of the total). Similarly, only 36 percent of those in this survey who believed that inflation was the most important economic problem voted for Carter, while fully 75 percent of those whose primary economic concern was over jobs and unemployment voted for the Georgian. It is also worth noting that 38 percent of the Democratic identifiers in this survey thought themselves worse off, while only 15 percent considered themselves better off than a year earlier. Among Republican identifiers, the proportions reversed: 13 percent thought themselves worse off and 35 percent thought themselves better off in 1976 than in 1975.

Given these divisions, the mystery would seem to deepen except for one basic fact of the 1970s: the heavy and growing class skew in electoral participation. Evidence is overwhelming that nonvoting is concentrated in the lower half of the American socioeconomic structure. The evidence is equally strong that, as turnout has decayed in both presidential and off-year elections from 1964 on, it has fallen more rapidly—both absolutely and relatively— among low-participation social groups than among those whose members participate most.[13]

Viewed at the aggregate level, the national presidential turnout of 54.6 percent in 1976 is mediocre enough, with lower rates of participation found in our past only in 1920, 1924, and 1948. But even this low turnout is masked, to a considerable extent, by the remobilization of the Southern electorate since 1948. If we exclude the Southern states from analysis, the 1976 participation rate of 56.8 percent is the lowest of all time, marginally lower even than 1920 and 1924. It stands 14.3 percent below the 1960 level, and 16.1 percent below the turnout rate achieved in the 1940 election. The detailed census survey of voting and nonvoting for 1976 is not yet available, but it is possible to obtain a sense of the magnitude of this problem by looking at the presidential turnout in selected New York metropolitan area counties. One, the Bronx, has become nationally infamous for the concentration of urban social problems within its borders. The other, Westchester County, is not

Table 2. The case of the disappearing voters; Turnout and partisanship in two New York metropolitan counties, 1940–76 (Percentage of potential electorate)

	1940	1952	1960	1972	1976
Bronx					
Voting	75.6	68.8	63.6	47.0	36.4
Nonvoting	24.4	31.2	36.4	53.0	63.6
Democratic	51.0	41.6	43.2	25.9	25.8
Republican	24.2	25.7	20.2	21.0	10.4
Other	0.4	1.5	0.2	0.1	0.2
Westchester					
Voting	81.8	81.6	78.6	69.1	61.0
Nonvoting	18.2	18.4	21.4	30.9	39.0
Democratic	30.7	26.3	34.0	25.5	27.5
Republican	51.0	54.9	44.5	43.4	33.1
Other	0.2	0.4	0.1	0.2	0.4

wholly free of these problems nowadays but remains an older and generally prosperous suburban bedroom for New York City. Table 2 graphically reveals what has happened to turnout in both counties since 1940.

The decay in participation even since 1972—hence, with the 18-to-20-year-old vote held constant—is considerable in both counties, but especially in the Bronx, where it was 22.6 percent of the 1972 base, compared with 11.4 percent of the 1972 base in Westchester. Had the 1940 turnout levels been reached in 1976, at least 350,000 more people would have come to the polls in the Bronx than actually did so, compared with an additional 131,000 in Westchester. The picture is similar in California: predominantly Chicano and black-slum congressional districts in Los Angeles, the 25th and 29th respectively, were carried overwhelmingly by Carter with turnout rates of 28.1 percent and 36.2 percent respectively. The suburban 20th and 27th, however, had participation rates of 59.9 percent. Both were carried by Ford; it was the raw-vote pluralities he received in affluent, high-participation districts like this that helped him win a statewide plurality of 139,960.[14]

In the concrete situation of 1976, excessive emphasis ought not, perhaps, to be placed on this massive decline in voting participation in attempting to account for the election's partisan outcome. For example, Everett Ladd insists that an "inversion" has occurred in the demographic structure of liberalism and conservatism since the New Deal era.[15] Even if something of a shift back again is discernible in 1976, it would be very misleading to assume that the old-class-ethnic polarities had been restored to anything like their pristine state. The data in Table 2, limited as they are, hint as much. The outcomes of 1940 and 1976 in New York State were similar in the aggregate, with narrow Democratic victories in both elections. Very broadly, the decline in the Democratic share of the potential electorate in places like the

Bronx has been matched by a decay in the Republican share of the electorate both in the party's traditional upstate bastions and in the rapidly growing suburbs.

Thus, in New York at least, the interaction of these two processes produced a narrow Carter victory, but one grounded quite differently from FDR's 1940 victory in that state. The general argument still stands, especially if states like Illinois or California are taken into account: had lower-status voters participated even at the rates of 1960 or 1964, the presidential election of 1976 would not have been very closely decided. But perhaps one reason why they didn't participate at those levels is the growing extent to which the Democratic party has come to appeal to some sectors of the better-off electorate, as Ladd indicates. It is the "majority party" of a declining fraction of the adult population. It is important to note in this respect that turnout for Congress was still lower (51.0 percent nationally). Yet Democrats won handily with 57.6 percent of the two-party vote, compared with Carter's 51.1 percent. Put another way, Carter won with the support of 27 percent of the adult population, congressional Democrats with the support of 29 percent.

It is, in fact, this divergence between outcomes at different election levels which is of central importance in analyzing the meaning of the 1976 election. Taken together, the evidence presented by Ladd and others, the narrow balance between likes and dislikes for each candidate in the 1976 CPS study, the high levels of suspicion and distrust of leadership and political parties found in all recent surveys, and the turnout pattern all suggest a profound blurring of political images in the public mind. The dissolution of the partisan link between electors and elected continues its course. One chief symptom of this has been the now abundantly documented divergence of presidential and congressional voting coalitions. The election of 1976 extended and perhaps reinforced this recent pattern. The historic decline of the Republican party at the grass roots continued. In the Northern and Western States, for example, the party's state legislature representation hit lows not even reached in 1936 or 1964.

The evolution toward distinct voting coalitions at different election levels has given incumbents of both parties an increasing personalized protection. The election for the House of Representatives showed almost exactly the pattern described by a number of recent writers.[16] That is, the partisan distribution of outcomes in seats without incumbents tended to have a single mode close to the 50 percent mark, while incumbents tended to score lopsided victories over their opponents. Of the 74 supposedly vulnerable Democratic freshmen elected in the Watergate landslide of 1974, only 3 lost their seats.[17] Indeed, of all 267 incumbents in the non-Southern states, only 11 (4.1 percent) lost their seats.

This literature on the increasing insulation of incumbents in congressional elections implicitly presupposes its opposite: a more or less tight partisan linkage in the electoral process. From such a linkage comes a constraint on

Table 3. Anticipated consequences of two models for congressional outcomes

Category	Incumbent-insulation model	Party-salience model
Vote in CDs with incumbents, 2 terms or more	Swings paralleling national, little effect on outcomes because of bimodality (already existing landslide margins for incumbents of both parties)	Swings paralleling national; significant effect on outcomes because of relative closeness
Voting swing in CDs with incumbents, one term (freshmen up for reelection)	Strong countercyclic swing favoring freshmen of both parties	Swing paralleling national, large-scale loss for freshmen of party with adverse nationwide swing
Vote swing, open seats (no incumbents)	Strong convergent swing toward 50% (greater than national swing for losing party, countercyclical swing against gaining party)	Swing paralleling national in *both* categories (D and R at last election) of open seats
Heterogeneity of vote swing across all CDs	Tend toward maximum	Tends toward minimum
Amplitude of seat swing (losses or gains for a given party in HR representation)*	Tends toward minimum	Tends toward maximum

*Another, somewhat more technical way of putting this is to say the swing ratio ("predicted" by the cube law to be 2.80 to 1 on the long run) approximates a minimum under the incumbent-insulation model and a maximum under the party-salience model. Empirical verification of this expectation is clear-cut. Thus in the five-election sequence 1884–92 the swing ratio reached an all-time high of 8.62. In the five-election sequence 1968–76 it reached an all-time low over the past century of 0.97.

the divergence of electoral coalitions along office-specific lines—what might be called, in short, the electoral foundation of party government. Table 3 presents schematically what each of these models implies for leading categories of aggregate electoral performance in congressional elections.

Considerations of space preclude as detailed an analysis as one might wish, but let us briefly compare the elections of 1874–76 and 1974–76 with these models in mind. (For obvious reasons, these comparisons incorporate all congressional districts except those in the eleven ex-Confederate states of the South.) An attractive feature of this comparison is the existence of certain rough comparabilities in the two sets: a modest Republican electoral upsurge after a preceding rout for the party (a Democratic swing in the two-party vote of −1.8 percent in 1874–76, and of −2.1 percent in 1974–76); an incumbent Republican administration with scandals in the immediate political background; and economic hard times. If we look first at the pattern of partisan outcomes by incumbency status in the elections of 1876 and 1976, we note immediately a startling difference in the two configurations (Figures 1 and 2). The move toward much greater heterogeneity of outcomes for incumbents is clear-cut. It should also be noted that the proportion of seats with incumbents is, of course, much larger in 1976 than in 1876 (60.6 percent in 1876, 86.7 percent in 1976).

Similarly, the swings reported in Table 4 reveal a strong conformity with the patterns to be anticipated as we move from a party-salience situation (1876) to a more personalized incumbent-insulation situation (1976). In 1876 the nationwide swing entailed the loss of their seats by 32 Democratic incumbents, 30 of whom were freshmen (24.4 percent of Democratic incumbents), as well as 7 Republican incumbents (12.5 percent). But in 1976 a slightly larger swing was associated with the loss of only 7 Democratic incumbents out of 241 running (2.9 percent), and 4 Republicans out of 122 running (3.3 percent). As is obvious from Table 4, movements among these categories of

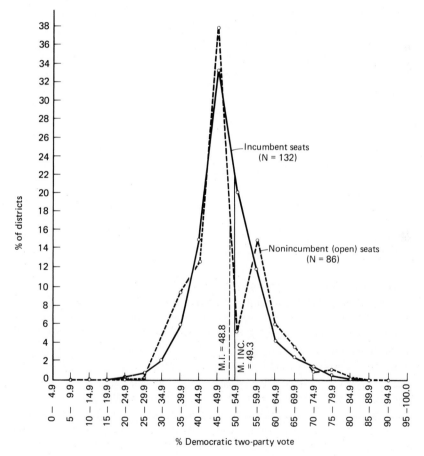

Figure 1. Distribution of contested seats, non-South, by incumbency status, 1876.

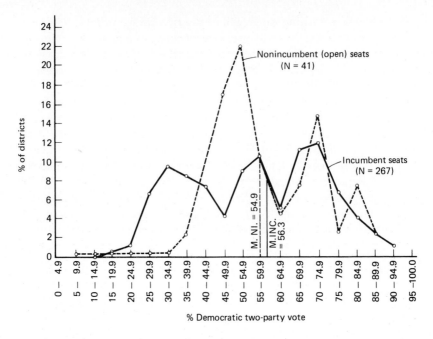

Figure 2. Distribution of contested seats, non-South, by incumbency status, 1976.

districts have a personalized heterogeneity which did not exist a century earlier.[18] A study of the heterogeneity of swing across all non-Southern districts
likewise produces results in conformity with the two models. Thus a 4.7 percent Conservative swing in the tightly partisan-constrained British electoral
universe between 1966 and 1970 was associated with a variance of 6.8; the
1874–76 Republican swing of 1.8 percent with a variance of 24.8; and the
1974–76 Republican swing with a variance of 76.7, a threefold increase over
the past century.

There is no question that the changing technology of political campaigning—the growth of congressional staffs, the increasing capacity and interest
of incumbents in concentrating on constituency service, and the steeply rising use of the congressional frank—has contributed heavily to this denation-
alization of electoral forces in the United States.[19] But the rapid and quite
massive recent shifts in this direction are also part of a general crisis of politics in which political parties have lost their salience as cue-givers for voting
among decisively large minorities of voters. This increasing personalization
of electoral politics critically affects not only electoral outcomes but the
power relations among elective institutions at the national level. As electoral
coalitions diverge and become more office-specific (actually, candidate-spe-
cific), the result is a profound behavioral reinforcement of the constraints

Table 4. Two models: The 1874–76 and 1974–76 congressional elections

Category of district	N 1876	1874–76 Mean % D 1874	Mean % D 1876	Swing to D Two-party	N 1976	1974–76 Mean % D 1974	Mean % D 1976	Swing to D Two-party
D incumbents, 2 terms or more, 76	18	58.9	56.5	−2.4	95	71.8	68.6	−3.2
D incumbents, 1 term (elected 74)	62	55.3	52.2	−3.1	60	58.2	62.7	+4.5
Open seats 76 won by D 74	38	58.1	54.7	−3.4	25	70.4	62.3	−8.1
Open seats 76, won by R 74	46	44.9	44.4	−0.5	17	40.1	48.9	+8.8
R incumbents, 1 term (elected 74)	17	44.9	44.7	−0.2	17	46.4	35.0	−11.4
R incumbents, 2 terms or more	28	42.2	42.4	+0.2	79	40.6	36.5	−4.1
Total H.R., non-South	209	51.2	49.4	−1.8	293	57.1	55.0	−2.1

*Non-Southern districts only. Based on percentage Democratic of two-party vote, in all districts where major-party opposition existed in both years of each pair.

which the separation of powers imposes on coherent policymaking and coordination. Put more bluntly, the emergence of these patterns implies a massive and, so far as we can tell, permanent reinforcement of institutional deadlock and the hamstringing of the presidency as an institution of national leadership.

To this we may add two other considerations. In the first place, we have Jimmy Carter himself: a man who owed little in his rise to the major interests surrounding the Democratic party, or to major leadership circles within the party itself. His style is technocratic and "above party." As he owed little to congressional Democrats in his rise to the presidency, so they owe little to him now. Second, there are abundant reasons to believe that the general crisis of capitalism into which we are now entering requires the formulation of the kinds of programs a Carter would tend to prefer: rational, comprehensive, highly complex, and with each major part crucially depending for its success on securing enactment of every other major part. Such is obviously the rationale behind the energy program. Other areas, notably taxation, will appear to require similar treatment. Yet if the history of our dispersed, nonsovereign political system teaches us anything, it is that Congress is unlikely to approve any such sweeping policy packages in any conditions other than those of overwhelming, immediate crisis. The syndrome of drift and mastery about which Walter Lippmann wrote more than a half-century ago was a crucial part of American experience, even when partisan links across separated branches of government were much more powerful than they are now.[20] A fortiori, the structural conditions favoring drift have become more deeply entrenched than ever.

Some years ago the historian Eric McKitrick wrote a seminal article which compared the effectiveness of Union and Confederate war efforts during the Civil War.[21] His analysis centered on a crucial comparative fact: the Union had powerful political parties, the Confederacy had no parties at all. The existence of these parties in the North meant two things for Lincoln's leadership. First, his own Republican-Union party gave him invaluable and predictable support at crucial times and created an essential institutional bond between him and Congress on the one hand, and between him and the war governors of the states on the other. Second, the existence of a powerful and structured Democractic opposition gave Lincoln essential information on the nature, scope, and effectiveness of the opposition to the war effort. In the Confederacy, on the other hand, the war effort was vitiated at every stage. The president and vice-president, coming from very different political traditions, were constantly at loggerheads; so were Jefferson Davis and the Confederate Congress. State governor opposition to the centralizing thrust of war was also endemic. The Confederacy was a political system in which opposition, unchanneled by party, seeped in here, there, and everywhere, to the detriment of its capacity to achieve the most elemental collective purposes.

The truth of the matter is that in a modern state political parties are required not only to give some substance to nominal democracy, but to act as channels of information and power through which political leadership can penetrate the society.[22] If the short-term interests of concentrated economic power and campaigning politicians are promoted by a disintegration of party linkages between rulers and ruled, their broader interests—both as individuals and as members of the hegeomonic class—may be very seriously compromised in the not-so-long run. In the American system, above all, the price for creating the support of mass electorates and cadres of activists—a support essential to the accumulation of power resources by executives—is the existence of some organized force able to overcome the fragmenting which the institutional structure mandates. Moreover, it may well be said that the essential job of the chief executive is to manage the capitalist political economy as a whole, both domestically and in its relationships with other political economies and states. It seems probable that today this management requires not only the existence of working instrumentalities for executive power accumulation, but the use of the power accumulated to secure assent to policies that will seriously affect both the position of organized interests and the daily habits of most Americans. If so, the progressive disappearance of party—a disappearance that successful candidates for office these days are capitalizing upon—would appear to favor political instability and ineffective performance on a scale without recent precedent.

Of course, one cannot have a resurgence of political parties (or a creation of new and more relevant ones) just by wishing for it. A price has to be paid by our various power elites to realize their purposes within the existing order. The shape of the 1976 election gives scant evidence that they are willing to pay it or even able to perceive that it must be paid if political stability and effectiveness are to be minimally realized. My own judgment, then, is that the outward symptoms of crisis have recently been transformed, but the crisis itself endures. How will the contradictions that feed it be resolved within the existing structure of our political system? Will they be resolved at all without a breakdown of the system at some point? These remain the unanswered riddles of the 1976 election.

NOTES

1. Probably the best single discussion of the tensions involved is Theodore Lowi, *The End of Liberalism* (New York: Norton, 1969), chs. 2–4. See also the discussion in Grant McConnell, *Private Power and American Democracy* (New York: Knopf, 1966).
2. For a useful discussion of these changes and their relationship to the nomination process, see Denis G. Sullivan, Jeffrey L. Pressman, Benjamin I. Page, and John J. Lyons, *The Politics of Representation: The Democratic Convention 1972* (New York: St. Martin's Press, 1974). See also Denis G. Sullivan, Jeffrey L. Pressman, and F. Christopher Arterton, *Explorations in Convention Decision Making: The Democratic Party in the 1970s* (San Francisco: W. H. Freeman, 1976).
3. Sullivan et al., *Politics of Representation*, pp. 116–34.
4. See, for example, Norman H. Nie, Sidney Verba, and John R. Petrocik, *The Changing American Voter* (Cambridge: Harvard University Press, 1976), chs. 12–15; and Arthur H. Miller, Warren E. Miller, Alden S. Raine, and Thad A. Brown, "A Majority Party in Disarray: Policy Polarization in the 1972 Election," *American Political Science Review* 70 (1976): 753–78.
5. Walter Dean Burnham, "American Politics in the 1970s: Beyond Party?" in William Nisbet Chambers and Walter Dean Burnham, eds., *The American Party Systems*, 2d ed. (New York: Oxford University Press, 1975), pp. 308–57; and David Broder, *The Party's Over* (New York: Harper & Row, 1972). For a somewhat more optimistic view, see John G. Stewart, *One Last Chance: The Democratic Party, 1974–76* (New York: Praeger, 1974).
6. See particularly the sobering data and report by Louis Harris Associates in a survey commissioned by the U.S. Senate, *Confidence and Concern: Citizens View American Government*, Committee on Government Operations, U.S. Senate, 93d Cong., 1st sess. (Washington: Government Printing Office, 1973). The Michigan Center for Political Studies 1976 study shows comparable findings.
7. Michigan CPS 1976 election study, frequency counts, Variable 3689.
8. The standard deviations of the presidential vote among the usual broad groups of Gallup's surveys increased for Party from 21.3 to 28.2; for Income from 4.7 to 9.0; and for Occupation from 2.9 to 6.5. For Ideology, on the other hand, the standard deviation declined from 23.3 to 18.0; for Region from 5.1 to 2.7; and for Age from 5.1 to 2.7.
9. For example, in the postelection survey the question was asked whether given groups had too much, enough, or too little influence. Big business led among twenty-four groups (77 percent said too much, 14 percent about right, 3 percent too little) in responses in the "too much" column. Next came labor unions (64 percent too much) and black militants (49 percent too much).
10. In fact, this is quite the largest single negative-mention item concerning Ford in this study.
11. See the evidence reported by Douglas A. Hibbs, Jr., "Economic Interest and the Politics of Macroeconomic Policy" (Cambridge: Center for International Studies, MIT, paper C/75–14, 1976), esp. pp. 24–40. This article was published in abridged form in the *American Political Science Review* 71 (December 1977).
12. Typescript, "NBC Election-Day Poll, 1976."

13. Since 1964, the U.S. Bureau of the Census has been asking questions as part of its extensive Current Population Reports, series P-20, about voting participation in presidential and off-year congressional elections. These reports have grown more detailed over the years. While the detailed reports for 1976 are not available at the time of writing, the longitudinal patterns for 1966–74 are clear-cut.

Category	Turnout 1968	Turnout 1972	Shift 1968–72	Normalized shift, 1968–72	Shift, 1966–74	Normalized shift, 1966–74
Propertied middle class (professional, managerial, farm owners)	81.7	79.7	−2.0	−2.4	−10.0	−14.3
Nonpropertied middle class (clerical sales)	77.4	72.8	−4.6	−5.9	−14.6	−22.5
Upper blue collar (craftsmen, service)	66.9	60.4	−6.5	−9.7	−14.7	−26.3
Lower blue collar (operatives, laborers)	57.5	49.5	−7.0	−11.3	−12.5	−27.1

14. Data from California *Statement of Vote,* relevant years; population base for 1976 districts, *Federal Register* 42, no. 79 (April 25, 1977): 21129; and *Congressional District Data Book,* 1974 California supplement. It is entirely characteristic that the absolute level of turnout declined from 1968 through 1976 by 9.3 percent and 17.4 percent in the poor/minority districts—CDs 25 and 29, respectively—but only 4.8 percent and 3.0 percent in the affluent suburban CDs 20 and 27, respectively.

15. Everett C. Ladd, "Liberalism Upside Down," *Political Science Quarterly* 91 (Fall–Winter 1977): 577–600.

16. For example, Burnham, "American Politics in the 1970s," pp. 322–33; Edward R. Tufte, "The Relationship between Seats and Votes in Two-Party Systems," *American Political Science Review* 67 (1973): 540–54; and idem, *Data Analysis for Politics and Policy* (Englewood Cliffs, N.J.: Prentice-Hall, 1974), pp. 96–101.

17. Allan Howe (1st Utah); Tim L. Hall (15th Illinois); Richard VanderVeen (5th Michigan). Howe was involved in a local personal scandal; VanderVeen ran in the solidly Republican district from which Gerald Ford came, at a time when Ford was at the top of the Republican ticket; and Hall represented an also normally Republican district in an exceptional year for the GOP in Illinois.

18. More formally, the variance across these categories of seats increased from 1.96 in 1876 to 49.0 in 1976, or by 25 times.

19. The best single discussion of this so far is Morris Fiorina, *Congress: Keystone of the Washington Establishment* (New Haven: Yale Fastback, 1976). The mechanisms which incumbents use to entrench themselves are spelled out thoroughly. See also David R. Mayhew, *Congress: The Electoral Connection* (New Haven: Yale University Press, 1974).

20. Walter Lippmann, *Drift and Mastery* (New York: 1914).

21. Eric McKitrick, "Party Politics and the Union and Confederate War Efforts," in Chambers and Burnham, *American Party Systems,* pp. 117–51.

22. Note the discussion of this in Giovanni Sartori, *Parties and Party Systems: A Framework for Analysis* (Cambridge: Cambridge University Press, 1976), 1:41–42. Sartori is much more preoccupied with the evils of hegemonic claims on the society by "monopolistic" parties (such as the Communists), and with problems of political order, than with the possible uses of parties as collective vehicles of democratization, but the point he makes is in any case important.

IV

The 1980s Come Next

8

American Politics in the 1980s
1980

I

The United States is clearly in the midst of a "conservative revival." As usual in such cases of "mood shift," the revival is spearheaded by intellectuals— particularly those who cluster around the Committee on the Present Danger and *Public Interest*. Symptoms include the California tax revolts of 1978 and 1980 (Proposition 13 and "Jarvis II," an initiative aimed at cutting the state income tax); the adoption by two dozen state legislatures of a constitutional amendment proposal requiring balanced federal budgets; the string of victories by very conservative Republicans in Southern and Western Senate elections; and, last but by no means least, Jimmy Carter's position as the most conservative Democrat to sit in the White House since Grover Cleveland.

Add to these signs of change another and far weightier one: the crisis in international relations unleashed by the seizure of the American embassy in Iran, reinforced by the Soviet invasion of Afghanistan. In early 1980 both these events are "swamping" the election campaign, scrambling all kinds of political calculations that seemed valid as late as November 1979. Whatever the long-term consequences, in the short term this international upheaval reinforces our domestic "conservative revival."

The thought occurs that the Republican party might capitalize on the situation, and perhaps even displace the Democrats as the majority party. It is not the first time in recent years that such speculations have arisen. Back in the late 1960s conservative intellectuals and politicians briefly thought they had discovered an "emerging Republican majority," arising from an imminent realignment. Kevin Phillips, arguing this thesis, traced the roots of this supposed counterrevolution to a white lower-middle and working-class backlash against the Great Society and its nonwhite beneficiaries, disagree-

From *Dissent*, Spring 1980, pp. 149–62.

able Vietnam War protesters, even more disagreeable and largely affluent college students, and, perhaps most disagreeable of all, conspicuous members of the counterculture.[1] This new Republican majority, then, would be created from the "silent majority's" rejection of these outgroups, coupled with Southern conservatism, patriotism, and racism.

Things did not work out that way. The realignment failed to realign; the supposedly emergent Republicans of 1968 became the submergent Republicans of 1974. The disruptive forces Phillips identified were real enough. They were quite sufficient to elect Richard Nixon twice, to give George Wallace his brief moment, and to rupture the Democratic party. In the end they could not prevent the election of Jimmy Carter or even his capture of all but one of the ex-Confederate states. Yet at no time did they fundamentally change the voters' underlying attitudes—whether one looks at the distribution of party identification, stances on economic questions, or propensity to vote for Democrats elsewhere on the ticket, even in the Nixon landslide of 1972.

It is not too difficult to see why. Some years ago Donald E. Stokes, then of the University of Michigan's Survey Research Center, identified two major types of issues in campaigns.[2] He spoke of "position issues" and "valence issues." "Position issues," Stokes said, are distinguished by the existence of two more or less clearly defined sides; on the other hand, what makes "valence issues" so useful to politicians of a minority party is precisely that they do *not* have two sides. An overwhelming majority of the population either supports or opposes a given valence issue. If these minority-party entrepreneurs can find a way to link the majority party to the "wrong" (i.e., unpopular) side of a significant valence (or "social") issue, the prospects of victory are much enhanced. By 1952, for example, Republicans had lost each of the preceding five presidential elections, defeated largely on the position-issue axis forged out of the Great Depression and the New Deal. Their successful 1952 strategy centered around their choice of a widely esteemed, essentially nonpartisan war hero and their skillful employment of such "valence issues" as "Korea, Communism, and corruption."

II

Attempts to build coalitions by using backlash against minority outgroups are closely linked to "valence-issue" strategies. Victory seems assured if the large majority under stress swings against blacks, Catholics, Communists, or other opportune targets. All this has a long history in American politics. The population is immensely heterogeneous by comparative standards. The rapid pace of socioeconomic change produces serious individual and group stress that, under the right circumstances and with skilled leadership, can provide intensely combustible materials for electoral warfare. Well-known examples include the Native Americans of the 1840s and 1850s, the Klan, Prohibition, and Catholic issues of the 1920s and, in part, the "great red scares" that followed American participation in both World Wars. Similarly, the Nixon-

Agnew campaigns of '68 and '72 hammered with great success at a variety of outgroup and negative-valence themes.

Yet there is a curious impermanence to such movements. They are, as it were, "luxury products" in electoral politics. They come to the fore either when the leading position-issue cleavages have been severely eroded with the passage of time, or when (as in the mid-1850s) they have not yet been conclusively defined. More than a century ago Horace Greeley pointed to this fragility in his analysis of the anti-Catholic, anti-immigrant Know-Nothings. Writing in 1855, when the Know-Nothing party seemed to be taking everything by storm, Greeley observed:

> It would seem as devoid of the elements of persistence as an anti-Cholera or anti-Potato-Rot party, and unlikely long to abide the necessary attrition of real and vital difference of opinion among its members with respect to the great questions of foreign and domestic policy which practically divide the country. These must soon dissolve its compact organization, distract its councils, "And, like the baseless fabric of a vision,/ Leave not a wrack behind."[3]

Greeley's forecast was right on the money. Scarcely a year after he wrote it, the Know-Nothing party had disintegrated as a national force, ruptured by the massive cleavage over slavery that ran right through both its organization and American politics as a whole.

Something of the sort also happened in the mid-1970s, as the Vietnam War receded into history and economic difficulties unprecedented since the Great Depression came into view. Confronting the issues, the candidates, and voter attitudes of 1976, one might well have thought oneself in a different country from the America of 1968 or 1972. Both foreign-policy and social-issue questions declined to the lowest level of intensity in public attitudes since 1948, perhaps even earlier. The national scene now was dominated by the malfunctioning of the economy and by such issues left over from Watergate as Ford's pardon of Nixon. This was not good news for Republicans. Ford was the first incumbent president to be defeated since Herbert Hoover in 1932. Moreover, at other electoral levels the Republican party had scarcely recovered from the near-record low to which it had fallen in 1974. Needless to say, it was particularly bad news for those conservatives who had hoped to build a new Republican majority on such flimsy and ephemeral foundations as those provided by "valence" or social-issue politics.

Once, very long ago, the Republican party enjoyed clear dominance. Decisively confirmed in the realigning election of 1896, this dominance lasted with little interruption until 1932. It lasted as long as the Republicans could command majority support on the main position-issue sets of the day. These issues centered on a sectional cleavage over modernization under industrial-capitalist auspices, though much else of importance was going on as well. Arrayed against each other were the country's "backward" primary-producer areas in the South and West, and the "advanced" industrial heartland, the

Northeastern and near-Midwestern metropole. Republicans projected them-selves to the electorate as the "modernizing party," the party of progress and affluence, and their Democratic opponents as backward-looking, not com-petent to manage the affairs of a modern state, and "responsible" for the Depression of 1893–97.

Laissez faire and the Republican majority perished in the flames of an even worse economic disaster. After 1932 both the dominant political ideology and the activities of government were sweepingly changed, though both remained well within our Lockean-liberal "tradition." A new, interventionist *political capitalism* was born. The state and its managers took responsibility for permanent involvement in the workings of the economy, the develop-ment of a welfare "safety net" for the victims of the capitalist business cycle, and the creation of new organizations of countervailing power, such as the industrial unions in the CIO. This displacement was not the work of a day: from beginning to end, the budgets from 1933–34 through 1941 classified all New Deal relief expenditures as temporary, and not until 1946 did Congress pass an Employment Act that made the federal government assume some for-mal responsibility for the elimination of mass unemployment.

The full-fledged claimant state we know today was a still later creation, of the 1960s and not of the 1930s. But the realignment of values, expectations, and political parties was striking enough, even in the early years. Out of these changes emerged the modal public-opinion structure described by Lloyd Free and Hadley Cantril in 1967.[4] American voters are arrayed not on one attitu-dinal dimension but two: ideological and operational. Ideologically they are predominantly conservative, regularly producing large majorities against "big government." Operationally they are liberals, just as regularly "voting" for specific programs that require government spending—particularly those from which they expect to benefit personally. However messy this may be from a logical point of view, such an opinion structure is admirably suited to support a nonsocialist but interventionist state that deals in the currency of particularized benefits.

The persistent Republican problem has been that its elites have never been able as a whole to accept this post–laissez faire dispensation. For many years the conservatives of this conservative party operated on the assumption that there was a "silent conservative majority" out there—somewhere—just wait-ing for a choice rather than an echo: a theory conclusively disproved in 1964 for reasons clear enough to public-opinion specialists. Nor did the Eisen-hower era and the politics of Nixon's 1968 and '72 campaigns make up for this *institutional* ill-adaptation to the modern political order. Presidential victories have come and gone, but support in the electorate for a Republican self-identification has fallen from nearly two-fifths in 1940 (only 4 points behind the Democrats) to little more than one-fifth (about 20 points behind) in the late 1970s.

One of the regular dynamics of politics seems to be that as a party shrinks, the influence of its hard-core elements grows within it. The hard-core Republicans are very, very conservative. Not surprisingly, substantial parts of the American business elite and the affluent upper-middle classes have recently been taking their political business elsewhere. Sophisticated Trilateralism has migrated from the Rockefeller Republicans to a Democratic administration headed by Jimmy Carter. Academic intellectuals now generally—if perhaps too easily—write off the GOP as representing only a parochial, declining-sector-plus-new-money brand of capitalist politics. Democratic presidents from John Kennedy to Jimmy Carter choose from the same pool of "serious men" that Republicans have tapped to manage the most pressing problems of the American elite, notoriously in the foreign-policy arena. No matter that this closed circle is still closed, no matter that these "serious men" have made a disastrous mess of things: you do not need Republicans in office to keep the circle closed and the range of policy choices narrowed. After all, the interests of the top elite as they perceive them will be protected well enough anyway—and perhaps all the better with Democrats running the transfer-payment state they believe in, and thus promoting social peace at the bottom of the system.

III

Until very recently, such reflections would have looked much more plausible than they do now. A further consideration of the problem is thus in order, beginning with two obvious facts.

First, both the domestic and world systems are dynamic and historically contingent, not static and eternally defined.

Second, the two are now so interpenetrated that neither politicians nor scholars can any longer profitably deal with them in isolation. The upheaval in the Middle East has resulted in Jimmy Carter's being born yet once again, so far as the Russians are concerned. Similarly, the whole 1980 election campaign will deal with foreign-policy and imperial-defense issues to a degree that scarcely could have been imagined before November 4, 1979. This and other signs of deep crisis involve important changes in the arena of conflict that may well reopen the question, "Who needs Republicans anyway?"

We have already considered the options generally available to the leaders of a minority party and dismissed two. Valence-issue politics can lead to ad hoc victory but—as the political economics of the Nixon years demonstrated—to no fundamental changes. Similarly, there are excellent reasons why a conservative minority party cannot simply lift itself up by its bootstraps and come forth with a set of position issues so compelling that a general political breakthrough ensues.

There is a third but relatively weak option, depending on two large "ifs." If the substructure of politics is to some extent dominated by class polarity, and if the basic issues of most campaigns are economic, then it would be possible for conservatives to score gains even without basic change with regard to the underlying position issues. This could happen if there was so large a withdrawal of lower-class elements from the voting population that the remaining, more affluent fragment could be won by traditional conservative appeals. As we shall see, some such disappearance has been going on during the past two decades. Congressional Republicans have been successful in thwarting efforts to liberalize electoral law and reduce procedural barriers surrounding the ballot box. Yet theirs is essentially a passive role. Declines in the participation rate depend crucially on what the *majority party* does or fails to do in order to mobilize the half of the electorate that now forms the "party of nonvoters."

Finally, the minority party can be propelled into majority status when a sharp crisis suddenly discloses how much the foundations of the domestic and international order have been changing and how irrelevant divisions over older position issues have become. Such events have been quite rare in American history. Still, they have formed the only historical route through which so sweeping a reorganization of the electoral market has occurred. Thus, if we inquire into the prospects for such a reorganization in the foreseeable future, we must attempt to evaluate changes in collective and individual existence whose effects could sweep away the accepted order of political capitalism and replace it with something else. To this we now turn.

IV

The state is primarily in business to promote capital accumulation and to maintain social harmony and legitimacy. Political capitalism was created out of the necessity to do both, following the collapse in 1929 of corporate self-regulation. Political capitalism was further consolidated when, with the suicide of Europe in World War II, a polar international confrontation—both in ideology and in *Realpolitik*—saw the launching of the Cold War and the creation of the American (quasi-) empire. Throughout the advanced industrial-capitalist world, "growthmanship" and the construction of expensive welfare states became the twin pillars of postwar public policy. The incentive for this was obvious: at all costs, the tremendous social conflicts and deadly international anarchy that erupted in the 1930s had to be avoided if the system was to survive. Assuming sufficient economic growth, the surplus thus generated could achieve military superiority over the Soviets, pay for the welfare state, lead to rising affluence in the private sector, and, moreover, do all these good things at the same time.

These objectives were achieved with spectacular success. An enormous, worldwide boom—unprecedented in both length and magnitude—unfolded from 1945 until about 1970. The secret of achieving sustained capitalist growth, increasing social harmony, and a relatively stable order in the world seemed to have been found. That persistent liberal utopia, the "end of ideology," seemed at last on the verge of realization.

This boom, of course, had its own contradictions; we can discuss only some of the more important ones here. First, the 1945–70 period was optimal for the growth of oligopoly capitalism from its national bases into a full-fledged multinational stage. In some respects, the multinational corporations may be said to have created the first true world economic market. Going wherever profit would be greatest, their operations in time led to a fateful compromising of essential national interests in the states of their origin—see the most obvious example in the energy-automotive sectors, where by 1980 the United States had become dependent on external sources for half of its oil supply. American leadership has done little or nothing to arrest this development, much less reverse it, even after the OPEC embargo price crisis of 1973–74.

The immense growth of competition between European and Japanese firms and domestic American corporations for consumer markets, especially within the United States, presents problems of comparable difficulty. Our free-trade policies form an integral part of imperial economic management. They were created and have been sustained by acute memories of the role that protectionism has played in wrecking the international economic and political order during the 1930s. A return to Smoot-Hawley tariff schedules would probably create comparable devastation, particularly in Japan and West Germany. Yet without protection, vitally important sectors of the domestic American economy were seriously undermined by the late 1970s. As this process continues in the 1980s, and it will, we can expect protectionist, national-capitalist candidacies to emerge, as indeed has already happened with John Connally—significantly, on the Republican side.

This brings us to another major legacy of the Great Boom of 1945–70. This boom was largely fueled by immense increases in consumer spending, and in the mass expectations of an affluent, middle-class living style. The ramifications of these changes have been exceptionally far-reaching; two are of particular interest here. Suburbanization spread on a massive scale. This involved a radical dispersion of urban populations and industry, made possible solely by the existence of cheap private transportation and cheap, apparently limitless energy. Dispersion was more extensively carried out in the United States than anywhere else in the Western world, and most completely in the rapidly growing Sunbelt regions. With time, this process gave rise to the famous "urban crisis." Yet far more formidable problems surfaced when the energy crisis hit in the mid-1970s. If and when it becomes necessary to impose World War II–style gasoline rationing, we will rapidly learn that the population is much less efficiently distributed than it was in 1941, and that public transportation will be far less capable of absorbing the load than it was in '41. Even without the necessity for rationing, the cost of transportation in so dispersed an environment will constitute a drag on overall economic performance far outweighing its counterparts in Japan or Western Europe—the more so as American fuel prices converge with those in Europe and in Japan.

V

Closely linked with the rise in consumer spending and expectations is the increasingly difficult relationship between debt and saving. Naturally, much public policy during the Great Boom was oriented toward promoting consumption and, at best, failing to encourage the saving from which capital accumulation is largely derived. In this favorable environment, the corporate sector went into debt on a scale rarely if ever seen in our economic history. It also promoted ever easier credit terms for consumers so that inventory could be profitably moved. The American economy is now a "debt economy," with the burdens of carrying this debt rising sharply since the death of the boom in the early 1970s. Profitable capital accumulation once again becomes visible as a key problem in capitalist economics, and in the 1970s it was joined by a dramatic slump in industrial productivity. In a no-growth situation, capital accumulation can be pursued only at the expense of consumption, both private and public. The contradictions here are obvious. They are reflected in part by the conservative intellectuals' rediscovery of the market, and of a parallel rediscovery by other intellectuals of Marxist critiques of capitalism.

If expectations of privately generated affluence were sharply raised by the Great Boom, still more were expectations involving the public sector's cornucopia-like output. One important aspect of booms is "boom psychology": the longer they last, the more people assume that they have become eternal and act accordingly. This psychology often takes extreme forms in the "go-go" years before the end—from the mad stock-market speculations of 1928–29 to the rushes into securities and real-estate speculation that marked the late 1960s and early 1970s. It is also in such eras that the state's leaders are likely to attempt to "complete" their task of promoting social harmony among the mass of the population.

During the heyday of laissez faire, this took such forms as Herbert Hoover's classically misplaced prediction of 1928 about chickens in pots and cars in garages—all, of course, to be provided by a beneficent private sector under cheerful Republican guidance. Of course, the quest for social harmony was pursued very differently during the boom of the 1960s than it had been forty years earlier: in the meantime, the interventionist state of political capitalism had come into being. There were more than enough problem areas, intensified by the boom's very existence—and these, like the plight of ghetto blacks, were now to receive "solutions" not only for domestic political reasons but sometimes for imperial ones as well.

Thus there was a fundamental change in the volume and, to a large extent, the very nature of domestic public-sector activities after the death of John F. Kennedy in 1963. In many respects this change was far greater in both its impact and implications than the New Deal had been. As even the most cursory inspection of pre-1963 budgets will show, the public sector of Eisenhower's time looks as nearly "prehistoric" to us as Hoover's. After 1963 new programs initiated the major redesigning and extension of such old ones as Social Security, and new claimants produced "entitlements" in a flow so powerful that it continued unabated right through the Republican presidential era of 1969–77. It continues still. Aficionados of the federal budget know that for all practical purposes a large part of it is out of anyone's reach on a year-to-year basis. These mandated expenditures, apart from servicing the national debt, are overwhelmingly in the domestic social-harmony and transfer-payment segments. These "uncontrollables" formed 65 percent of the whole budget a decade ago, in Nixon's time. The "uncontrollable" share of Carter's 1981 budget now stands at 76 percent.[5]

Forgotten along the way—because it was assumed that skilled Keynesian economists could "fine-tune" the economy while the interventionist state guaranteed eternal growth—was the bottom line. Promises could only be kept on condition that the boom did continue forever. Without a constantly growing surplus, the modern welfare state cannot be supported at levels that the public has been taught to expect. If the surplus stops growing—even more, if it disappears—retrenchment is as *practically* inevitable as it becomes *politically* hazardous for those in office.

VI

Well, the surplus did stop growing. It stopped especially after the revolution in the world political economy that was initiated with the oil embargo and OPEC price control in 1973–74. At the beginning of the 1980s, real discretionary income per worker had *declined* 9 percent since 1970 and 18 percent since 1973. The cost of basic necessities had risen 110 percent from 1970 and 80 percent from 1973.[6] Household and corporate debt stood close to, if not above, the record highs of 1974. Increases in American productivity fell to the lowest in the capitalist world, even lower than those in Britain. The inflation rate reached double digits, and as the 1980s begin there is little or no relief in sight. One might think that this dismal situation would produce some leftish reaction against an economy that, after all, has not lost its capitalist character. But we find instead the "conservative revival," which represents the best structurally based hope the Republicans have had in many decades. How to explain this?

First, there is the immense growth of the state since the Great Depression. And this virtually ensures that the public sector, much more than the private, will be widely held responsible for increasing discomfort—and for two reasons, ideological and operational. On the ideological plane, there still is no room for a coherent socialist opposition to the liberal individualism that has always dominated American political culture. This "liberal tradition" stacks the deck against the public sector in its domestic role, and it creates serious problems of legitimating that role in direct proportion as it expands. Accordingly, its intrusions and still more its mistakes will be much less tolerated than will those of any part of the private sector—except, perhaps, in the case of the oil companies. This quasi-permanent bias is an inevitable by-product of hegemony in the cultural-ideological domain, particularly when—as in the United States—that hegemony is to all practical purposes uncontested. In the 1930s the state came to be seen as an indispensable vehicle for national economic salvation; in the 1980s it is seen as a central part of the problem.

Operationally, the public sector's interventions for the sake of maintaining social harmony tend cumulatively to produce the opposite. This is not just an American problem; these days it is present in most Western political economies. But it is peculiarly exacerbated in the United States. The American Constitution of 1787 was explicitly designed to prevent the emergence of an internally sovereign and accountable political power structure. James Madison quite properly referred to it as "our feudal constituion."

We do not have space here to describe all that happens when a hugely increased load of demand and output is processed through such a power centrifuge, but merely for a few brief propositions. The traditional partisan channels connecting rulers and ruled, across the great fault lines at the center known as "separation of powers," decay. As they wither, the fragmented pluralism this constituional structure encourages keeps growing. Political executives, legislators, and judges have every incentive to go into business for themselves and for their clients, and they do. This becomes increasingly obvious to the public, especially to the unorganized "great middle" sectors, as more and more groups organize and get their pieces of the public pie. In consequence, a generalized crisis of legitimacy develops while the surplus declines. The "cultural contradictions of capitalism"—its persistent tendency to corrode general-interest bonds and to emphasize short-run personal gain—become the core of an explicitly political problem.[7] Finally, we find a public sector that really does share some features with feudal regimes. It becomes widely, and not inaccurately, perceived as being both feeble and oppressive at the same time. If the contemporary American state is under such widespread attack, one important reason is that it is the kind of state it is.

The growing "conservative revival" is no less fueled by permanent inflation of socially damaging proportions—an inflation virtually inevitable so long as the forces that simultaneously promote economic decay and an elective political feudalism continue to exist. For declines in the general level of well-being must occur so long as these forces predominate. This is particularly true of the enormous transfer of wealth to the OPEC countries, akin to the reparations exacted from Germany by the Allies after 1919, but of vastly greater magnitude.

Granted the electoral incentives, it is politically almost impossible to "disentitle" any significant part of the population. After all, more than 50 million government checks are at stake. It is also highly undesirable from the government's own perspective to have a situation in which, at the private-sector level, there would be actual, visible, and widespread pay cuts. Were the currency stabilized, both would be required under present and immediately foreseeable economic conditions. So the promises are kept, but the people who do the promising have the unique power under the Constitution to regulate the value of the units of account used to fulfill those promises. It also seems that people are much more quiescent when their real disposable income declines while nominal income increases than they would be if they received pay cuts totaling, say, the 18 percent by which discretionary income has declined from 1973 to 1980. Finally, and most important, the inflationary "solution" permits the government to enjoy the benefits of "tax creep," as increases in nominal income push people into higher income-tax brackets. This becomes even more important for our rulers as they contemplate the military spending a Cold War II would entail. For unlike the situation in the last Cold War, this additional expenditure on arms will have to be added to the top of a budget, now well over $600 billion, three-quarters of which is locked into domestic "uncontrollable" expenditures.

This solution would be nearly perfect from the state's point of view were it not for certain unfortunate consequences of inflation. One is that inflation tends to rot the social order of capitalism in direct proportion to its magnitude. It can, for example, seriously compromise the pursuit of capital accumulation strategies by diverting funds from investment in securities into speculation in securities and Krugerrands, and by even further discouraging savings and encouraging immediate consumption. More generally, inflation stimulates an ever more bitter struggle among organized fragments of the population to use whatever market power or political clout they have to shove the burden of declining real income onto others.

Carried to its logical extreme under conditions of hyperinflation, this can lead to truly Hobbesian social conflicts, a war of all against all such as historians of the Weimar period have described in vivid detail. *Sauve qui peut* becomes the order of the day, the only principle of social intercourse. Even at the far lower inflation rate that prevails in the United States today, inflation coupled with declining real income prompts an increasingly desperate and widespread search by the middle classes for ways out. As California real-estate owners found with Proposition 13, cutting back on the public sector can give immediate relief. But the local public sector is much more accessible to such revolts than the national. The federal government continues insistently to pursue "tax creep" fiscal strategies to keep its own entitlement structure more or less intact—ensuring that these efforts at relief will be redoubled in the years immediately ahead. The victims, at all levels, will be the poor and other dependent populations, particularly schoolchildren; but schoolchildren do not vote, and the poor don't vote in anything approximating their true numbers.

VII

We now turn to the matter of electoral participation, since its dynamics give additional support to the "conservative revival." Some recent studies, notably by Douglas Hibbs, have identified a clear class dimension affecting the public-policy choices governments have made in the recent past.[8] Two broad conclusions are relevant here. First, the greater the strength of the organized left in the electoral market, the larger the postwar growth in the welfare state has tended to be, though somewhat less so in the United States than elsewhere. Second, opinion data reveal that working-class people are more worried about losing their jobs than about inflation, while worry about inflation is much stronger among middle-class people.

It follows that a political system with *no* organized working-class left will be marked by heavy abstentions among the lower classes; that the active electorate will thus be significantly more middle-class than the population at large; and that there will be, comparatively, a tendency for the electoral market and public policy to reflect a higher tolerance of unemployment and a lower tolerance of inflation than if the lower classes organized and participated. Obviously, as inflation has become a much more acute problem than it was when these studies were conducted, a different set of findings might well exist. Even so, we suspect that if a "conservative revival" were to occur under more acute inflationary pressure, the quite extreme class skew in the electoral market would maximize the chances for this revival to prevail.

We may not be far from a description of actual conditions in the United States today. Throughout this century participation among the American lower classes has been significantly lower than in virtually all other Western polities. This class differential, not surprisingly, was notably reduced during and for some time after the New Deal realignment of the 1930s. But since 1960 turnout (outside of the very special case of the South in the aftermath of voting-rights legislation) has sharply plummeted from the relatively mediocre levels of the 1940–60 period. Geographically, participation has fallen especially severely in metropolitan areas, and most of all in their cores. In 1978, for example, turnout for the gubernatorial election in New York State reached a 150-year low, 38.5 percent of the potential electorate, even lower than it had been in the gubernatorial election of 1810, when nearly three-fifths of the potentially eligible population was disfranchised for failure to meet a property qualification! In 1978 it was, of course, lowest of all in New York City. Sociologically, this decline in post-1960 participation levels has been particularly concentrated among working-class Americans who already vote least. Thus by 1976 blue-collar and service workers constituted only 48.5 percent of the active electorate, but fully three-quarters of the "party of non-voters."[9]

Nor can the abysmal turnouts of the 1970s be discounted by the enfranchisement of the 18- to 20-year-old age group in 1971. True, participation rates in this group are more than 50 percent lower than among their West German counterparts. But this age cohort added only between 6 and 8 percent to the U.S. population base. It is trivial to show that even if they were altogether excluded from that base in computing turnout rates, this would increase the nominal turnout rate by less than 2 percent. The post-1960 decline and its concentration among the lower classes of American society are thus real enough. And if blacks have a generally lower turnout rate than whites, this too is overwhelmingly a matter of class.

To repeat, conservatives profit from this huge participation skew in a variety of ways, but theirs is not the active role in creating it. Proximately, it is what the supposed representatives of non-middle-class Americans, the Democratic party and the trade unions, have done and especially what they have failed to do that has made the larger contribution. More remotely, this huge "hole" in the active electorate is largely the artifact of uncontested hegemony and its effects on political consciousness.

One can hardly conclude a discussion of the postboom situation and its effects on our domestic politics without reference to the so fundamentally changed international situation. The boom and the Cold War were inseparably linked to America's ascendancy, and both created and maintained American quasi-sovereignty in an essentially Hobbesian international arena. But the 1970s began with the unilateral American abrogation of key features of the 1944 Bretton Woods monetary agreement. This was required by the deterioration of American economic supremacy in the non-Communist world. It ended with the seizure of American embassy personnel by Iranian revolutionaries and with Soviet aggression in Afghanistan, following a long period of growth in the USSR's military capacity. The latter in particular appears to confirm the acute misgivings among conservatives about American power and "national will." The invasion has served as an admirable accelerator for these misgivings. Both upheavals are linked to the utter dependence of the industrial-capitalist world on Persian Gulf oil. To a much greater extent than almost anyone realizes, the United States, dependent on external sources for half its oil, has lost the capacity to pursue a genuinely independent foreign policy this side of a nuclear war. Conservatives are the least likely people to attempt a rigorous inquest as to how this fatal dependence was created, by whom, and for whose benefit. But they will profit nevertheless, at least in the short term.

The probable consequences of this decay in the American empire's control over international economic and political life now are visible. We may well expect the reinstitution of the draft on both economic-cost and geopolitical grounds, and very large "catch-up" expenditures on the military. Both will probably have unsuspected price tags. The draft will be more widely resisted than it was during the Korean War. Above all, the addition of huge supplemental defense expenditures to the present budget structure, with all its "uncontrollables," will also trouble a much more problematic American economy. To the extent that these expenditures go up without major cuts elsewhere, the danger of hyperinflation becomes more real than during the Korean or Vietnam wars. Not only does this suggest that the "guns versus butter" stage may be reached surprisingly soon; it also suggests that, if the international situation continues to deteriorate, something very akin to "war capitalism" is likely to emerge as our dominant survival strategy.

VIII

The reader at this point may justifiably say that even if all the foregoing is an accurate assessment, it still doesn't tell us very much about how a "conservative revival" will help the Republicans. After all, the absence of organizable oppositions in American politics has meant that the Democratic party is a remarkably protean beast. Its leaders can and do have freedom to wander widely over the policy landscape. Jimmy Carter is not only the latest example, but one of the purest. He can and will, so far as possible, preempt issues that his primary opponents and the Republicans might use in their efforts to unseat him. So will many in the congressional Democratic party. So what, fundamentally, will change in and around the electoral arena?

Not very much will change, unless or until the survival needs of the dominant mode of production require it. But the foregoing analysis raises the lively prospect that we may at last be reaching such a condition. In the short term, the Republican candidate of 1980 may or may not succeed by tapping public reactions to high inflation and the administration's ineptitude abroad. But it is hard to imagine how runaway increases in the Social Security levy, rapid and heavy increases in defense spending, high inflation, possibly substantial increases in unemployment, and persistent declines in the mass standard of living can be sustained indefinitely without producing a political explosion.

A rediscovery of the economic market's virtues comes to seem, under these conditions, both appropriate and nearly inevitable. So do arguments and policies favoring capital accumulation over consumption and a conversion of as much as possible of the domestic public sector's activity to a pay-as-you-go (and pay-as-and-if-you-can-afford-it) basis. For in the absence of an organizable socialist critique of political capitalism, the operational code of American politics can only fluctuate between neo-laissez faire and the "entitlement"-oriented welfare state. Both the capital accumulation needs of the regime at the top and the personal survival needs of a dominantly middle-class voting population now point toward disentitlement as the order of the day. If so, then the initiative in political discourse and action will continue to move toward the neo-laissez faire pole. There is an obvious implication in all this: the American Lockean-liberal tradition, as capable in the past of unexpected revitalizations as the capitalism it supports, may at last have run out of historically creative options.

But dismantling the welfare state—even placing it at zero real growth for any length of time—would involve speeding up the impoverishment of the poor and even of significant parts of the middle classes. If the Republicans today enjoy the identified support of little more than one-fifth of the electorate, there are very good reasons for this. Given their incapacity to develop a credible political formula appropriate to an age of retrenchment, their success in curbing the entitlement state would inevitably, sooner rather than later, produce an electoral reaction in favor of the Democrats. Large-scale sacrifices from the American people are already being called for and are almost certainly on the way. They will require legitimation. If a model of the state's "proper" role based on the conditions of 1936 no longer seems adequate to the difficulties we now face, the 1886 variety seems even less suitable so long as a relatively free electoral market continues to exist.

This sort of dead end at least suggests that some sort of socialist (or at least laborite) alternative vision of our human potential might possibly be seriously organized. Perhaps so; but while the future ought not to be foreclosed, such an eventuality seems remote. Uncontested hegemony is not easily overthrown. Among other things, the organization of world politics since the Bolshevik Revolution has served to displace the struggle between such alternatives onto the international arena.

Insupportable ideological demands are placed on Americans *as a whole*, along with threatening *Realpolitik* demands, by the so-called Socialist bloc. Equally insupportable material demands are being made on Americans *as a whole* by the Third World "southern" nations in the ongoing north-south dialogue, and especially by the OPEC cartel. Domestically, such external pressures notoriously tend to weld a "coalition of the whole." They narrow seriously organizable alternatives in domestic politics—very much as, in 1896, the Democrats with their free-silver campaign united and rallied against themselves overwhelming majorities of voters in the industrial Northeast. In such situations of threat, people who regard themselves and are regarded by others as "haves" (no matter how illusory the perception may be) will unite on this issue, while they divide on everything else. We may be left for the foreseeable future with an ongoing, increasingly sterile debate between laissez faire conservatives and "statist" liberals.

In the last analysis, what we have been describing is a chronic but now escalating crisis of rule—a crisis, therefore, in the foundations of the constitutional regime that now exists in the United States. Such a crisis is not likely to be resolved within the framework of that regime, nor, *a fortiori,* within the traditional Republican-Democratic choice structure. The vacuum of power and policy created along the way is patently dangerous to the long-term survival of the American state and of capitalism itself. The problem of *authoritatively* defining a collective national purpose becomes as intractable as it is necessary.

This point is now clearly understood by those intellectual conservatives who go virtually as far as they dare in their proposals to curb what John Adams, in simpler times, once called "the democratical element."[10] For the decay of the American empire's economic and international foundations returns us squarely to the dilemma so visible in the 1930s and forgotten during the postwar boom. There is inevitable conflict between accumulation and legitimation strategies, visible precisely to the extent that surplus does not grow. There is, therefore, an even deeper conflict, when ultimate push comes to ultimate shove, between capitalism and democracy—made deeper, if possible, by the feudalized governmental machinery through which democratic impulses are processed in the United States. It has been the first principle of the "liberal tradition in America" to deny the existence of any such conflict. But it has now become irrepressible. It is a mark of the times that only those on the right seem to have become aware that this is so, and to draw the appropriate conclusions (for them) as to the future ordering of politics.

Antonio Gramsci once observed that "the crisis consists precisely in the fact that the old is dying and the new cannot be born; in this interregnum a great variety of morbid symptoms occurs."[11] If you seek a monument to the truth of this statement, look around you. Such a blocked context would seem to require of people who do not owe allegiance to the "conservative revival" that they pay more rather than less attention to what is going on in the 1980 election and its successors. They also have more reason to concentrate their attention on the course of the "conservative revival" in both the ideological and operational spheres, and to shape their own actions accordingly. The task, one would suppose, is to find a way out that liberates rather than destroys man's capacity for self-realization not only as an individual but as a social being. There will be work enough for anyone who feels the call to undertake it.

NOTES

1. Kevin Phillips, *The Emerging Republican Majority* (New Rochelle: Arlington House, 1969).
2. Donald E. Stokes, "Spatial Models of Party Competition," in Angus Campbell, Philip E. Converse, Warren E. Miller, and Donald E. Stokes, *Elections and the Political Order* (New York: Wiley, 1966), pp. 170–74.
3. *Tribune (Whig) Almanac for 1856* (New York, 1855), p. 7.
4. Lloyd Free and Hadley Cantril, *The Political Beliefs of Americans* (New Brunswick, N.J.: Rutgers University Press, 1967).
5. See the analysis of the 1981 budget in the *New York Times,* January 20, 1980, sec. 3, pp. 1 and 14.
6. These estimates are taken from the account of the problem of declining real income in *Business Week,* January 28, 1980, pp. 72–78.
7. For a full discussion, see Daniel Bell, *The Cultural Contradictions of Capitalism* (New York: Basic Books, 1976). Bell's argument is close to, and partially follows, that presented a generation earlier by Joseph Schumpeter in *Capitalism, Socialism and Democracy* (New York: Harper, 1942, 1947, 1950).
8. Douglas A. Hibbs, Jr., "Political Parties and Macroeconomic Policy," *American Political*

Science Review 7! (1977): 1467–87. For a provocative account of these and related issues, see also Edward R. Tufte, *Political Control of the Economy* (Princeton: Princeton University Press, 1978).

9. For a fuller discussion of this problem see my essay "The Appearance and Disappearance of the American Voter," in Richard Rose, ed., *Political Participation* (London: Sage, 1980) [Chapter Four in the present collection].

10. Most notably, so far, in Samuel P. Huntington's essay, "The Democratic Distemper," in *The American Commonwealth 1976,* ed. Nathan Glazer and Irving Kristol (New York: Basic Books, 1976), pp. 9–38. Kristol has made an interesting effort to evade the point in his essay "The Worst Is Yet to Come," *Wall Street Journal,* November 26, 1979. Kristol attempts this by restating the classic Cold War synthesis: democracy based on an ever-expanding economic pie, capitalism, and a militarized response to external challenge. It is one thesis of my present essay that the historic conditions for the success of that synthesis have expired.

11. Antonio Gramsci, *Selections from the Prison Notebooks of Antonio Gramsci,* ed. Q Hoare and G. N. Smith (New York: International, 1971). p. 276.

9

Into the 1980s
with Ronald Reagan
1982

I. Introduction

The 1980 election resulted in the easy triumph of Ronald Reagan over the incumbent Democrat, Jimmy Carter. Along with the defeat of Carter, the first elected incumbent president to lose since Herbert Hoover in 1932 (and, one may add, by a comparable margin), the Democrats also lost control of the Senate. Moreover, as the events of 1981 were to demonstrate, their unexpectedly heavy losses in the House elections proved large enough to turn over effective control of that body to a conservative coalition. Very probably all that saved Speaker Thomas O'Neill's job in the end was the incumbent-insulation effect in congressional elections.[1] Even more striking than the election itself has been the aftermath so far. Ronald Reagan and his allies have wrought one of the largest and most comprehensive policy changes in modern times. The growth of the domestic public sector, a fixture of American politics for nearly two decades, has been truncated, and a long-term tax cut primarily benefiting the wealthy has been adopted. Moreover, there are potentially important straws in the wind of public opinion since the election.[2] They sum up to the proposition that the white electorate, at least, supports these changes generally and Reagan personally, and that a shift of party identification toward the Republicans may be going on as well.

The 1980 election elevated to the presidency the most ideologically conservative man to hold that office since the 1920s. Since his inauguration, he and his coalition have already enacted into law a substantial part of the economic program on which they campaigned. Much else has happened as well. The Democratic party, its intellectuals, idea brokers, and activists are in near-total disarray, a condition that has in fact been underway for years. Equally long term, and now at fruition, has been the development—organizationally and intellectually—of their opposite numbers on the right. Keynesian economics, once so mainstream that even Richard Nixon proclaimed himself a

Keynesian, is now widely discredited. The inheritance is being claimed on one side by monetarists, on the other by supply-siders—both intensely conservative and both concerned with "liberating" the private-capitalist sector to "reindustrialize" the American economy. While the liberal left is now in severe disarray, the right is well organized, enthusiastic, sure of itself.

It goes without saying that in the American past, critical realignments have been fashioned out of just such materials. Their chief contribution to American politics has been to bring order out of a preceding disorder that has grown to unacceptable proportions. They decisively "clear the air." When all is said and done, this happens because the victorious coalition is able to revitalize the basic American political formula by apparently bringing it up to date, giving it a "relevant" variant in accord with the politically and cognitively organizable needs of the times. Conversely, the former majority coalition tends to fall into disarray in large part because—like ruling classes in prerevolutionary France or Russia—they have lost confidence in their right to rule, and they have lost as well any convincing intellectual rationale for their activities. Certainly both Republicans and Democrats are behaving very much as we would expect participants in a critical election to behave during its aftermath. Indeed, the Reagan administration can lay claim to being the most intellectual administration of modern times, as it is surely the most ideological; the two are not unrelated. We may note with Charles Beard that men in politics are largely ruled by ideas—true and false—and with Karl Marx that the dominant ideas are likely to have a close relationship to the needs of the dominant mode of production at any given time. Rarely have these truths been more obvious than they are now.

Yet there are important reasons for being very careful to avoid any simplistic analogies with the past, especially where old-style critical realignments are concerned. For one thing, there are technical requirements of such events that are not visible in the behavior of the American electorate, at least not yet. For another, the criterion of durability obviously cannot be satisfied until a good deal of time passes. And these days, in sharp contrast with our isolationist and self-sufficient past, the world economy and power structure are not likely to leave us alone for fifteen or thirty years. But quite apart from all these considerations, the present political situation in the United States peculiarly demands a careful analysis of the conditions that preceded and produced the political revolution through which we are now passing. This includes considerable attention to the longer-term roots of crisis as well as the disjunctions that have widened over the past generation among levels of the political system. There seems to be as much reason for supposing that the events of 1980–81 have shifted the American regime crisis onto a higher (and more dangerous) plane as for supposing that an old-style critical realignment has occurred. What follows, therefore, will be a discussion of the longer-term and shorter-terms factors leading to the Democratic debacle of 1980; the leading features of the election itself as a moment of our political history; and the implications of the events that have crowded in after Ronald Reagan's victory and inaugruation.

II. The Background of Crisis

One sure sign of a genuine "crisis of the old order" is the development of intellectual and operational stress among those who guide the destinies of state and, parallel to that, the growth of self-confidence and élan among those who are in opposition. The pattern is common enough in the prehistory of genuine revolutions. It has been frequently remarked by Crane Brinton, Leon Trotsky, and other analysts of revolution that a considerable time before the outright collapse of the ancien régime, the intellectual and political elites that support and attempt to reproduce it somehow lose their vitality, self-confidence, and effectiveness as persuaders and actors.[3] The defenses of the regime seem to become increasingly ritualized and barren. Conversely, the opposition raises a new standard (or standards, since there is often more than one current of opposition at work) that becomes increasingly persuasive, increasingly dominant in defining reality. The opposition therefore comes to dominate action agendas in politics. Thus there is the classic case of the *philosophes* in France under Louis XV and Louis XVI, and the devastating intellectual and moral destabilization of the regime in the wake of the Enlightenment. It seems to be an axiom that before revolutionary change is possible, the legitimacy of the ruling class must first be thoroughly undermined—not least importantly in the minds of the ruling class itself.

Something very analogous has clearly happened in the United States during recent years. The rupture between the Cold War and "new-liberal" wings of the Democratic party in the late 1960s contributed importantly to the subsequent intellectual paralysis that has gripped the organic intellectuals of the political-capitalist state, and for good reason. However matters may be in other advanced industrial-capitalist societies, the sheer necessities of permanent competition between capitalism and Soviet-style Marxism for the future of the world have required that the American state become—as it has in very large part—a garrison state. The old Cold War Democratic left permitted its obsessions with Communism to bring about the moral and material disaster that was the American misadventure in Vietnam. On the other hand, the newer Democratic left never developed or maintained a position that appeared to face the realities of international politics with sufficient realism. Things might have turned out somewhat differently in the late 1970s, or even in 1980, had the Soviets' behavior been modified toward the less aggressive stance that many Americans had believed that détente, the Helsinki accords, and the SALT I treaty implied. That it was not, in places like Angola and Afghanistan, simply underlined the crucial dilemma that tore at the vitals of the Democratic coalition. To a much greater extent than is commonly realized, the permanent international struggle has been decisive in paralyzing and undermining the cohesion of the left in this country. This is a very long-term process. It began in and just after World War I at about the time of the "great red scare" and has continued to unfold ever since, though by no means smoothly. It speeded up again, after a temporary hiatus following the Vietnam excesses, in the late 1970s.

But important as foreign-policy considerations have been in the undermining of the Democratic coalition and the partial dispersal of its intellectuals, the crisis gripping the political-capitalist state and those who have created and supported it runs quite across the board. It is reflected in a fourfold way: in the political economy; in the empire created after 1945, and the relationships between it and its Soviet and Third World opponents; in the cultural system; and, not surprisingly, in the political system itself. Accordingly, a vast and growing gap has gradaully emerged between the inarticulate major premises of the "American dream" and emerging realities. This in turn has set the stage for just the kind of reactionary revitalization movement that has elevated to power in 1980.[4] This movement's cardinal article of faith is that the "American dream" and the world of reality can be somehow rejoined. Evaluation of the extent of the crisis can best begin with a brief—if no doubt idealized—view of the situation prevailing in the "good old days," i.e., in the "American century," which actually lasted from 1945 to about 1965.

The economics of these "good old days" were chiefly dominated by the overwhelming ascendancy of American capitalism in the world economy following the mass destruction overseas wrought by World War II. This lasted approximately from the signing of the Bretton Woods agreement in 1944 until its forced abrogation by Richard Nixon in 1971. In this period the dollar enjoyed a hegemonic position in international relations, effectively replacing the gold standard of the *really* "good old days" and faithfully reflecting the creation of an American imperial economic order out of the lethal economic disorder of the 1930s. Not surprisingly this was a period of boom, in which the rate of profit—however calculated—achieved high levels that were both historic and sustained. Indeed, this quarter-century boom—the longest in the history of the business cycle—actually accelerated as it matured. During the 1960s, prosperity veritably "took off."[5]

It is hardly surprising that this period was also one in which Keynesian consumption (or "demand-side") economics became the standard wisdom among professionals in that trade. Toward the end of the boom, under Democratic administrations, much was even said about how these professionals had acquired the ability to "fine-tune" the economy. By this they meant the final flattening out of the capitalist business cycle by a judicious blend of fiscal and monetary policy. But preoccupation with mass consumption as a major goal of public policy far antedates Great Society programs or the fine-tuners of the mid-1960s. The postwar "G.I. Bill of Rights" involved direct federal payments to or on behalf of veterans' higher education, with enormous and permanent effects on upgrading the labor force from prewar levels. More than this, taxation and banking policy were shaped so as to stimulate mass purchasing of major consumer goods—houses (often with VA or FHA loan guarantees), automobiles, and the whole range of consumer durables.

The material means were thus provided for the vast explosion of suburbanization that was forever to alter the American landscape after 1950. Further incentives for "average" Americans to acquire assets like houses were provided by the deductability from income taxes of interest on mortgage (but not rent) payments. The basis for the burgeoning "debt economy" was thus laid—at the same time also laying the ghost of underconsumption that had plagued the 1930s and given rise to Keynes's revolution in economics. As for saving or investment, the tax codes in this period were of course frequently biased to encourage them. Yet it is fair to say that, apart from such encouragements, relatively little attention was paid to savings as a subject of public policy, because savings was not at that time a significant macro-economic problem.

In this supposed golden age, the imperial system that grew out of the ruins of World War II appeared to function relatively efficiently and under American control. To be sure, the immediate postwar years were punctuated by loud noises as these ruins continued to collapse, down through the fall of China to the Communists in 1949. Events such as this and the Soviets' absorption of all Eastern Europe into their political, economic, and military system mightily contributed to the forging of an intense anti-Communist consensus in the United States. (This consensus spilled over into the absurdities of McCarthyism, but once this scare subsided, the consensus remained.) Order was established in Western Europe through the Marshall Plan (1947) and the North Atlantic Treaty Organization (1949). Recalcitrants in other parts of the "free world" were eliminated or controlled by actions of the CIA (Iran, 1953; Guatemala, 1954) or direct military intervention (Lebanon, 1958; Dominican Republic, 1965).

To be sure, this control was not perfect. Its most conspicuous failure—with long-lasting and still-ramifying effects, not least on the behavior of American political elites—was the Cuban Revolution of 1959, and the CIA's spectacular failure to overthrow Fidel Castro at the Bay of Pigs (1961). But the imperial *lines* were successfully defended from attack in Korea (1950–53), and an apparently viable partition solution was imposed with the end of French colonial rule in Vietnam (1954–56). Above all, this was a period of formal decolonization in the Third World, but one in which the excolonial powers and the United States retained impressive economic and military control, including control over the decolonization process. This also meant control over the supply and price of critical raw materials (notably oil, of course). Equally significantly, it was both economically and politically possible for American defense expenditures to absorb about 10 percent of the gross national product even during periods of relative peace (e.g., 1953–65), and for the nation to launch a massive and expensive space program in direct competition with the Soviets. Public support for these burdens (and others, such as the extraction of human resources for imperial purposes, for instance the military draft) was overwhelming in general, even if there were some political conflicts at the margins.

The cultural system was molded accordingly. The generation of World War II veterans had been socialized to a certain regimentation in a common cause against enemies who were universally seen as sinister and evil. The transfer of the Manichaean enemy image from Fascists and Nazis to Communists was achieved remarkably smoothly and completely within five years after that war had ended. Without doubt, the Korean War was the final catalyst in this process. It was, broadly speaking, an era of consensus and cultural conformity. Traditional American values appeared in the ascendant, along with a wholesale celebration of American political, economic, and social institutions—often enough, as Will Herberg has shown, a celebration enveloped in explicitly religious forms.[6] There was essentially no such thing as a "counterculture" in the Eisenhower era; the civil rights movement was in its infancy, and at that time was overwhelmingly concentrated in the South; women's liberation hardly existed in consciousness, much less as an organized dissenting political movement; Marxism in any form was essentially taboo in or out of university-intellectual circles; and, needless to say, such matters as abortion, generally available pornography, or out-of-the-closet homosexuality were simply invisible as national issues.

Finally, the American political system presented few if any problematics either to political scientists or to the larger public. Until well into the 1960s, the direct role of the federal government in the domestic arena was severely circumscribed by later standards. Budget-balancing was commonly accepted by Presidents Truman and Eisenhower as a matter of course, as it has been historically (except for major crisis periods).[7] Poverty had not been discovered as a pathology on which war had to be declared. Detailed environmental, safety, and other federal regulations on business, its processes, and its products were almost nonexistent compared with what they would become under the aegis of the Occupational Safety and Health Administration and the Environmental Protection Agency in the 1970s. There was next to no federal role in health (apart from very limited, and traditional, public-health programs) and no Medicare program. Similarly, there was next to no federal role in education as such, though—especially after the Sputnik scare of 1957—increasing federal money for contract and general research began to flow into university coffers. In short, the domestic loads on the state were very small by comparison with post-1965 development. This meant, among other things, that there were that many fewer claimants, lobbyists, and Washington lawyers mobilized to get their piece of the action. At the same time, public support for and trust in political institutions and leaders were very high.

This was the age of "the civic culture," allegiant, participatory, *civiste*.[8] It was also a time in which the traditional parties had not yet begun to disintegrate.

As we have said, this whole picture has no doubt been overidealized somewhat, but it still presents a credible benchmark for evaluating the development of a major crisis after this "golden age" began to wane. Limitations of space preclude an elaborate documentation of each aspect of that crisis, but certain main themes seem clear enough. In the economy, the key changes have been associated with the rise of formidable industrial competition from other capitalist nations (Western European and especially Japan); the aging of the American industrial plant, along with declining investment in research and development, declining productivity, and declining profitability; the explosion of the "debt economy," at the governmental, corporate, and consumer levels; and, finally, the impact of dramatic upward shifts in energy costs after the consolidation of OPEC as an effective price cartel (1973–74).[9] There is no question that, whatever the specific weight of international, public-sector, and private-sector causal factors, or whatever remedies are proposed to deal with these ills, the American economy has shifted from rapid growth toward an inflationary stagnation within a time period that, historically, has been very short. There have been three stages to this painful transformation so far: an upheaval around 1969–71, associated with a major collapse in the stock market and the abrogation of Bretton Woods; the inflationary semidepression of 1974–75, triggered by tbe impact of the Arab oil embargo and price increases of 1973–74; and the 1979–80 inflationary upsurge, importantly linked with the collapse of the Shah's rule in Iran and its sequel.

One set of consequences is spelled out in Table 1: a 1968–1981 decline of *one-fifth* in the real standard of living among "typical" factory workers with three dependents. It is not to be supposed, of course, that the decline in earning shown in Table 1 has been the typical plight of the entire American labor force. On the contrary: in certain regions of the country, among certain occupations—especially among many portions of the middle and upper classes—business has almost never been better, and the money is rolling in. A crucial part of the economic picture in the most recent period is the rapid escalation of *uneven development* along such lines, rather than a universal bankruptcy of the Great Depression variety. But as Lester Thurow has convincingly pointed out, this uneven development now takes place in increasingly zero-sum conditions, as assured, sustained economic growth has disappeared.[10] The losses of one group, increasingly, become the gains of another. The nature and political consequences of this uneven development significantly change. Expectations under such conditions acquire an altogether different form from those to which we had become accustomed. As we shall see, ideology, far from ending, makes a decisive comeback—not merely into the active political arena, but into the very seats of power.

Table 1. Declining standard of living for some: Spendable weekly earnings for a worker with three dependents, 1965–81

Year	Spendable earnings, 1967$	Index: 1965 = 100	Index: 1973 = 100
1965	$102.41	100.0	107.0
1966	102.26	99.9	106.8
1967	101.26	98.9	105.8
1968	102.43	100.0+	107.0
1969	91.07	88.9	95.1
1970	89.95	87.8	94.0
1971	92.43	90.3	96.6
1972	96.40	94.1	100.7
1973	95.73	93.5	100.0
1974	90.97	88.8	95.0
1975	90.53	88.4	94.6
1976	91.79	89.6	95.9
1977	93.48	91.3	97.6
1978	92.50	90.3	96.6
1979	89.34	87.2	93.3
1980	83.51	81.5	87.2
1981	81.85	79.9	85.5

*January–May average.
Source: U.S. Department of Commerce, *Survey of Current Business* (section titled "Labor Force, Employment and Earnings"), various years.

This remarkable transformation required not merely a rapid movement toward zero-sum conditions in the most recent past, but development of the domestic political-capitalist state over the preceding generation. This, in turn, cumulatively emerged as a preempted, if creative, response to the shattering economic collapse of 1929 and its consequences, both domestic and international. In view of the multiple cataclysms of the 1930s and 1940s, it was hardly surprising that postwar political elites throughout the Western world were overwhelmingly preoccupied with ensuring that such ghastly conditions never returned. Nothing less than the survival of capitalism and democracy alike required this. Accordingly, it was not surprising that Keynesian doctrines came into their own as official policy. Maintenance of social harmony was paramount, and the way to achieve this was by stimulating mass consumption.

For a long time a major difference between the United States and Western Europe lay in the much greater stress Americans placed, ideologically and in policy terms, on stimulating consumption in the private sector that on social-consumption expenditures based upon the activities of the state. This reflected both long-term ideological commitments in the United States and the vastly superior macroeconomic position that it enjoyed in the years immediately after World War II. But the objectives were essentially the same: to preserve capitalist democracy through mass consumption and welfare programs. As this program could only be sustained through economic growth rising faster than all consumption claims on it, "growthmanship" was an essential ingredient in all "forward-looking" government policies during this period. Without such growth, contradictions between the accumulation-savings needs of capital and the political and economic necessity of effective mass consumption could not be avoided. With such growth these contradictions would seem to disappear, and with their disappearance the great goal of social harmony and the "end of ideology" could be achieved.[11]

At some point during the mid-1960s a momentous transition occurred in the United States toward a much more active and intrusive federal role in the maintenance of this social harmony. This transition was to some extent made possible by the rout of Barry Goldwater and the Republicans in the 1964 elections. But underlying it was a remarkable transformation of official perspectives concerning the federal budget, a broad shift toward a concept of "full-employment balance" that, along with other complementary changes in macro-economic ideas, led to a condition of more or less permanent federal deficits. Still further, racial and other problems had shifted from the rural South to the urban North in the preceding two decades, and poverty had been established as an officially recognized fact. What seemed more logical— as well as humane—than a war on poverty and the development of public-sector programs of regulation and expenditure in areas formerly outside the purview of the federal government? For underlying all else was the enormous boom of the 1960s coupled with the rapid accumulation of very real, not to say grave, domestic social problems that neither the development of the private-sector economy nor the warfare state of the "golden age" had addressed.

With the conversion even of Richard Nixon to Keynesianism came the rapid discovery of the uses to which the "political business cycle" could be put by politicians concerned to maximize their election chances.[12] Unfortunately, this rapid development of political capitalism turned out not to promote and maintain social harmony as its architects intended, but cumulatively to do exactly the opposite. This was evident even before the economic conjuncture began decisively to change after 1969—particularly in the massive escalation of racial tensions around which George Wallace's national career was largely organized. It became far more so in the 1970s, as the network of regulatory and expenditure programs became fully established. We shall explore some plausible reasons for this remarkable development in the following section of this essay. For now it is enough to suggest that the terms of debate began decisively to change in the 1970s as growth was replaced by zero-sum inflationary stagnation.

By the mid-1970s the business press was viewing developments in the political economy with open alarm. For example, special issues of *Business Week* in the latter half of the decade included articles with such titles as "The Debt Economy," "The Shrinking Standard of Living," and, in 1980, "Revitalizing the U.S. Economy."[13] At the same time, intellectuals (political scientists and others) were discovering the existence of a "governability crisis" based upon the very attributes of pluralism and brokerage that their predecessors of the "golden age" had celebrated. Nothing less than "the crisis of democracy" was involved: in a nutshell, too many demands from the wrong quarters, too little respect or support for authority, increasingly insupportable "overload" on the political system.[14] Very unsurprisingly, this chorus of alarmed voices swelled to a roar as a climax of economic stress was approached during Jimmy Carter's administration. By the late 1970s it had become obvious that consumption would somehow have to be curtailed if capitalism's crisis was to be eased, and that accumulation would have to receive much more detailed support from government policy than it had hitherto. In a word, the economic contradictions of capitalism, which the "golden age" had supposedly banished for good along with "ideology," dramatically reappeared. So did the enduring tension between the underlying needs of the economic system and the norms and practices of political democracy, especially as the latter had been concretely worked out in public policy after 1965.

By the same token, liberal economists and other intellectuals fell increasingly into disarray, uncertainty, and even confusion. With inflation veering more and more often into double digits and the competitive position of the American economy declining in the world market, it was increasingly obvious that the received economic orthodoxies of the preceding period were inadequate at best, irrelevant at worst. Moreover, the whole of the political-capitalist formula had been predicated on both the possibility and reality of sustained economic growth, and hence upon the preservation of a class peace (or social harmony) dependent on that growth. Where does one turn when this disappears, and the hard choices that Thurow and others have identified have to be made?

Broadly speaking, there are two positions that accept, if not insist upon, the necessity of making such choices: socialism and neo–laissez faire capitalist ideology. Whatever may happen in the future, socialism has been a conspicuous nonstarter in American politics. The immense residual strength of neo–laissez faire right-wing ideology in America, on the other hand, was first fully revealed by the nomination of Barry Goldwater as the Republican candidate in 1964. Unfortunately for him, he arrived on the scene fifteen years too early. But the underlying point remains.

From Tocqueville's time onward, a tremendously broad commitment to capitalist values has been understood to be a uniquely developed property of American political culture.[15] To all intents and purposes political debate and action in the American political mainstream have been organized around a common belief among political-capitalist centrists and the laissez faire right alike that the fundamentals of capitalism as such are sound and beyond

Table 2. Tax revenues as a percentage of gross national product: Some comparative data

1966		1976		Change, 1966–76	
1. Netherlands	37.0	1. Sweden	50.9	1. Luxembourg	+18.4
2. Sweden	36.4	2. Luxembourg	50.5	2. Sweden	+13.9
3. Austria	35.5	3. Norway	46.2	3. Denmark	+12.1
4. France	34.8	4. Netherlands	46.2	4. Norway	+11.6
5. Norway	34.6	5. Denmark	44.7	5. Finalnd	+10.1
6. Belgium	32.9	6. Finland	42.2	6. Switzerland	+10.1
7. Denmark	32.6	7. Belgium	41.9	7. Netherlands	+9.2
8. United Kingdom	32.2	8. France	39.5	8. Belgium	+9.0
9. West Germany	32.2	9. Austria	38.9	9. Ireland	+8.9
10. Luxembourg	32.0	10. Ireland	36.8	10. Turkey	+7.9
11. Finland	31.8	11. West Germany	36.7	11. Portugal	+7.4
12. Italy	29.0	12. United Kingdom	36.7	12. Italy	+6.8
13. Ireland	28.0	13. Italy	35.8	13. Australia	+5.9
14. Canada	27.6	14. Canada	32.9	14. Greece	+5.7
		15. New Zealand	31.8	15. New Zealand	+5.4
15. UNITED STATES	26.9				
		16. Switzerland	31.6	16. Canada	+5.3
16. New Zealand	26.4	17. Australia	30.0	17. Spain	+5.3
17. Australia	24.1			18. France	+4.7
18. Greece	22.2	18. UNITED STATES	29.3	19. West Germany	+4.5
19. Switzerland	21.5			20. United Kingdom	+4.5
20. Portugal	19.1	19. Greece	27.9		
		20. Portugal	26.5	21. Austria	+3.4
21. Japan	17.6			22. Japan	+3.3
22. Turkey	17.0	21. Turkey	24.9		
23. Spain	15.0	22. Japan	20.9	23. UNITED STATES	+2.4
		23. Spain	20.3		
Means	28.1		35.8		+7.7

Source: Tax Foundation, Inc., *Facts and Figures on Government Finance,* 20th biennial ed. (Washington: Tax Foundation, 1979), p. 35.

debate. The recurring question, even in crisis periods, is not over whether it should be replaced by something else but over ways and means to protect it and perfect its operations. As Douglas Hibbs and others have pointed out, this has had a concomitant in the postwar world that, as a whole, the American political system is more tolerant of unemployment and less tolerant of inflation and of "big government" in nondefense areas than any other system with which it may be readily compared.[16] This may account in substantial part for a truly remarkable comparative datum revealed in Table 2. When total tax revenues are evaluated as percentages of gross national product (in the United States as elsewhere, this includes taxes raised by all levels of government), the United States ranks very low indeed among twenty-three selected countries for the most recent year, 1976—in fact, lower than any other industrially developed country except Japan.

Moreover, the comparative position of the United States in this regard actually *declined* between 1966 and 1976, in the very midst of the political-capitalist era. In 1966 the United States ranked fifteenth among the twenty-three countries. A decade later it ranked eighteenth, and in terms of the change in the tax "take" of gross national product between 1966 and 1976 it ranked last! The absolute growth of 2.4 percent, from 26.9 percent to 29.3 percent of GNP, is truly very little across the space of a decade. The last percentage, 29.3, is evidently crushingly large to many Americans; it surely is compared with 1929, when about 10 percent of GNP was extracted by taxes at all levels of government. On the other hand, even in 1976, at 6 percent of GNP, the United States extracted more for defense from its population than any other of these countries except Greece (its nearest colleagues being Greece at 6.3 percent, Turkey, Britain, and Portugal at 5 percent, and France at rather less than 4 percent). Even were we to return to a defense expenditure of 10 percent of GNP, and assuming that the 4 percent difference was fully absorbed by increased taxes, this would bring the United States public sector thus defined to exactly one-third of GNP, 33.3 percent. Even this, very slightly ahead of Canada and substantially behind Italy, still leaves the United States in fourteenth place, i.e., in the bottom half of the whole distribution.

Moreover, there is something rather considerably less than meets the eye so far as federal budget deficits are concerned. It should be recalled—but somehow very rarely is—that state and local government budgets run very substantial surpluses in the aggregate. Naturally these surpluses do not wholly cancel out the federal deficit. But when the receipts and expenditures of all levels of government are evaluated together, we find that a $6.3 billion surplus was achieved in 1973, a $11.9 billion surplus in 1979, and that the budget was almost precisely in balance in 1978, with a deficit of $0.2 billion. The great aggregate deficits on this consolidated basis since 1970 were run in 1975 ($63.8 billion), 1976 ($36.5 billion), and 1980 ($34.8 billion).[17] The predominant explanation for this is quite simple: there were years when the private economy was in, or just emerging from, serious difficulty and the Keynesian machinery for coping with recession was doing its basic job of damping down the capitalist business cycle.

The basic truth of the matter is that the federal government's share of the gross national product has risen very little over the past quarter-century, which is why the United States ranks twenty-third of twenty-three on this indicator for the 1966–76 period. Similarly, the parade of horribles concerning federal budgetary deficits carefully conceals the fact that total government budgetary deficits present a much less dramatically alarming picture. It is worth observing in this regard that the $34.8 billion 1980 deficit constitutes little more than 1 percent of the gross national product. In fact, the overall deficits in the 1977–80 period constituted less than .5 percent of the GNP accumulated during this period. It is little wonder that most economists believe that, by itself, public-sector deficit cannot remotely explain the extent of inflation, declining productivity and profitability, and stagnation that has dominated the economic landscape of the United States and most other advanced industrial-capitalist countries since the early 1970s.

To an important extent, then, one must look less to data than to mythology to explain so extraordinary a general hostility to the interventionist state in America. We have attempted to suggest several reasons for this, for example, the increasingly serious problems—not mythical at all—of capital accumulation in an economy that decays over time compared with its competitors. It is possible that, *given the current organization of the American economy, society, and electoral-political market,* these burdens of state expenditure are somehow objectively more onerous than they would be in other and differently organized capitalist political economies. But much of the uproar must, I think, be traced back to the continuing vitality of the American liberal-capitalist political tradition, and the enduring hostility to the state, to organized labor, to the poor that arises from this kind of uncontested cultural hegemony. Of course, to suggest that American propertied-class mythology has had an important role in recent events is by no means to underrate that mythology's empirical importance to our politics. What people believe about their world (however poorly grounded empirically) is almost certainly more important to what happens in politics than the actual structure of objective reality. Accordingly, the facts just cited are not officially recognized facts, nor are they likely to be so recognized in the near future.

We will say very little about the unfolding crisis of empire, punctuated as it has been by the spread of Marxist regimes into Southeast Asia, Africa, and Latin America, as well as by the Soviet empire's thrust into Afghanistan and the Iranian and OPEC crises. The point here is that the United States has clearly and cumulatively lost the power to control events beyond its borders, to a degree that has made substantial portions of the American elite (and public) very uneasy. The overall crisis, rooted in a manifest loss of control of events by American political and economic elites in each of our four realms, has become overwhelmingly visible in imperial-defense matters as well during the 1970s. Finally, the actions of the Soviet Union itself have provided an important stimulus to defense-oriented revitalization in the United States. One aspect of mainstream American political culture has been a theme of near paranoia about Marxism in general and the Soviet Union in particular, a theme that goes right back to the Bolshevik Revolution of 1917 and the "great red scare" of 1918–20. On the other hand, it must be said in all candor that sometimes paranoids have real enemies. The need for greater competitive military effort vis-à-vis the USSR cannot be regarded under present circumstances as exclusively a figment of fevered right-wing imagination. Finally, it may be noted that developments in this dimension, perhaps more than any other, prompt the evaluation that—whatever happens politically to

Ronald Reagan and his administration—an important and probably durable sea-change has occurred in the balance of *domestic* political forces in the United States.

The most conspicuous short-term feature of America's culturally related politics from 1978 onwards has been the rapid emergence of religious-based interest groups and their linkup with the economic right under Ronald Reagan. Until quite recently most analysts of American political culture have been insensitive to the pervasiveness of religious symbolism and values within it. Actually this culture has been less "secularized" than development-modernization models in Europe and elsewhere would lead us to expect. I have recently analyzed this aspect of American politics in another place.[18] It is enough to say here that pietistic religious values, ultimately derived from the dissenting Protestantism that saturated the founding culture, have retained exceptional scope and importance in shaping popular consciousness. The impact of the Moral Majority and other elements of the social-issue New Right have made people aware of the importance of religion in American culture. This upsurge may be regarded as the latest in a series of wavelike conservative-revivalist upheavals since colonial times,[19] episodes that have sensitively reflected the cumulative pressures of untrammeled cultural change on Americans. And thus today the Moral Majority is in rebellion against the manifold cultural revolution that has swept over America during the past twenty years.

More generally, as Daniel Bell has brilliantly demonstrated, the contemporary cultural crisis may also be read as arising from capitalism itself, especially in its mass-consumption, political-capitalist phase.[20] Basic to Bell's argument is the cultural counterpart of the economic tension between accumulation and consumption. The dominance of the former implies a cultural configuration that may broadly be said to be Victorian in flavor. The primary cultural themes are bound up with themes of self-discipline as an adjunct to accumulation—not merely in behavior on and related to the job, but the self-discipline required for the preservation of the nuclear family and the voluntary renunciation of immediate gratification that is essential to the individual's propensity to save rather than to spend. Needless to say, this self-repression in its classic form and time is reinforced by externally defined carrots (the Horatio Alger "rags to riches" myth) and sticks (permanent social insecurity as a goad to appropriate action by the individual). Such self-repression serves as an appropriate cultural pattern for accumulation. Yet to the extent it dominates, it also undercuts another need of advanced capitalism: that for broadly based mass consumption, created in large part by the stimulation of advertising.

In an age of high mass consumption, the cultural balance is altered in a fundamental way. Gratification of wants, spending rather than saving—indeed, cheerfully and unashamedly going into debt to finance consumption—are of the essence. The economic stability and profitability of capitalism in this stage come to be dependent upon this mass consumption and the culture that goes with it. Unfortunately, the hedonism that also and necessarily results has unacceptable consequences in the longer run. Divorce and breakup of the nuclear family, the spread of countercultures, the growth of demands for more income through government action, problems of labor "indiscipline," soaring crime rates, and much else follow in its train. If sex and violence sell soap in television, it is sex and violence we will see on television: the point of the enterprise is to make a buck. And so it goes.

More generally, from a political point of view the development of hedonic rather than Victorian-accumulation values in the culture implies the development of hedonic expectations from government too. Under conditions of political capitalism it has been very easy for politicians to meet such demands and buy off possible trouble by new programs or expansions of older ones. It goes without saying that if the dominant needs of American capitalism now lie in the accumulation (or "reindustrialization") sphere, the culture must be very substantially modified accordingly. Cultural neo-Victorianism and economic neo–laissez faire would seem to have important and perhaps mutually indispensable complementary functions. The proliferation of atomistic hedonism comes to be seen as subverting the foundations of the social order and of Christianity itself. It is therefore not surprising that religious sects of all sorts have proliferated in recent years, that evangelical Protestant denominations in particular have had major growth, and that on the political front the Moral Majority has emerged as a reactionary-revitalization cultural movement, increasingly linked with other elements of the Reagan coalition.

As we have suggested, there is evidently a certain logic in all this. But whether, even in America, quite so much toothpaste can be thus shoved back into the tube remains to be seen. Whatever the outcome of that particular struggle, the cultural contradictions of capitalism are likely to endure in rather acute form. As old-fashioned European conservatives and Marxists have agreed, they arise from the logic of the system itself, a logic of atomistic utilitarian calculation that tends to dissolve all collective bonds of responsibility between one individual and another.[21] No less than the enduring tension between what might be described as "accumulation cultures" and consumption cultures," this inherent tendency may be modified, but it is not likely to be repealed by the results of the 1980, or any other, election. But the mazeway deformations now involved in these tensions—the mounting sense of unacceptable cultural decay—can produce abundant kindling for right-wing political explosions in America.

III. The 1980 Election

Standing as it did under the sign of acute crisis in economy, empire, and culture, the 1980 election was destined to be a most unusual one. Jimmy Carter had originally won the Democratic nomination in 1976 against a splintered opposition, in the context of a party whose bitter internal struggles of the preceding eight years had created a genuine vacuum. With the decay of organization, the ascendancy of the media in nomination processes, and the open-primary rule structure adopted by the party in 1971, Carter won nomination in very much the same way—though on an infinitely larger stage—he had won the Democratic nomination for governor of Georgia in 1970. He confronted Gerald Ford in the general election, a man who three years earlier had been but a congressman from Grand Rapids; who owed his elevation to the disgrace of the former leaders of his own party, first Spiro Agnew and then Richard Nixon; and who, while widely considered to be a nice man, was almost equally widely believed to lack important qualifications for the office. In retrospect, Ford was a *locum tenens*. His bid for election was particularly hampered by the fallout from Watergate (more specifically, his pardon of Nixon, which continued to plague him throughout his presidency) and the aftermath of the major 1974–75 recession. With all these disabilities, Ford came within a hair's breadth of victory, in what was in actuality the third closest presidential election since Jackson's time. This near miss, by the way, reminds us again of the peculiar problem of the Democrats, the so-called majority party. Since Franklin Roosevelt's death in 1945, its candidates have won only four presidential elections to five for the Republicans, and only in 1964 was the Democratic victory of convincing proportions.

Carter thus was in a real sense the artifact of a political vacuum arising from the general American crisis. His tenure in office greatly deepened this vacuum and directly led to the victory of Ronald Reagan at what was, for Reagan personally, the last possible historical moment. Jimmy Carter's administration was plagued by misadventure and bad luck. But it was also, and decisively, plagued by his lack of any clear objectives and the resultant incoherence of his administration. He did not really know what he wanted to do with the power he had so narrowly won (a problem which, for better or for worse, is certainly not shared by his successor).

To some considerable extent, this lack of a coherent point of view—which was coupled with what events demonstrated to be naiveté in his handling of American foreign policy—was his own personal contribution to defeat in 1980. A moderate Southern Democrat who was a curious blend of technocrat and evangelical preacher, Carter appointed many quite leftish Democrats to important positions. But at the same time, when the economic crisis accelerated in 1979–80 it seems never to have occurred to him to question received, conventional, and profoundly conservative-capitalist doctrine about the economy as a whole. It was he, and not Ronald Reagan, who appointed Paul Volcker to be chairman of the Federal Reserve Board. His reward was (in conjunction with the oil crisis of 1979 and rapidly growing inflationary pressure) a combined credit crunch and recession in the presidential election year itself. This entailed the highest "discomfort index" to occur in any presidential election year since 1932 (see Figure 1), when

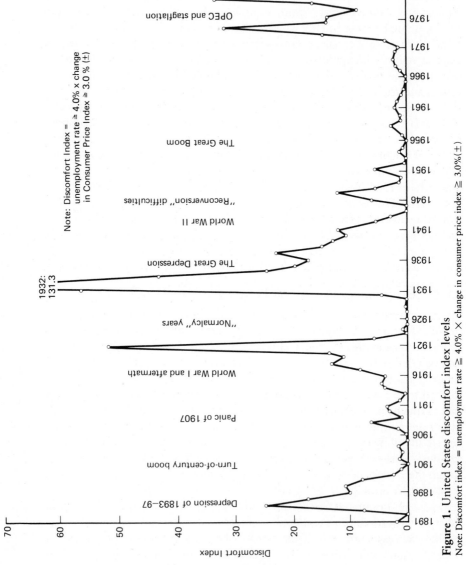

Figure 1. United States discomfort index levels

Note: Discomfort index = unemployment rate $\geq$ 4.0% $\times$ change in consumer price index $\geq$ 3.0% ($\pm$)

284

another incumbent seeking reelection was repudiated by the voters. In no small measure Carter was defeated in 1980 because he handed the economic issue on a silver platter to the Republicans, the first such occasion since FDR created the modern Democratic party a half-century ago.

It was the fact that Carter was for his time the most conservative Democratic president on "core" economic issues since Grover Cleveland left office in 1897 that impelled Edward M. Kennedy to enter the nomination race against him in late 1979. Kennedy's bid was as ill-starred as Carter's presidency, for his announcement occurred simultaneously with the eruption of the Iranian hostage crisis. In the resulting "rally 'round the flag" atmosphere Kennedy could only be damaged by commenting truthfully on the nature of the Shah's regime in Iran. He was hurt even more, at crucial early stages, by Carter's almost shameless exploitation of the hostage crisis and his pursuit of the "Rose Garden strategy" of campaigning from the White House. Of course, at the outset virtually no one had expected this crisis to continue to the last day of Carter's presidency. What made a crucial difference in his bid for renomination turned out, equally crucially, to be an albatross around his neck in the general election. Even at that, the remarkable thing about Kennedy's campaign is that it was resurrected after his disastrous rout in Illinois in March. Thereafter he did remarkably well against the president in the industrial cities and states that were the bedrock of the modern Democratic coalition. The issues were quite fundamental. Kennedy's continuing race made it clear that Carter had lost vital support within the party's very core in the industrial states. If Iran and his own inadequacies had torpedoed Kennedy's candidacy, the credit crunch and recession of 1980 resurrected it. When Kennedy charged that Carter was "Ronald Reagan's clone" he exaggerated greatly, in the fashion of politicians on the stump. But he had something; indeed he had a large part of the 1980 puzzle. The later 1980 primary returns ratified this.

Just the same, the general election was only in retrospect a walkaway for Ronald Reagan. Reagan had not been the first choice of the corporate boardrooms: that honor had gone to John Connally.[22] Of course, they rallied nicely in the end. And from the beginning there was a strong core of support among social-issue conservatives and among imperial-policy "hawks," along with very large parts of the business and lower middle classes in the Sunbelt states. Even so, and even with Carter's own negative image problems, the emergence of John Anderson as an independent candidate for the many dissatisfieds in the electorate and the multilevel systemic crisis, by then in full swing, made the election seem—and it probably was—extremely close until the very end of the campaign.

The evidence is now very strong that Jimmy Carter's support simply dribbled away across most demographic and issue groups in the American electorate. At least until the debate, however, dissatisfaction with Carter was counterbalanced by unease over Reagan. Of course the dissatisfaction predominanted in the end. The postelection Gallup survey shows that, among all Reagan voters, 38 percent gave as their main reason an explicit dissatisfaction with Carter (22 percent said they were "dissatisfied with Carter," 6 percent said that "Carter has not accomplished anything"). Only 31 percent of these voters gave as their reasons a liking for Reagan's economic policies (or policies in general). The remainder were divided, with 21 percent for that time-honored response, "It's time for a change," while another 12 percent said that they believed that Reagan would make a better leader.[23]

Connected with all this were debits for Reagan too. A very important negative factor for him was the war-peace issue. Prior to the debate, 43 percent of respondents in the *New York Times*–CBS poll expressed fear that Reagan would get us into war, compared with only 19 percent making that judgment about Carter. After the debate the figures had shifted to 35 percent and 12 percent respectively, one indication of the debate's considerable—if marginal—influence on the final result. Similarly, 66 percent of voters believed that Reagan said things too carelessly, compared with 46 percent making this judgment of Carter. Fully 70 percent believed that Carter understood the complicated problems a president has to deal with, compared with only 51 percent believing this of Reagan. And 60 percent chose Carter as most likely to be steady and patient in a crisis, while only 28 percent chose Reagan.[24]

Since the election ended in a decisive victory for Reagan, we must assume that his strengths vis-à-vis Carter considerably outweighed his weaknesses. These included a preponderance of belief that the economy would get better (43 percent so believing with Reagan elected, only 21 percent with Carter elected); that Reagan would be much more likely to maintain at least military parity with the Soviet Union (66 percent for Reagan on this issue, 27 percent for Carter); that Reagan would be much more likely (77 percent "yes") than Carter (55 percent "yes") to see to it that the United States is respected by other nations. Similarly, on questions of leadership quality Reagan came out decisively ahead of Carter in offering a vision of where he wished to lead the country (67 percent to 48 percent) and as a strong leader generally (62 percent to 32 percent). It is worth noting that in a 1980 Gallup comparison of Gerald Ford and Jimmy Carter's performance as president, fully 46 percent of respondents judged Carter as "below average" or "poor," compared with 21 percent so evaluating Gerald Ford.

In truth, at the time of the election there was far and away the greatest negative affect against *both* of the major candidates ever recorded in the entire 1952–80 period covered by the Michigan election studies.[25] Clearly associated with this was a continued decline in *turnout* outside of the Southern states, now beginning to approximate the all-time lows reached in the 1920s and, in some major states like New York, even to fall below that former rock-bottom level. Obviously the very existence of John Anderson's independent candidacy was also closely bound up with this pervasive negativism. Not surprisingly, then, there was an exceptional "softness" of support for the presidential candidates of 1980, even by contemporary standards of party decomposition in the electorate. Gallup, for example, reports that fully 35 percent of voters did not make up their minds until after watching the debate at the end of the campaign. Moreover, vote switchers—those saying that they had at some point intended to vote for another candidate—constituted fully 27 percent of the total sample, compared with 21 percent in 1976, 12 percent in 1972, and 17 percent in 1968 (the last also an election with an important third "protest" candidate).[26]

Most contemporary survey-research analysis of the American electorate has pretty severely discouraged the notion that very large numbers of voters make their decisions on ideological grounds. On the other hand, there seems to be conclusive evidence that voters pay much more attention to issues and perceived candidate stands on issues in some elections than in others. So far as can now be determined, the 1980 record would seem to confirm both generalizations. For example, the *New York Times*–CBS survey found that only 11 percent of Reagan's voters gave as their primary reason for support that "he's a real conservative," while fully 38 percent of them gave as their reason that "it's time for a change."[27] Similarly, as we have seen earlier, the Gallup survey revealed that only 31 percent of Reagan's voters gave as their primary reasons for supporting him that they liked his economic policies (17 percent), or that they just liked his policies in general (14 percent). There is some limited evidence to suggest a gradual shift toward conservatism in general. The National Opinion Research Center's data on this point are summarized in Table 3.[28]

The ABC exit poll of November 4, 1980, found a rather greater conservative tilt: 24 percent of their respondents identified themselves as liberals, 37 percent as moderates, and 39 percent as conservatives. Of still greater interest, perhaps, are the preliminary signs that *since the election* the Republicans may have considerably narrowed the party identification gap—which, were this process first to continue and then stabilize, would provide strong support for the view that the nation is undergoing a critical realignment.

It may well be wondered what this "conservatism in general" really amounts to in concrete policy terms. For example, when NORC asked people in 1980 whether they thought that present levels of public services should be provided even at the cost of maintaining present spending levels, or that

Table 3. Conservative gains?

Year	Percentage of respondents:			L − C
	Liberal	Moderate	Conservative	
1974	31	40	30	+1
1976	29	40	31	−2
1978	28	38	34	−6
1980	26	41	34	−8

services (even in health and education) should be cut in order to reduce government spending, 47 percent agreed with maintaining spending, 21 percent took a midroad position, and only 33 percent selected the cut-services-and-spending alternative.[29] Even among self-identified conservatives, moreover, 36 percent were for maintenance and only 44 percent for cutting. As is well known, moreover, there are a large number of issue areas where consensus more or less exists across such self-defined "ideological" groups. Thus, for example, there is consensus that too little is being spent on the problems of crime and drugs, and too much on foreign aid and welfare. Not surprisingly, when Communism as a form of government is assessed, there is a difference only in degree between liberals and conservatives. For both groups, only 1 percent would agree that it is a good form of government, while 44 percent of liberals and 63 percent of conservatives think it the worst possible form of government (the national total for the latter is 59 percent). In view of all this, it is not surprising that the AIPO–Gallup report on the election was unable to detect any significant shift to the right on such specific issues as abortion (modal position moderate), the death penalty (heavily for, but declining support), ERA (substantially for), gun control (heavily for, but also declining support), or a constitutional amendment to require a balanced budget (overwhelmingly and stably for, from the mid-1970s onwards).

What does all this add up to? Essentially we are dealing here with an electorate that is very strongly (and consensually) anticollectivist, hence "conservative," especially by comparative standards. But this is nothing new, and when evidence for a massive ideologically oriented breakthrough toward the right is sought it cannot conclusively be found. On the other hand, this is an electorate that is seriously bothered by national issues. In 1980 these issues, pretty clearly, dealt with the quality of presidential leadership; the economic situation, particularly as it impacted on the respondent's family finances over the year, creating, as we have said, the highest "discomfort index" of any year since 1932–33; the international situation, so far as both a perceived decline in our military posture vis-à-vis the Russians and issues of our "national honor" were concerned; and surely in a number of quarters a perceived decline in religion and morality as well. All of these cut directly athwart Jimmy Carter's ill-fated reelection bid, and—doubtless with the help of the debate and the last-minute upswing in the visibility of the Iranian hostage crisis—eventually overwhelmed the many reservations that "swing" voters appear to have had about Ronald Reagan.

Carter's losses from 1976 to 1980 were particularly concentrated among what could be generally described as white "middle Americans." The profiles of this loss offered by the Gallup poll results and those of the *New York Times*–CBS survey differ slightly, but there is pretty substantial agreement between them.[30] Carter's losses were particularly concentrated among men, Catholics and Jews, blue-collar workers, members of labor-union families, high school graduates and/or those with some (but not complete) college education, Democrats, Westerners, Southerners, voters of age 50 and older, and those whose family finances were worse off than a year ago (the last particularly concentrated among the lower-middle and working classes, and much larger a proportion of the electorate—34 percent—than in 1976). Conversely, Carter made gains or suffered only minimal losses among women, blacks, younger voters (18–29), Republicans (especially liberal Republicans, the few that are still left), college graduates, white-collar workers (a very substantial minority of whom work in the public sector), Easterners, Midwesterners, and those whose family finances were better than a year ago (only 16 percent of the sample).

Perhaps one of the most telling bits of information from the *New York Times*–CBS survey has to do with the attitudinal question concerning whether the respondent was more worried about unemployment or inflation. For one thing—despite the near-total preoccupation among politicians, intellectuals, and the media with the inflation problem—only 44 percent of this sample chose it as the most serious problem, while fully 39 percent of the sample were chiefly worried about unemployment. For another, Carter's 1976–80 losses among people concerned with unemployment were, on a two-party basis, a full 19 percentage points, compared with a loss of only 2 percent among inflation-preoccupied respondents. Of course in 1976 he received only 35 percent of the candidate vote among the latter, compared with 75 percent among unemployment-preoccupied respondents. Still, it is worth noting that while Carter's share of the former fell only to 33 percent, it fell to 56 percent of the latter.

Now this is a question with a strong and obvious class resonance: unemployment is concentrated among manual and lower-status white-collar workers. For Carter to suffer so large a decline in this category reveals quite a bit, in conjunction with our other information about this election, as to what went wrong for him. The crucial swing groups were concentrated among those whites who formed the core of the non-Southern Democratic coalition. His loss of support among such groups, foreshadowed by Kennedy's showing in the primaries, was especially decisive. What happened in 1980 was not, at the mass level, a sudden burst of ideological right-wing clarity, but a vote of no confidence in an incumbent administration that had stumbled into an inchoate economic conservatism, and in a party that had fundamentally lost its way.

This becomes particularly evident as one examines recent trends in voting turnout. The Democratic party, the carrier and initiator of post–Great Society political capitalism, might have been thought to have a permanent and very strong interest in doing whatever was necessary, organizationally and "educationally," to stimulate voter turnout to the maximum possible degree. By now it is surely not lost on anyone that a crucial factor in the New Deal revolution, which displaced the Republicans from their generation-long voting majority, was a huge influx of former nonvoters.[31] In the 1920s as in the

1970s and in 1980, the overwhelming majority of these nonvoters were concentrated in the lower classes. They came pouring into the active electorate between 1928 and 1940—especially in 1936 and 1940—because they were pushed by acute personal economic distress and pulled by a combination of FDR's personal charisma and the relief policies over which he presided. Not surprisingly, to judge from analyses extending from Paul Lazarsfeld to Robert Alford, 1940 was the most class-polarized election in American history.[32] It was also, especially in the great industrial-urban centers, the election with the highest voting participation since well before World War I.

One of the more notable discoveries arising from recent historically oriented analyses of turnout change is the extent to which Republican ascendancy in the "era of normalcy" was in fact based upon a de facto oligarchy under the nominal forms of universal suffrage. It is clear in any case that this ascendancy rested upon a huge vacuum at the bottom of the social structure.[33] Astonishingly enough, by 1980 a very similar pattern had reemerged (outside of the South, of course) after a half-century in which the Democratic party, the supposed "party of the people," had enjoyed majority status. In 1980, for example, Ronald Reagan won the presidency with 28.0 percent of the established potential electorate voting for him, a figure comparable to the 28.3 percent that Adlai Stevenson garnered in losing to Dwight Eisenhower in 1952, or that Wendell Willkie secured while losing to FDR in 1940. Of course, Jimmy Carter was only able to win 22.6 percent of the potential electorate in 1980—in itself the fourth worst showing for *any* Democratic candidate since Andrew Jackson's first election in 1828, and of course the worst for an incumbent president running for reelection since Herbert Hoover's identical percentage in 1932.

In a state like New York, the 1980 results are yet more remarkable by historic standards. Carter won only 22.1 percent of the state's potential electorate, less even than George McGovern in 1972. Across thirty-nine presidential elections, only Cox in 1920 and Davis in 1924 did worse. But Reagan was not overwhelmingly popular in this state either. His 23.4 percent was quite enough to carry New York under the circumstances. But it was hardly larger than the 21.2 percent Barry Goldwater got in his landslide defeat at Lyndon Jonson's hands in 1964, and is in fact the *third lowest share* of the state's electorate won by any Republican since the creation of that party in 1856. (Willkie, running against FDR in 1940 and losing in New York as he did nationally, won 36.4 percent of the potential electorate, i.e., 13 percent more than Reagan did in winning the state in 1980.) Overall, by 1980 the "party of nonvoters" in the Empire State had risen to exactly half of its potential electorate, the highest in history. It was more than 6 points higher than in the nonparticipant 1920s when, we are told, immigrant women were a chief factor in low turnouts. In 1940, on the other hand, abstainers constituted less than one-quarter (24.3 percent) of New York's potential electorate.

There is superabundant evidence that when people ask "Where have all the voters gone?" they should really be asking where all the working-class Democrats have gone.[34] And this, of course, requires that we ask ourselves what it is about this party that has evidently contribted to the disappearance of so much of its "natural" voting clientele into the "party of nonvoters" over the past two decades. For underlying and preceding Jimmy Carter's dramatic failure in 1980 has been the fact of turnout decline since the early 1960s—and the concentration of this decline among core working-class groups that already voted least, and whose mobilization in the 1930s was crucial to bringing a national Democratic majority into being.[35]

We may begin our more extened discussion of the "party of nonvoters" with the obvious point that "political alienation" is linked to nonparticipation in some way: the more of the first, the more of the second. To be sure, "alienation" covers a great deal of analytic territory, including among other things distrust of politicians, cynicism about major sociopolitical institutions and elites, lack of internal efficacy (for example, feeling personally "out of it," or dominated in one's life by Lady Luck rather than by the effects of one's purposive activity), and lack of external efficacy (for example, believing that one has little or no influence on what politicians do). Most mainstream survey-research analysis has come to concentrate upon the relationship between a respondent's level of external political efficacy and nonparticipation, since this relationship appears much stronger than those involving cynicism or distrust. In other words, all other things being equal, a person with a high sense of external efficacy but low political trust is more likely to vote (very probably as an "alienated voter") than one who has higher political trust but a low sense of external efficacy. On the other hand, Arthur Hadley's 1978 study, *The Empty Polling Booth,* has concluded on the base of the survey he employed that the sharpest single point of cleavage between two sets of otherwise similar respondents is the degree to which they believe that fate or chance determines what happens to them.[36] "Fatalists," *ceteris paribus,* have a much greater likelihood of abstaining than purposive actors. But of course such a variable must be regarded in terms of this analytic scheme as one of *internal* rather than *external* efficacy. Nor can one doubt, upon further reflection, that a "synthetic" rather than "analytic" or reducing model of alienation would stress that when *all* of its components dramatically increase at the same time, interactive "contaminating" effects are probably at work across all the variables. This seems to have been the case over the past twenty years in all categories except, according to the studies by the University of Michigan Survey Research Center/Center for Political Studies (SRC/CPS) the internal-efficacy measures. Yet even here the presumption is very likely that, if we have far more abstainers now than in 1960, then some crisis-related process is at work to produce an ever larger number of adults who think that Lady Luck decisively governs their lives.

Another vantage point may be gained in thinking about the abstention problem and its concentration in the social structure by accepting Anthony

Downs's arguments about the causes and consequences of rational abstention.[37] His model stresses that individuals operating under conditions of more
or less extreme uncertainty and imperfect information seek to calculate a
differential between competing parties when casting their votes. If party A
clearly promises more positive utilities to such individuals, then they will
vote for it, and similarly with party B. But when this partisan differential
approximates zero, then it becomes irrational for them to expend scarce
resources to achieve a null result. They will therefore abstain. The inference
is obvious that, if there is a secular trend toward growing abstention in segments of the electorate, something essential to the calculation of these utilities has disappeared or has become severely compromised in some way. In a
Downsian framework, this "something essential" would include at the least
a breakup of partisan cue-giving as a means of calculation and, associated
with that, a growing indeterminacy of relationship between what politicians
promise to win office and what they actually do once in office. The chronological coincidence of the growth of abstention—especially among those
social classes whose participation was already least—with acute party
decomposition after 1960, tremendous negative shifts in "allegiant" or "supportive" attitudes and in external-efficacy measures, and the full emergence
of the political-capitalist state seems altogether too complete to permit the
view that the relationships among them are casual or accidental. From an
inner-subjective point of view, a growing Downsian indeterminacy of relationship between voter choice and the subsequent actions of elected politicians would surely be expressed in terms of declines in levels of external efficacy, and would spill over into antipartisan, antipolitician attitudes as well.

The inference from all this is that "alienation," both in its subsets and in
its totality, is in essence socially determined. To put the horse before the cart,
the psychological-attitudinal characteristics that survey researchers identify
are responses or adaptations to a reality given by the social milieu in which
the individual lives. They are, therefore, very largely a function of the class
structure. This point, essentially Downsian in implication, seems strongly
supported by works such as James D. Wright's *The Dissent of the Governed*,
which largely rely on surveys but ask questions only intermittently raised by
the survey-research mainstream.[38] The basic argument of this work can be
summarized and extended here, since it seems to permit a reasonably orderly
entry into the relationship between voting, nonvoting, and the results of the
1980 election.

Viewed in sociological terms, a set of "alienative" responses that is concentrated among working-class adults corresponds to a relatively rational
view of their environment. This is very probably true not only in America
but more generally. Wright finds little evidence—at least around the 1970
date of his references—for assuming that alienative responses are much
greater in the United States than in countries like Canada, West Germany,
and the Netherlands. If we assume that such responses are learned as part of

the socialization process in working-class subcultures—just as their opposites are learned as part of those of the upper-middle classes—then one would expect to find that the density of such opinions across national lines, but in societies with broadly similar class structures and so on, would be relatively uniform. The crucial difference is *political:* it lies in the activity, or lack of activity, in the electoral market, which sharply differentiates American working-class politics from those of other advanced industrial-capitalist societies.

There is almost no satisfactory evidence to support the fears of academic "democractic elitists" that these American working-class abstainers are potential "runaway masses," easily capable of being stampeded by authoritarian-populist or even Nazi-type demagogues. As we now know, such a model fundamentally misdescribes political events in the latter Weimar era as well, since Nazi "contagion" was overwhelmingly concentrated among farmers and the middle classes who were neither Catholic nor Marxist.[39] Still less does it apply here. Such work as has been seriously done on this problem suggests that in America such "contagion" is pretty much an across-the-board phenomenon.[40] In fact, there is some reason for supposing that anti-democratic attitudes and behavior can under certain stress conditions be particularly visible at the *higher* levels of the American class structure.[41] So far, at least, the chief measurable political effect of alienation in American politics is *abstention* rather than either demagogue-led *mobilization* or the *channeling* of alienation into the active participatory forms provided by mass working-class parties of the European type. This brings us squarely up against the chief political difference between the United States and other advanced industrial-capitalist countries: the near-total absence of organizable alternatives to individualist, capitalist liberalism in the political culture and therefore in the electoral market.

As is obvious, all modern class society is characterized by great inequalities in life chances based directly upon relative proximity to the ownership of the productive assets of capitalism. Broadly speaking, there are three possible psychological responses open to the losers in this struggle. First, they may project rather than introject their relatively poor life situation onto "the system," "the bosses," or other external negative reference groups. Historically, working-class parties have come into existence in most non-American environments because, among other things, severe social inequality has been mandated and defined not only for members of this generation but also for their offspring. There is a second and parallel kind of projection. This has primarily involved lower-middle-class elements, especially those suffering from a poor and deteriorating life situation. Here the out-groups identified are organized labor, Marxism (especially Communism), "international finance capital," ethnoracial minorities such as the Jews, and so on. The modal response under stress here may be populist or, in this century, fascist or Nazi in character. Such movements correspond to a specific context of economic stress and perceived social disintegration, with the classic case that of Eastern and Central Europe after 1918. Yet not dissimilar projection politics has erupted time and again in the United States, underlying such events as regularly recurring right-wing religious revivalism, "great red scares," and McCarthyism. These reflect not the politics of "hysterical masses" of workers nearly so much as the politics of a *middle class under stress.*

The third and overwhelmingly normal American response has been introjection rather than projection of a relatively poor life situation, and acceptance of (or at least acquiescence in) the individualist-liberal norms that are densely articulated and believed in at the top of the social structure. The individual takes blame on himself for his situation and supports the inalienable right of property owners to own and enjoy property.[42] It follows that in such a setting of political consciousness and culture, the existing order comes over time to be accepted nearly universally as "given," and psychological adaptations to its disciplines and inequalities follow as a matter of course. This means that the electoral market is essentially defined by a "politics of excluded alternatives." In such a politics, doctrines of social harmony and of capitalist fundamentalism tend to dominate political discourse and outcomes. Leftist political movements whose minimal objectives must include redistribution of life chances through political action are conspicuous by their absence. Correspondingly, abstention rather than "left" voting is a major response among the lower classes to the political environment.

What this adds up to is that most organized, active politics in the United States concerns issues and agendas among various contending racially, ethnically, occupationally, and otherwise defined parts of the American middle classes. To the extent that general affluence prevails among those who actually vote, and if other political conditions are specified and present, this interplay of contending middle-class elements will tend toward a maximum of fragmentation. The Democratic party, much more heterogeneous in its social composition than the Republican, will be particularly affected by the political consequences. In particular, it will be guided both by its underlying doctrines of natural social harmony and by its leadership's responses to well-organized sectoral interests within its heteroclite coalition to embrace public policies for which general mass support is at best doubtful.

Changing the focus, then, from the problem of abstention to the problem of maintaining coalitional integrity, we turn to an evaluation of how the former majority party and the interventionist political capitalism it has created are related to certain aspects of contemporary public opinion in the United States. It has long been axiomatic among left liberals that the state serves as an important balance wheel, whose policies are designed not only to eliminate the "inhuman face of capitalism" and its devastating social effects but to provide a counterweight to the power wielded in the private sector by those who own and control it. A parallel axiom, advanced by many party theorists such as V. O. Key, Jr., is that political parties form an essential mechanism by which the many individually powerless can collectively aggregate their voting power and thus ameliorate their social condition.[43] It seems evident that in order for these views to be remotely relevant to the real political world, at least two conditions would be necessary. First, the state should so act, and should be widely accepted as so acting. This means that there should be a strong *and enduring* support for it among those, at the least, who could be expected to gain from its policy interventions. Second, by the same token, parties acting as agents for achieving some positively evaluated purpose through elections and control of government should retain relatively robust support, especially among those who are supposed to need them most.

Longitudinal survey evidence makes it absolutely certain that neither condition has been fulfilled in recent years. To the contrary. The massive and very real growth of domestic state activity in regulation and apportioning benefits since the early 1960s has occurred alongside a precipitate decline in levels of trust for government and a massive shift toward the view that government is too powerful. These changes are among the largest ever recorded in opinion surveys. Moreover, they are clearly connected with a general "conservative" shift on a number of policy questions between 1964 and 1978 such as support for a governmental role in ensuring comprehensive health care, or support for government guarantees of a job and a good standard of living. Overall, there are clear signs that the population has been tending (or perhaps creeping) rightward.[44]

Still further, there is much reason to suppose that these swings have been of a sociologically across-the-board character, with the result that by the late 1970s there was wide consensus for the view that government had grown too powerful. This implies as well that the old class and liberal-conservative anchorings have been seriously compressed, i.e., that the relative swing toward a more "conservative" position has been even more marked among liberals, Democrats, and lower-class elements than among conservatives, Republicans, and upper-middle to upper-class elements. Two cases in point from the Center for Political Studies survey files are shown in Table 4.[45]

Three overall impressions are formed from the data array in Table 4. First, the movement away from support for government power and from trust in government is huge. Moreover, it cannot be said that by 1978 we are any longer in the direct wake of the Vietnam War or even Watergate; more general processes appear to be at work. Second, the initial figures reveal political, ideological, and sociological stratifications in the expected directions, particularly so far as the government-power index is concerned. However, these are remarkably flattened out by the effect of differential movements, even within an overall across-the-board movement. The variances across these items are indicated in Table 5.

The specifically political implication of this is that the negative swings against government power and away from trust in government are *particularly concentrated among just those elements in society that could have been*

Table 4. Declining support for "big government," 1964–78

	Power-of-federal-government index			Trust-in-government index		
	1964	1978	Shift	1964	1978	Shift
Income percentiles						
95–100	−22	−37	−15	44	−9	−53
68–94	−3	−36	−39	45	−29	−74
34–67	13	−35	−48	45	−33	−78
17–33	14	−24	−38	49	−29	−78
0–16	15	−17	−32	28	−36	−64
Social class						
Professional	−17	−38	−21	43	−30	−73
White collar	−2	−30	−28	42	−31	−73
Blue collar	19	−23	−42	44	−34	−78
Unskilled	42	−24	−66	59	−32	−91
Farmers	15	−58	−73	45	−30	−75
Housewives	6	−32	−38	38	−42	−80
Party identification						
Strong Democrat	33	−13	−46	57	−23	−80
Weak or leaning Democrat	20	−25	−45	46	−31	−77
Independent	0	−33	−33	31	−35	−66
Weak or leaning Republican	−22	−41	−19	33	−40	−73
Strong Republican	−52	−50	+2	17	−41	−58
Liberal-conservative index						
Liberal 1	32	2	−30	44	−48	−92
2	31	−18	−49	49	−33	−82
Middle 3	11	−33	−44	53	−30	−83
4	−12	−41	−53	31	−25	−56
Conservative 5	−59	−68	−9	6	−38	−44
Total population	6	−29	−35	42	−33	−75

expected to benefit from and support government the most. The most striking comparative feature may be that by 1978 the lowest-income, lowest-occupation strata in American society had developed negative attitudes toward governmental power that were about as substantial as those shown by the most affluent and upper-occupation groups in that society in 1964. Such

Table 5. Differentials in relative support for big government, 1964–78 (Variances in PDI)

	Government power		Trust		Reduction in % variance, 1964–78	
Variable	1964	1978	1964	1978	Power	Trust
Income percentiles	205.04	62.96	53.36	89.76	69.3	68.2 (inc.)
Social class	494.25	35.69	48.50	2.19	92.8	95.5
Party identification	917.76	163.04	186.56	53.20	82.2	76.8
Liberal − conservative	1145.84	545.84	289.04	61.36	52.4	78.8

swings form an integral part of the deep background to Ronald Reagan's decisive victory in 1980.

It is patently obvious that developments of this sort raise particular problems of aggregation and effectiveness for Democratic (rather than Republican) party elites. For the Democrats in this system are the pro-state party, while the Republicans are, if not wholly antistate, oriented toward private-sector solutions to practically all collective problems except for law enforcement and national defense. A number of studies have shown, not surprisingly, that the gap between party elites over issues and ideology is substantially larger than it is among the mass of party identifiers. In 1960 McCloskey demonstrated that this was so, and also that Republican elites at that time were much further to the right of center than Democratic elites were to the left.[46] Since then, of course, the content of "liberalism" has undergone very profound changes.[47] In particular, racial, minority, and social issues are very much more important to political conflicts within the Democratic party than they were then. This changing content reflects, and reinforces, the growth and recent degeneration of the political-capitalist state over which Democrats have often presided, and which has largely reflected Democratic (and liberal) initiatives and perspectives. At the same time, enduring liberal-conservative cleavages remain much more important than Ladd seems to recognize.

To pursue the matter of Democratic party vulnerability more fully, some opinion data stratifying party elites and registered party voters are needed. Fortunately, these are provided in a survey of California elites and voters performed by the *Los Angeles Times* in the summer of 1981. Six issue questions were asked, and a profile of the ideological positions of Californians was also included. Using the same kind of position-differential index (PDI) employed by the CPS studies, we find that these profiles are as portrayed in Table 6. (See the Appendix on PDI Scores for a brief description of this mea-

Table 6. Issue differences between party activists and registered voters in California, 1981

Variable	Democrats Elites	Democrats Mass	Republicans Elites	Republicans Mass	Public (Mass)
Attention paid to blacks, other minorities, by California government	24	−10	−30	−30	−15
Opinion of Proposition 13	43	−13	−62	−49	−12
Equal Rights Amendment	71	41	−10	18	32
Death penalty	20	−45	−80	−74	−52
Busing for integration	9	−56	−84	−76	−53
Prayers in public schools	34	−36	−27	−35	−30
Ideology (liberal − conservative)	56	−5	−77	−49	−18

Table 7. Proximity of elites to mass in California, 1981

Variable	Democratic elites		Republican elites	
	Party mass	Public	Party mass	Public
Attention paid to blacks	+34	+39	0	−15
Opinion of Proposition 13	+56	+55	−13	−50
Equal Rights Amendment	+30	+39	−28	−42
Death penalty	+65	+72	−6	−28
Busing for integration	+65	+62	−8	−31
Prayers in public schools	+70	+64	+8	+3
Ideology	+61	+74	−28	−59
Mean proximity, all items	+54	+58	−11	−32

sure.) The relative proximities on these issues that this yields are shown in Table 7.

Indeed, on most of these issues Republican elites actually stand *closer to Democratic registered voters than Democratic elites do.* The latter almost literally stand by themselves in left field, as Table 8 indicates. The mean proximities to Democratic registered voters are: Democratic elites, +54; Republican elites, −35, which indicates an overall (wholly unweighted) Republican proximity advantage of 19 percentage points. By contrast, proximities with Republican registered voters are: Democratic elites, +78; Republican elites, −11—a Republican advantage of 67 percentage points.

Needless to say, this poll does not indicate the relative salience of these questions, and others to which we would have liked answers were not asked. Moreover, the poll does not tap the more than one-third of California's electorate of over 17 million who are not registered at all. But for reasons we have already indicated, it seems unlikely that there are radical position-issue differences between the nonregistrants and the registrants. (If there were, it might be possible for a determined entrepreneur to organize them, and then they would of course register.)

The overall thrust of this poll is clear enough. The only one of these items

Table 8. Proximity of party elites to registered democrats in California, 1981

	Democratic elites– Democratic voters	Republican elites– Democratic voters	Closest and margin
Attention paid to blacks	+34	−20	R 14
Opinion of Proposition 13	+56	−49	R 7
Equal Rights Amendment	+30	−51	D 21
Death penalty	+65	−35	R 35
Busing for integration	+65	−28	R 37
Prayers in public schools	+70	+9	R 61
Ideology	+61	−72	D 11

for which the public as a whole (including Republican registrants) has a "liberal" position is the Equal Rights Amendment. All the others show conservative positions in this survey, and in the "big three" social issues—death penalty, busing, and prayer—there are very large conservative leads among rank-and-file Democrats and Republicans alike. These issues are known to be salient, and to be identified with important differences between Democratic party activists and the large bulk of the American population. It goes without saying that to the extent of their salience they contribute to a situation in which the Democratic party cannot possibly maintain its integrity. Relationships of attitude between Republican elites and the party rank and file are, as we would expect, much closer.

Indeed, it would seem reasonable to suppose that the general—if far from generally ideological—rightward creep in the white American electorate of recent years has increasingly narrowed the overall gap between Republican elites and the "mainstream," while widening the gap between Democratic elites and that "mainstream." Again, we call attention to the fact that organized American electoral politics these days is overwhelmingly middle-class politics, pitting middle-class liberals, with all the "new liberalism" of the post-1968 period, against middle-class conservatives who are both economically *and* socially conservative. One obvious action implication of all this is that left liberals who wish to get anywhere in American politics during the 1980s can only succeed, in all probability, to the extent that they are able to change the subject of *primary* political discourse away from social-issue politics and toward economic issues.

Reviewing all this material, we are still confronted with the underlying question: why should the rapid post-1960 growth of the political-capitalist state at home be associated with a collapse of support for it, particularly among those strata who would presumably benefit the most? Part of the answer lies in the discrepancy just reviewed between the newer social- and racial-issue liberalism of Democratic party elites and the more conservative attitudes of the general population on such dimensions. This discrepancy is considerably reinforced by being coupled with the relatively high salience of such issues—indeed, they are definers of "liberalism"—in the recent politics of the Democratic party. Connected with these realities are several others that are intimately linked to the way the political-capitalist state works in fact and the way it is perceived to work by many Americans. For in the first place, many of the Great Society programs were targeted to specific clientele subgroups whose needs and discrete (often racial) status marked them out for what others not so favored could only regard as preferential treatment. As these programs developed, a very important consequence was that people among the working and lower-middle classes again and again found themselves in a profoundly irritating situation. The irritations took a variety of forms—clearly enough racial in many cases such as busing (judicially rather than legislatively imposed, to be sure), but also economic. Often enough necessary goods they could not afford to purchase in the private market (medical care, for example) were provided—after a fashion, of course—to those who could pass the means tests required.

Second, one of the basic principles of this state was the maintenance of social harmony through public-sector intervention. In American political, cultural, and constitutional conditions, the preferred strategy was one of syndicalist cooptation of leaders and potential leaders of newly mobilized interest groups into the governing coalition. But inevitably this also led—sooner rather than later—to the rapid development of "iron triangle" relationships among such leaders, public-sector bureaucracies with large regulatory discretion, relevant actors in Congress, and other Washington "insider" lawyers and lobbyists. Under such conditions, programs tended to take on a life of their own within the feudalities of political power thus created. And it is hardly surprising that, over the years, the public as a whole (including those in the lower end of the social structure) should increasingly believe that "parties are interested in votes, but not in your opinions," and that "those we elect to go to Washington tend to lose touch with the people back home." One important consequence of this proliferation of feudalized power islands in the political process was spelled out by "Mr. Great Society bureaucrat" himself, Joseph Califano. In his discussion of his responsibilities in the medical field, Califano observed that he could not now support the creation of a nationwide comprehensive medical-care program of the sort that Senator Kennedy and other liberals have advocated. Why not? Because any such program would bankrupt the country under the political conditions in Washington that he describes. It would do so because of the immense resources of the medical lobby "iron triangle" and, ultimately, because Congress lacked the integrity that would be necessary to keep the costs of such a program within reasonable bounds.[48]

This brings us to a third feature of the political-capitalist state: its mobilized interests and leadership. They tend to pursue nonpopular strategies to get what they want through the political process when broad public support is not available and is known not to be available. Such strategies have preeminently involved litigation and judicial descision an integral part of group struggle, but bureaucratic linkages, administrative decisions, and "insider" low-visibility congressional legislation with major subsequent impact on the public have also played important roles. Such strategies, whenever they can be linked to constitutional rights, can obviously be defended on the merits. More than forty years ago Justice Stone was among the first to point this out in a famous footnote to his opinion in *United States* v. *Carolene Products Co.*[49] Where one is dealing with discrete, insular, and "unpopular" minorities, the normal judicial presumption in favor of letting the political process do its work may have to be waived. On the other hand, such waivers have grown enormously in volume and impact in recent decades: in general, the use of elite-insider strategies to pursue liberal ends in the absence of demonstrated popular support for them has gone far beyond anything Stone or his contemporaries could have imagined. One hardly wishes to adopt the formula "public opinion, right or wrong." Yet there can be no doubt that any political construction that is founded upon bypassing such opinion, in a country founded upon it, is sooner or later going to run into serious legitimacy problems.[50] For reasons that we hope have been made clear, these legitimacy problems came to a head in the Carter administration. In the absence of overwhelming support for a national agenda, politicization of issues has led to fragmentation both of policy and of political coalitions. The search for social harmony through public-sector activity thus *negates itself* so long as political consciousness and organization are in their present condition.

Much of the foregoing analysis makes most direct operational sense at the level of elites and of the well-informed publics just below them. It is indeed at the elite level that the political-capitalist state, its economic and social assumptions, and its workings have most clearly come to a dead end in the early 1980s. At the broader mass level, the setting for the 1980 election was fashioned one stage at a time from the mid-1960s onward. The efflorescence of George Wallace a decade ago was one part of this setting; so too were the overwhelming public rejection of George McGovern in 1972 and the creation of Jimmy Carter pretty much from whole cloth in 1976. The deeper context of Carter's defeat in 1980 involved the processes by which, through a series of stages, the "party of nonvoters" absorbed a larger and larger share of adults who, under other circumstances, would mostly have voted Democratic. Even more to the point, it involved rejection of his leadership by those who were still left in the active electorate—that is, by voters who were in the aggregate considerably more middle class than the population as a whole. Economic and imperial-management issues were central to the 1980 result, but there was something else as well. Notably in his so-called "malaise" speech of summer 1979, but also elsewhere, Carter appeared fundamentally to doubt that the old freewheeling, growth-oriented, boundlessly optimistic "American dream" could ever be restored. The first article of Ronald Reagan's faith, on the other hand, is that it must be and can if only the proper policies are followed. Reagan's faith is shared by those many Americans who want to believe this proposition, and by the many others who never doubted it in the first place.

There are important reasons for supposing that geographical factors played a major part in the 1980 election, a part not readily accessible to broad-sweep survey analysis, and that these factors may well have been related to the relative density of belief in the conventional pre-political-capitalist "dream" as a realizable ideal. Such geographical partitioning becomes rather easy to pick up by examining the magnitude of the swing away from Jimmy Carter between 1976 and 1980 at the level of the congressional district. Were we to ask where, geographically, we would tend to find concentrations of American voters for whom the traditional American dream lives, where would we look? Clearly to affluent-suburban small-town America in general and to the very rapidly developing (and prospering) Sunbelt states of the South and West in particular. Conversely, appeals for reactionary revitalization in terms of this dream would be least effective among blacks—tenacious supporters of the political-capitalist state for obvious reasons—and among residents of major metropolitan centers, especially in the aging, gradually decaying Snowbelt. Now, everyone knows that ecological and geographical relationships cannot shed light directly upon individual attitudes. But they are often quite suggestive, and nowhere more so than in this 1980 case.

We begin with John Anderson's percentage of the total vote. Here we find a sharply defined set of regional polarizations. Of the seventy-seven congressional districts in which Anderson received 10 percent or more of the total vote, twenty-four were concentrated in New England, and another twenty-five were scattered through the Mountain and Pacific states. There were ten districts in the country giving Anderson 15 percent or more of the vote, compared with his national total of 6.6 percent. Nine of these were in New England (one in Connecticut, Vermont at large, and seven in Massachusetts), while the other was in California—the prosperous 12th, centered in the vicinity of Stanford University. Conversely, of the 110 congressional districts where Anderson received less than 4 percent of the total vote, fully 82 were located in the eleven states of the old Confederacy, and another 13 were in the immediately adjacent border states from Maryland to Oklahoma. It is evident that Anderson did particularly well in university-dominated communties. Thus he received exactly one-third of the total vote in the town of Hanover, New Hampshire (Dartmouth College), just under one-fifth of it in Johnson and Story counties, Iowa (University of Iowa, Iowa State), 22.1 percent in the town of Orono, Maine (University of Maine), 32.0 percent in Durham, New Hampshire (University of New Hampshire), and 24.3 percent in Amherst, Massachusetts (University of Massachusetts, Amherst College). Anderson's essentially moralistic, liberal-Republican appeal in the center of the 1980 political spectrum struck relatively responsive chords in New England and certain other areas (upstate New York, northern Illinois, Minnesota, certain affluent suburban districts, across the country, and much of the West) where such traditions form important parts of local political culture.[51] His penetration was essentially nil where another third-party protest—the Wallace movement of 1968—reached its greatest extent. Not surprisingly, this candidacy of the center suffered extreme attrition during the campaign. It contributed significantly to certain local outcomes (e.g., Reagan's tiny plurality in Massachusetts). But overall the "Anderson difference" turned out to be of very little importance to the 1980 result.

If we turn to the anti-Carter shift, expressed as a differential in his 1976 and 1980 share of the two-party vote, a very different picture appears. Obviously there was an across-the-board character to this movement, but there are some very revealing differentials at the extremes of the 1976–80 swing distribution. Most suggestive of all, when the extremes are arrayed together in terms of the classic sectional partitioning of the 1896 election, we find striking evidence of its being *turned inside out*. But before we examine this striking singularity, let us briefly describe the kinds of congressional districts in which the Reagan (or anti-Carter) "surge" was greatest and those in which it was least. Of the 436 districts in the nation (including the District of Columbia), 89 showed a negative Democratic swing of 10 percent or greater in the two-party vote (The nation as a whole had a −6.4 percent swing), while at the other end, 80 districts showed less than 1 percent negative Democratic swing, with Carter actually showing a gain over 1976 in 59 of them.

In a subset of fourteen districts, the pro-Reagan swing was 15 percentage points or more. These districts were located in Jewish areas of New York City; the Illinois hometown district of both Ronald Reagan and John B. Anderson (the 16th); a group of five agricultural districts of the far Midwest; two districts in metropolitan Miami, Florida; the "great basin" areas of the desert Mountain states (including the adjacent and very similar 1st California); and the special case of Arkansas, three out of whose four districts showed at least a 15 percent swing away from Carter. In all of these cases except the Illinois 16th and the rather anomalous situation in Arkansas, specific populations had exceptionally intense grievances against the incumbent administration: Jews upset by the March 1, 1980, condemnation of Israel in the United Nations and by other evidences of possible pro-Arabism in the Carter administration; south Floridians impacted by a Cuban immigration-refugee crisis; citadels of the "sagebrush rebellion" in the Mountain West, perhaps reinforced by administration proposals to build MX missile sites in their territory; and farmers objecting to paying the price for post-Afghanistan economic boycotts of the USSR.

The much broader group of districts with an exceptionally large but not extreme Reagan "surge" in 1980 constitutes, not surprisingly, a more heterogeneous collection. Overall, however, most of them divide into five types: (1) the Snowbelt states of the Northeast and near Midwest, Jewish districts in New York City and Philadelphia, joined by a general pattern of concentration among relatively high-growth and especially high-status suburban districts; (2) agricultural and small-town districts in the Midwestern states and in the western border states (Missouri and Oklahoma); (3) the Mountain states, 9 (or 47.4 percent) of whose total districts showed a Reagan swing of 10 percent or more; (4) a group of thirteen districts in California, located in the Central Valley and the area from suburban Los Angeles to San Diego; (5) small-town and rural white districts in the Southern states, with particular concentrations in Florida and Texas, the most solidly Sunbelt and rapidly growing of them all. This pattern clearly suggests something: the concentration of Reagan's particular appeal (or Carter's lack of it) in areas that are affluent, dominantly upper-middle class, and, in the Sunbelt, undergoing exceptional growth and mushrooming prosperity. In most such districts we may assume that the continuing "American dream" based upon capitalist growth and prosperity was never seriously in question.

The eighty districts at the other end, relatively or absolutely pro-Carter, may also be divided into several groups. (1) Districts with black and Hispanic majorities. Fully eighteen of the eighty districts fall into this category (including two urban districts in the South, the Texas 20th and Tennessee 8th), and in the overwhelming majority of these cases Carter actually increased his two-party percentage over 1976. (2) Another group of thirty-nine districts, nearly half of the total, have central cities at their core, but are dominated by white (or, in Honolulu, Oriental) rather than black or Hispanic voting populations. Strikingly, they do not include districts from three very large metropolitan areas—Boston, New York, and Philadelphia. With these important exceptions, however, they are to be found from Providence and Hartford westward all the way to Detroit, Chicago, Minneapolis, Kansas City, and beyond to Seattle, San Francisco–Oakland, and Honolulu. (3) There is a cluster of five nonmetropolitan districts in Michigan, reflecting the disappearance of a clear "friends and neighbors" effect on behalf of Gerald R. Ford in 1976. (4) Finally, we find a group of "Yankee" and upstate New York districts, extending from Vermont to the western end of New York, where Anderson did relatively well and where Reagan's share of both the total and the two-party vote declined compared with Ford's in 1976. These four groups taken together add up to seventy out of the total of eighty districts with a relatively or (usually) absolutely pro-Carter swing between 1976 and 1980. The story they appear to tell is one of relative lack of aggregate Republican appeal in 1980 among blacks, central-city whites, and certain marginal, probably traditionally liberal Republican groups in the far Northeast quadrant of the country. In many urban districts, to repeat, Carter's major problem lay not in defending his share of the vote cast compared with 1976 but in failing to mobilize support among the "party of nonvoters," which tends to be concentrated in such areas these days.

What does this all add up to? Viewed most broadly, the magnitude of the relative slide to Reagan in 1980 appears closely related to the physical location of winners and losers in the contemporary evolution of demographic and class structure in the United States. As is now well known, the 1980 census returns reveal two population shifts of exceptional magnitude: from Snowbelt to Sunbelt generally, and from densely populated metropolitan areas to small-town and rural nonmetropolitan areas. The first represents an acceleration of a trend that has been under way for at least several decades. The second is a (if possible) more dramatic and (until recently) unexpected shift, the like of which has not been seen since the beginnings of industrial urbanization more than a century ago. These transfers of population involve relative and, to some extent, absolute transfers of wealth from areas of low or negative growth to those of rapid growth. To the effects of these shifts are added, of course, not only the general tendency of most affluent suburban

districts to produce Republican presidential majorities, but also their tendency to show a pro-Reagan shift in 1980 that was greater than the national
swing.

The evolution of the physical base of the American political economy
since the early postwar era has entailed a now rapidly accelerating inversion
of the classical patterns of concentration in wealth and population emerging
a century ago with the onset of full industrial capitalism. In that period,
industrially and financially generated wealth was concentrated in an area that
I have referred to elsewhere as the "metropole."[52] This metropole had a
rather precisely demarcated geographical boundary, defined by the outer limits of the railroads' "scheduled territories," i.e., those with preferential
freight rates. These territories incorporated the Northeast, the East North
Central heartland of the Midwest, West Virginia with its coal, and Maryland. In addition, the preferential rates carried just outside these limits to
include Duluth and the Mesabi iron-ore range, Minneapolis, St. Louis, and
Louisville. The area enclosed within these limits comprised 53.7 percent of
the total national population in 1900, some 87.2 percent of people living in
cities of 50,000 or larger, 77 percent of gross value of manufacturing products, and majorities ranging from two-thirds to three-quarters of the national
supply of such items as university library holdings and endowments and bank
capital. But it occupied only 15 percent of the land area of the continental
United States. The density of populaton within the metropole in 1900 was
ninety-two per square mile; in the areas outside of it, only fourteen. These
external areas were in essentially colonial relationships, economically and
otherwise, with the metropole. They were largely based upon primary-sector
production: ranching, mining, lumbering, and cash-crop agriculture in the
trans-Mississippi West and cotton monoculture throughout large parts of the
South.

As is well known, the epochal 1896 "battle of the standards"—silver versus
gold currency—pivoted on this regional geography, based as it was upon a
clear-cut division of economic labor in an imperial-sized country. McKinley
carried only four states outside its borders, losing none within. The subsequent "system of 1896," which lasted until it was smashed by the 1929
Depression and the New Deal, was based upon Republican hegemony
throughout the metropole and much of the West and, on the other hand, a
solid one-party Democratic South. While political conditions changed fundamentally after 1932 with the emergence of class-ethnic politics in the
metropole and the alliance of the West with the Democrats, the economic
pattern of concentration showed only marginal change until after World
War II. The accumulating transformations since have included not only the
full industrialization and urbanization of a once-rural South, but vast westward transfers of population as well. Thus in 1932 California had twenty
congressmen and Florida five, while Pennsylvania had thirty-three and New
York forty-five. Beginning in 1982, California will have forty-five, Florida

Table 9. Geographical factors in the 1980 election: Location of extremes in anti-Carter swing, 1976–80

Area	Anti-Carter swing −10 and over	Anti-Carter swing −0.9 to positive	Total	% of total districts in regional category −10 and over	% of total districts in regional category −9 to positive
Ex-metropole					
Northeast	12	23	105	13.5	28.8
East north central plus Minneapolis, St. Louis	5	34	92	5.6	42.6
Eastern border	0	5	15	0	6.4
Total ex-metropole	17	62	212	19.1	77.5
Ex-"colonies"					
West (inc. west north central)	36	9	98	40.4	11.3
Western border	7	1	18	7.9	1.3
South	29	8	108	32.6	10.0
Total ex-"colonies"	72	18	224	80.9	22.5
Grand total	89	80	436	100.0	100.0

nineteen, Pennsylvania twenty-three, and New York thirty-four. Needless to say, this process of transformation has been greatly accelerated by the sudden rises in energy costs that marked the 1970s and have the most fateful implications for the longer-term economic viability of the former metropole as a whole.

In view of what we have said so far, we would expect to find a relationship between this geographical partitioning and the relative density of pro-Reagan swing in the 1980 election at the extreme ends of our congressional district distribution. As Table 9 shows, this relationship is, in fact, very strong.

To the rather stark regional profile offered in the table it perhaps only needs to be added that, as of 1982, some 16 congressional seats will shift from the ex-metropole to the ex-"colonies," leaving totals of 196 and 240 respectively. Moreover, while detailed analysis must await final reapportionment decisions by the states, it seems very probable that a great deal of the shift will be concentrated on taking away seats from the pro-Carter end of the distribution and transferring them to the pro-Reagan end. That, after all, is what a significant relationship between partisanship and growth (or stagnation) is all about. And it is the existence of relationships of this sort that reinforce the sense that, whatever specifically happens in the short run to the Reaganite attack on the fourfold crisis of decay we have been describing, a very important turning point in our modern history was reached in 1980. Among other things, the Republican capture of the Senate represents a power shift that is not likely to be soon overcome this side of a disaster of the 1929–

32 sort. The profound shift in the direction of public policy that occurred in 1981 will, therefore, not easily be reversed in the immediately foreseeable future. As always in our modern history, an enormous amount will depend on the future interchanges between the Democratic voting coalition and the "party of nonvoters." But, as this essay has attempted to stress, an enormous amount of intellectual and policy rebuilding will have to be done within that party first, or—a far more immense task—a new party will have to be built on the ruins.

IV. The Reagan Revolution: 1981 and Beyond

As the election of 1980 was most unusual by the conventional rubrics of American politics, so its aftermath has been even more so. Recent conventional wisdom about the workings of the American political process has stressed the acute decay of the political parties at the grass-roots level and in Congress, coupled with the ongoing proposition that the establishment of decisive centralized control over policy agendas is not normally to be had. American politics, we have all been taught, is a messy "slow boring of hard boards," in which all sorts of interest groups engage in interplay, and in which policy change is incremental rather than comprehensive. This conventional wisdom is by no means without its force even in the early 1980s. The Reagan administration's struggle to get its way with Congress on the sale of AWACS aircraft to Saudi Arabia is one striking evidence of this. So, too, the success of the administration's continuing struggles to bring government spending into line with its ideology is obviously and importantly dependent on the perceived success of its earlier program. It will depend even more importantly on the extent to which legislators who have supported the new administration come to smell a greater likelihood of political defeat in and after 1982.

Nevertheless, the magnitude of the change and the nature of congressional behavior associated with it are staggering by normal standards. They are particularly so when it is recalled that neither the extreme stimulus of acute crisis (as in 1861 or 1933) nor exceptionally large majorities for the president's party (as in 1965–66) are present in the current political situation. This administration came to power with an ideological point of view. To a considerable extent this ideology is conformant with American business ideology about government spending, deficits, and the like, but it also differs in important respects from that variant of American business ideology that dominates the East's financial markets and their managers. Basic to Ronald Reagan's version of capitalist revitalization are several components whose capacity to hang coherently together is doubted even by many Wall Street business conservatives. According to this view, it is possible and necessary to increase defense spending toward its pre-Vietnam level; to cut the budget sufficiently to bring it into balance; and to cut federal taxes over the next several years by as much as $700 billion. The rationale offered for the last—which is the true key to what will happen to the state and to a very large proportion of the American people—is that offered by Arthur Laffer and his school: cutting the taxes paid by the rich and corporations will "liberate" so great a volume of creative investment energy from the thralldom of state extraction that business will boom. Indeed, in the pure Laffer version it will boom so much that at least as large an aggregate volume of revenue will come to the federal government under much lower tax rates as could have been expected under the higher ones. It was this, of course, that George Bush criticized during the 1980 primary season as "voodoo economics." In all probability this approach will duly be revealed over time as a classic example of wish fulfillment (always an important ingredient of ideological politics), of trying yet another time to get something for nothing.

Obviously, if the Lafferite, Kemp-Roth approach to federal taxation policy proves justified by events—or if some overwhelmingly favorable economic change develops elsewhere that makes it seem so—then the political forces that came into the ascendant in 1980 may well be unbeatable for the foreseeable future. Doubts as to whether this scenario will materialize are, however, very widespread even among orthodox economic conservatives. One need only recall not only George Bush's remarks but Senate Majority Leader Howard Baker's uneasy observation following congressional approval of the initial Reagan program that it was "a riverboat gamble." What is politically astonishing is that this "riverboat gamble" was adopted almost intact by Congress, despite a rather narrow Republican lead in the Senate and a Democratic majority of fifty-three in the House of Representatives.

The conventional wisdom of American politics points to the kind of fragmentation in the policymaking process, the kind of "dynamic deadlock," that admittedly reached pretty extreme form under Jimmy Carter. Another basic element in this wisdom is the accepted proposition that ideological movements in American politics lose: consider Goldwater in 1964 and McGovern in 1972. Pragmatism rather than ideology is therefore almost always the rule in this politics, and especially in the White House. Yet here we have an ideological administration in which the chief right-wing "mole" is the president himself. What is more, this administration was able to work its will on Congress and secure the largest and clearly most comprehensive set of policy changes to have occurred in the past half-century. Whatever happens to later portions of the Reagan policy, this change will in all probability endure for some time to come. The budget and tax victories of spring and summer 1981 were the outgrowth of two complementary political facts: an iron, monolithic discipline among congressional Republicans and the merger with them of conservative Democratic "boll weevils" in the House, about one-sixth of the total Democratic strength there.

The history of politics, elsewhere as well as in America, makes it clear that parties come into being to amass a collective program. They tend to be strong to the extent that the power blocs they represent need to change the foundations of public policy in some comprehensive way. There is a collective agenda that requires some sort of state action, and hence the collective discipline necessary to achieve it. Historically, this sort of thing has usually developed from a left that is in opposition to an incumbent right that sees no particular reason for any major change in a status quo from which it benefits.[53] In very much the same way, the politics of 1981 reflect the amassing of the political resources necessary to achieve change. On this occasion, however, the initiative has come from the right. It has built up throughout the 1970s, as right-wing Political Action Committees, single-interest groups and others have learned how to "wire" the American electoral market through a wide variety of ultramodern, media-oriented techiques. Reagan in power was a man with a program for comprehensive change, which could only be realized through state action. The change intended represents nothing less than the complete liquidation of the political-capitalist state brought into being since the end of Eisenhower's administration. That this goal could be seriously pursued by *any* president, *any* administration, is by itself conclusive proof of the gravity and generality of the economic, imperial, cultural, and political crisis now afflicting the United States. That a very large part of it could be achieved—in large measure through a party discipline which would do credit to American legislative politics in the Civil War era— is the most striking commentary of all on the narrowness of organizable political alternatives in this country. It is also yet another measure of the profound disarray into which the Democratic opposition has fallen.

As for the tax-cut segment of Reagan's program, the whole point of neo-laissez faire capitalist revitalization is to shift the fulcrum of state action from promoting mass consumption to promoting accumulation in the private sector. This inevitably means that clear-cut choices will be made that grossly favor the rich, are more or less neutral for the upper-middle classes, and adversely affect groups below that level, in ascending severity as one travels down the class structure. But the probability that Lafferite tax economics lacks realism vastly multiplies the stakes in this zero-sum game. For if the increased volume of revenues on a lower tax base does not materialize, the only way to balance the budget (another part of the ideological program) will be to cut ever deeper into the expenditures of the state. To the extent that the imperial-sector objectives of the administration require that no substantial cuts come from defense—indeed, that defense expenditures go up very substantially before the mid-1980s—this means that virtually all of them must come from domestic-sector activities of the federal government. There is a very clear possibility here of a kind of public-sector "doomsday machine" that could take one of at least two forms. The first, under conditions of continuing stagnation and high interest rates, points toward continuing and large "overruns" in the federal budget deficit, leading to increased budget cuts, and so on, in a scenario that has some characteristics of infinite regression. The second, presuming continued commitment to large defense increases but an eventually decisive political resistance to any further domestic budget cuts, points toward the prospect of large and escalating budget deficits, with limited contribution to the rate of inflation but much more decisive contribution to the indefinite maintenance of damagingly high interest rates. To the extent that Lafferite economics comes very widely to be judged wrong, the only alternative at some point will be to rescind a very large part of the tax cuts that were a centerpiece of the 1981 Reagan policy program. This in turn will collide with the president's ideological commitments, with an outcome that cannot now be predicted.

Politics, like economics, is anything but an exact science. But it does not take much perspicacity to see that the longer the administration's ideological program remains in place, the greater the probability that something approximating class warfare will occur in America—quite apart from its international economic ramifications. Very broadly, there have been three major approaches to the problems of late capitalism in recent decades. These are socialism, one or another variant of political capitalism with a more or less elaborate welfare state, and neo–laissez faire capitalism revitalization. In the United States, of course, socialism or even social democracy is an excluded alternative a priori. The choice has been narrowed therefore to political capitalism and business-corporate conservatism: in short, to the Democratic and Republican parties and the power blocs they represent in the electoral market. Crucial to the ideology of political capitalism is the gospel of social harmony among classes and other definable population groups that are capable of political organization. As we have seen, political capitalism at work, under specifically American institutional and cultural conditions, has failed to accomplish this objective. But the dominant ideology of its leadership remains dead set against any philosophy that accepts the necessity of organized disharmony in politics, i.e., the "class struggle" or something like it. Its policy recipes, accordingly, are so organized as to avoid wherever humanly possible the making of hard choices, no matter what the circumstances. But for effectiveness, credibility, and legitimacy, this requires the reality of assured, sustained economic growth, precisely the condition that no longer exists.

The other two traditions are, on the other hand, quite unafraid to make hard choices, though of course diametrically opposite ones. In the case of right-wing capitalist ideology, this is of course veiled in various ways—by appeals to patriotic emotion, by insistence upon the general well-being that will arise as the private sector trickles down its benefits to the public, and so on. But the reality is that of class struggle on behalf of the owners and managers of the country's major economic assets. When the business press and others talk, therefore, of "reindustrializing America" through the "liberation" of corporate enterprise from bondage to government regulation and extraction, they are talking about the repeal of relative social harmony in

America, both in theory and in practice. Obviously the class struggle thus proclaimed is at present wholly one-sided in its political impact. This fact helps us to understand how the huge Reagan program could be pushed through Congress with such apparently slender political resources. It is worth noting in this regard that a *New York Times*–CBS poll, reported on September 29, 1981, comments that "there was a strong class tone in the variations in approval. For example, 70 percent of those with family incomes of $40,000 or more, but only 34 percent of those with incomes under $10,000, said they approved of his [Reagan's] handling of his job."[54]

Considering the posture of the Democratic party, the one-sidedness of the struggle is inevitable at this stage. For in the country of the blind, the one-eyed man is king—a major theme of this essay, and a leading reason for supposing that the transition of 1980 was much more fundamental than many political scientists appear to believe.

Assessment of the future is often a bootless enterprise at best. It is especially so now, when so many unprecedented things have happened. Still, it does seem much more likely than not that issues of class struggle will become more central to American politics during the 1980s than they have been at any time since Pearl Harbor. By the same token, it seems unlikely that this class struggle will continue to be so hopelessly one-sided as it was in 1980 and 1981. It is entirely possible that the repeal of social harmony that Ronald Reagan pushed through Congress as a silent rider to his budget and tax packages will prove to be irreversible, whatever happens in the political short run. If so, we will have been present at the creation of the most momentous change in American politics since the Civil War.

Another major, and final, question is likely to be given at least the beginnings of an answer during the 1980s. We have made the obvious point that, at least down to the present moment, the socialist (or social-democratic) approach to the problems of late-capitalist political economy has simply not been a politically organizable option in the United States. Nor is it likely to be in the foreseeable future unless some unforeseen circumstances compel a revolutionary transformation of political consciousness and political culture first. This leaves us with the other two. If our analysis of the generic and specifically American difficulties facing political capitalism in the present era is correct, it has suffered an intellectual and operational bankruptcy which will not soon be repaired. After all, only in some such vacuum as Jimmy Carter inherited and intensified could Ronald Reagan have come to power, ideology and all. So for the moment we are left with only one leading alternative which has public plausibility and intellectual vitality: Reagan's specific variant of the American Business Creed.

The question that follows is now obvious: where do Americans turn if this too fails? If relative stagnation, accelerating regional and class triage, and damagingly high rates of inflation and interest continue to unfold as they have in the past, what then? Obviously, everything would depend upon the concrete extent of this failure, which could well range from continuing but very gradual long-term deterioration of the economic foundations (and our other crisis variables) to events as dramatic as a great depression or a major war. If we leave aside the latter two, with their literally incalculable consequences, and focus upon a more middle-of-the-road but still rather depressing scenario, a number of specific shorter-term results become imaginable, including defeat of the now-incumbent coalition, but also including its retention of power. The former would presumably be based upon standard short-term political-economy considerations: if the discomfort index is high enough, the incumbents will be voted out of office. Even if this were to occur, the prospects for a re-reversal of macro-policy—back toward the political capitalism much beloved of Democratic leaders—would be politically poor and would probably remain economically very dubious.

On the other hand, it is very easy to imagine the incumbents doing surprisingly well even if the general economic situation remains rather poor (but not universally desperate). We have touched upon some of the reasons for such a judgment: for instance, the extreme class skew in voting participation, the continuing commitment to doctrines of social harmony in a conflict situation by the leadership of the Democrats and much of organized labor; and the political culture itself. It seems to be axiomatic in the study of revolutionary change that people simply will not abandon their traditional ideological mazeways even though they suffer severe discomfort. A comprehensive alternative to the present order must first come into being, must achieve organizational and intellectual institutionalization, and must begin to penetrate the mass of the public. Failing the emergence of such an alternative, what is much more likely to happen is that the rallying around the traditional mazeways and the collective symbolisms linked to them will become more and more frenetic, more and more extreme, as the situation deteriorates. It is becoming increasingly obvious that the long-standing class skew in American political participation is becoming energized on one side—in the upper half of the American class strucutre. Reagan personally, with the traditional cultural symbols that he so masterfully evokes, is uncannily well positioned to carry out a perimeter defense of the American Business Creed, the "American dream," and American capitalism. And who knows? Perhaps he can do so the more effectively, among the American bourgeoisie, to the extent that the gap between ideology and reality actually widens.

In a nutshell, there is some reason to suspect that in the shorter term the new coalition can expect to win if things go well and perhaps even if things go poorly, so long as they don't go too poorly. A zero-sum society is, pretty much by definition, a society whose dynamics and characteristics are best explained by models of conflict rather than consensus, ideology rather than pragmatism. In such circumstances, political opposition to a hegemonic coalition cannot fail to be ineffective if it refuses to reorganize its political perspectives and organization into a mode that accepts this conflict as the inescapable core of political action. But such acceptance of conflict by an opposition must also entail the radicalization of that opposition. Only when that point is clearly understood, and when life-and-death issues are at stake, will the one-sided struggle now going on become more competitive. Probabilities would appear to favor this development over the longer run. Like nature, politics abhors a vacuum, and those now in power must depend upon a continued vacuum on the other side in order to retain it. This was also the case in the 1920s, as subsequent events proved. But one of these subsequent events was the largest economic catastrophe in the history of capitalism, and it is a very nice question indeed as to whether anything less cosmic could have destroyed that earlier, vacuum-supported "system of 1896."

Viewed in the longer term, the question retains its oppressive character. If the two alternatives possible in this system both prove hopelessly inadequate to provide the political and economic means of a collective well-being for most Americans, then a full-scale regime crisis—wholly equal in its importance and implications to the greatest such crises in modern world history—will erupt into plain view at some point. Many potential conflict lines of such a crisis, which would surely include bitter and polarized struggles over state power, are already clearly illuminated across the entire electoral map of the 1980 election—very importantly including the now-silent "party of nonvoters" on that map. What will happen politically if the perimeter defense of American capitalism that was launched in 1980–81 becomes ever more desperate? In particular, what will happen if, for the first time in our modern history, that defense comes to be seriously challenged from within the country, especially by a large-scale social-democratic opposition?

At this point readers may well conclude that the author of this essay suffers from a fevered imagination, if not worse. Fair enough: many generations of critics and dissenters have gone broke waiting for *der Tag* to strike American politics, and so it may well be again. At the same time, the apparent exigencies of our overall situation do not appear likely to be easily amenable to cure by reciting Professor Coué's mantra, "Every day, in every way, I am getting better and better." Many or most of them can be traced back to a fundamentally changed world from which America can no longer escape into isolation. That prescient cultural historial Louis Hartz worried more than a quarter-century ago about what would happen to the United States when it was forced to rejoin world history after a 150-year vacation from it.[55] We are now in the process of finding out, as we move into the 1980s with Ronald Reagan.

Appendix on PDI Scores

As the compilers of *American National Election Studies Data Sourcebook* point out, the PDI (position-differential index) is a convenient summary measure. It is typically, though not always, made up by subtracting the extreme positive value from the extreme negative value in a distribution of three or

Table 10. Democratic elite and mass: Positions on the death penalty in California

Position	Elites	Mass	Difference (E − M)
Strongly approve	24	59	−35
Approve	15	11	4
Disapprove	12	6	6
Strongly disapprove	44	14	30
Don't know, no answer	5	10	−5

more possible responses to a question. In the case of all six of the issue questions in the *Los Angeles Times* poll, the scale is a four-point one (usually "strongly approve, approve, disapprove, strongly disapprove"), and the "liberal-conservative" anchoring in each case is well known and not in doubt. A specific distribution, the position of Democratic politicians (elites) and registered Democratic voters (mass) on the death penalty, is shown in Table 10. The PDI score, subtracting "strongly approve" from "strongly disapprove," is +20 for Democratic elites and −45 for registered Democratic voters.

So far as proximity is concerned, the closer the value is to zero the closer the "fit" in attitudes measured by PDI between any two groups (usually elite versus mass). Thus we find on this particular issue that Republican elites, while more intensely conservative (pro-death-penalty) than Democratic registered voters, are still 30 points closer to the position of the latter than are Democratic elites on the more liberal (anti-death-penalty) side (−35 to +65 respectively).

With regard to the seventh measure in the Los Angeles poll, "ideology," PDI involves subtracting the two conservative responses ("somewhat" and "very conservative") from the two liberal responses ("somewhat" and "very liberal"). The explicit distributions of Californians on this measure in 1981, shown in Table 11, may be of some interest.

Table 11. California party elites and masses: Ideology

Ideology	Democrats Elite	Mass	Republicans Elite	Mass	Total public (mass)
Very liberal	24	6	1	2	5 ⎱ 21
Somewhat liberal	41	20	3	9	16 ⎰
Middle-of-road	24	40	14	27	37 ⎱ 37
Somewhat conservative	8	21	52	41	28 ⎰
Very conservative	2	10	30	20	12 ⎱ 40
Refused, no response	2	3	..	1	3 ⎰
PDI (liberal − conservative)	56	−5	−77	−49	−18

NOTES

1. Take the case of California, for example. Reagan carried thirty-four congressional districts to Carter's nine, but Democratic House candidates won twenty-two seats to twenty-one for the Republicans. Thus, there were thirteen districts in which Reagan and a congressional Democrat won (30.2 percent of the total), though none with Carter and a Republican. Twelve of these thirteen cases involved Democratic incumbents running for reelection. It is also worth noting tha Alan Cranston (D), running for and winning reelection to the Senate in a landslide over Paul Gann (R), carried thirty-six of the state's forty-three congressional districts, including twenty-seven in which Reagan won a plurality, and eleven in which Republican incumbent congressmen were also reelected. For the general "party decomposition" phenomenon involved, see my essay, "The 1976 Election: Has the Crisis Been Adjourned?" in Walter Dean Burnham and Martha W. Weinberg, eds., *American Politics and Public Policy* (Cambridge: MIT Press, 1978), pp. 1–25, and the literature there cited. [Chapter Seven of the present volume].

2. A Wirthlin poll taken in March 1981 indicates a sudden narrowing of the gap between Democratic and Republican party identifiers from the Michigan SRC/CPS levels of the fall of 1980. See *National Journal* 13 (1981): 681; and R. B. Wirthlin, "The Republican Strategy," in S. M. Lipset, ed., *Party Coalitions in the 1980s* (San Francisco: Institute for Contemporary Studies, 1981), pp. 235–66. Similarly, the *New York Times*–CBS poll reported in the *Times* on August 28, 1981 (p. 1), reveals a notable drop in Democratic identification, and on the other hand, other poll data from the first quarter of 1981—notably the Gallup survey—indicate no significant change at all from 1980 levels. See *Public Opinion*, April–May 1981, pp. 29–31. As of this writing, the jury is still out; by the time of reading, we should have a clearer idea of whether the shifts reported by some are real or a flash in the pan.

3. See Crane Brinton, *The Anatomy of Revolution* (New York: Vintage, 1965), especially "The Desertion of the Intellectuals," pp. 41–52; and Leon Trotsky, *The History of the Russian Revolution,* 3 vols. (New York: Simon & Schuster, 1932), especially 1:78–100.

4. The term "revitalization movement" is Anthony Wallace's. For an analysis of processes leading to such movements, based upon Wallace's model of response to system stress, as well as of the movements themselves, see Chalmers Johnson, *Revolutionary Change* (Boston: Little, Brown, 1966).

5. Between 1950 and 1959, per capita disposable income (PCDI) in 1972 constant dollars rose from $2,386 to $2,696, a gain of 13.0 percent. Between 1960 and 1969, PCDI rose from $2,697 to $3,515, for a gain of 30.3 percent over the decade. For that matter, despite the stagnation of the mid-1970s there was also a quite substantial rise in PCDI from 1970 to 1979: from $3,619 to $4,509 in constant dollars, a gain of 24.6 percent during this admittedly troubled decade. See *Economic Report of the President, 1980* (Washington: Government Printing Office, 1980), p. 229. The latter statistic is the more remarkable in view of the decline in spendable weekly income among factory workers reported in Table 1, not to mention the massive fall in rates of profit and in stock prices after 1969.

6. Will Herberg, *Protestant, Catholic, Jew* (New York: Doubleday, 1960).

7. The intellectual and operational transformation among policy elites that was necessary to develop the political-capitalist state is analyzed in Herbert Stein, *The Fiscal Revolution in America* (Chicago: University of Chicago Press, 1969).

8. Gabriel Almond and Sidney Verba, *The Civic Culture* (Boston: Little, Brown, 1963).

9. Two very illuminating discussions are Manuel Castells, *The Economic Crisis and American Society* (Princeton: Princeton University Press, 1980); and, on the international side, Fred Block, *The Origins of International Economic Disorder* (Berkeley: University of California Press, 1977). The oil shocks inflicted by OPEC and the Iranian crisis in 1974 and again in 1979 were real enough contributors to economic malaise. But it seems worth stressing that the decisive fall in profit rate, whichever of several possible ways it can be approximated, and in the income data shown in Table 1 occurred in 1969 (workers' income, stock prices)

and 1970 (profits)—i.e., *not* at the time of the OPEC crisis in energy, but at the time when the Bretton Woods agreement and the hegemony of the dollar associated with it collapsed.

10. Lester Thurow, *The Zero-Sum Society* (New York: Basic Books, 1980).

11. Cf. Daniel Bell, *The End of Ideology* (Glencoe, Ill.: Free Press, 1960); and also another period piece, Robert E. Lane, "The Politics of Consensus in an Age of Affluence," *American Political Science Review* 59 (1965): 874–95.

12. The best and most persuasive analysis of this—particularly with respect to Nixon's 1972 reelection campaign—is Edward Tufte, *Political Control of the Economy* (Princeton: Princeton University Press, 1978).

13. See, for example, *Business Week*'s special issues: "Capital Crisis: The $4.5 Trillion America Needs To Grow," September 22, 1975; "The Slow-Investment Economy," October 17, 1977; and "Revitalizing the U.S. Economy," June 30, 1980. These and other cries of alarm reflect the underlying point—which I think largely explains a suddenly increased intolerance of state burdens that have not in fact grown very much, and despite a very substantial decade-long growth between 1970 and 1980 in per capita disposable income. The underlying point is that American capitalism has grown steadily less profitable, and it is exceptionally easy under cultural conditions of uncontested hegemony to propagate the view that the state is the primary culprit. The aggregate rate of profit can be measured in a variety of ways, and none of them seem to be without inextricable difficulties. Three are presented below, with the third about midway between the other two and probably coming closest to the "real" rate of profit.

Estimated rate of profit for nonfinancial corporations

	Rate of profit 1	Rate of profit 2	Rate of profit 3
1955	19.6	10.1	14.4
1960	15.9	7.9	11.2
1965	19.7	9.1	15.8
1968	17.8	7.9	13.6
1969	14.4	6.7	11.9
1970	11.8	5.1	9.4
1971	12.1	5.5	9.7
1972	12.6	5.9	10.7
1973	12.9	5.7	10.3
1974	10.4	4.4	7.8
1975	10.5	5.8	8.3
1976	12.0	6.8	9.3
1977	11.9	6.8	9.7
1978	11.5	6.6	9.7
1979	11.4	6.3	9.2
1980	9.9	5.2	8.7

Note: Rate of profit 1 is the volume of profits with inventory valuation adjustment and capital consumption adjustment plus net interest, divided by fixed residential and nonresidential capital stock (nonfinancial corporations). See Thomas E Weisskopf, "Marxian Crisis Theory and the Rate of Profit in the Postwar U.S. Economy," *Cambridge Journal of Economics* 3 (1979): 341–78.

Rate of profit 2: Gross profits of nonfinancial corporations divided by current-dollar gross national product.

Rate of profit 3: Rate of return on depreciable assets, nonfinancial corporations. See *Economic Report of the President 1981* (Washington: Government Printing Office, 1981), p. 331.

The obvious point is that, whichever of these metrics one employs, aggregate profitability has been cut very substantially, with a first turning point occurring in 1969–70 and a second in 1974–75. Cumulatively, by 1980 these rates of profit had all been cut by substantially more than one-third from their mean 1960–68 levels.

14. Samuel P. Huntington et al., *The Crisis of Democracy* (Report to the Trilateral Commission) (New York: New York University Press, 1975). See also Nathan Glazer and Irving Kristol, eds., *The American Commonwealth, 1976* (New York: Basic Books, 1976), especially (but not exclusively!) Professor Huntington's essay. For a fascinating industrial-sector analysis of the classical horrible example of "crisis of democracy," stressing the collision between the growth of welfarism and the needs of business for accumulation, see David Abraham, *The Collapse of the Weimar Republic: Political Economy and Crisis* (Princeton: Princeton University Press, 1981).

15. The *locus classicus* is, of course, Louis Hartz, *The Liberal Tradition in America* (New York: Harcourt, Brace, 1955). Hartz's basic themes plus the religious variable that he neglected are developed in Donald J. Devine, *The Political Culture of the United States* (Chicago: Rand McNally, 1972).

16. Douglas A. Hibbs, Jr., "Political Parties and Macroeconomic Policy," *American Political Science Review* 71 (1977): 1467–87.

17. *Economic Report of the President, 1981* (Washington: Government Printing Office, 1981), p. 318.

18. Walter Dean Burnham, "The 1980 Earthquake: Realignment, Reaction or What?" in Thomas Ferguson and Joel D. Rogers, eds., *The Hidden Election* (New York: Pantheon, 1981), pp. 98–140, esp. pp. 132–40. See also my fuller discussion of the subject in my *The Dynamics of American Politics* (New York: Basic Books, forthcoming).

19. William McLoughlin, *Modern Revivalism* (New York: Ronald Press, 1959).

20. Daniel Bell, *The Cultural Contradictions of Capitalism* (New York: Basic Books, 1976). Certain parts of his argument were formulated a generation ago by the eminent conservative economist Joseph A. Schumpeter in his *Capitalism, Socialism and Democracy* (New York: Harper, 1950). Self-renunciatory neo-Victorian accumulationist culture is by no means found solely in capitalist societies. As many students of the Soviet scene have remarked, it was very characteristic of Stalinism—a movement infinitely more counterrevolutionary in culture than in economics or politics. This Stalinist cultural pattern of course reinforces the view that neo-Victorian mores have, excuse the expression, some "objective" utility to the accumulation process and those who control it.

21. Of course, the *Communist Manifesto,* published by Marx and Engels in 1848, is the classic Marxist text on the subject. "The bourgeoisie, wherever it has got the upper hand, has put an end to all feudal, patriarchal, idyllic relations. It has pitilessley torn asunder the motley feudal ties that bound man to his 'natural superiors,' and has left remaining no other nexus between man and man than naked self-interest, than callous cash payment. It has drowned the most heavenly ecstasies of religious fervor, of chivalrous enthusiasm, of philistine sentimentalism, in the icy water of egotistical calculation." Karl Marx and Friedrich Engels, *Selected Works* (Moscow: Foreign Languages Publishing House, 1962) 1:36. Pope Pius XI was scarcely less critical, and on very similar grounds, in his major encyclical, *Quadragesimo Anno* (1931). The pure capitalist blueprint achieves its purest of all expressions in Ayn Rand, *Atlas Shrugged* (Boston: Houghton Mifflin, 1957).

22. See the discussion of this in Ferguson and Rogers, *Hidden Election,* pp. 42–47.

23. For a fuller discussion, see ibid., pp. 109–10.

24. *Public Opinion,* December–January 1981, pp. 28–30.

25. The proportion of the electorate with negative evaluations of both major-party candidates shows three clear stages of development. Between 1952 and 1960 those with such evaluations were 13, 16, and 15 percent of the SRC/CPS samples. This figure rose abruptly to 30 percent in 1964, and from then through 1976 remained in the thirties (30, 32, 35, and 31 percent). Then in 1980 it climbed sharply again, to fully 46 percent of all respondents.

26. *Gallup Opinion Index,* December 1980, p. 30. Quite characteristically—and, I think, accurately—the analysis of the election had the headline "Dissatisfaction Motivated Voters."

27. *New York Times,* November 9, 1980, p. 28.

28. *Public Opinion,* February–March 1981, p. 20. However, it should be noted that, as with the question of party identification shifts after the 1980 election, different polls yield different conclusions. Gallup, for instance, ibid., shows a decline of 8 percentage points for both liberals and conservatives, and a very large increase (from 34 to 49 percent) among "middle of the roaders" between 1972 and 1980.

29. Ibid., p. 27.

30. See Burnham, "1980 Earthquake," pp. 104–7, 129–31; and *Gallup Opinion Index,* December 1980, p. 70.

31. An excellent analysis of this point is given in Kristi Andersen, *The Creation of a Democratic Majority, 1928–36* (Chicago: University of Chicago Press, 1979).

32. Paul Lazarsfeld et al., *The People's Choice* (New York: Duell, Sloan & Pearce, 1944); Robert R. Alford, *Party and Society* (Chicago: Rand McNally, 1963), pp. 219–49, 352.

33. See my discussion of this point in Walter Dean Burnham, "The System of 1896: An Analysis," in Paul Kleppner et al., *The Evolution of American Electoral Systems* (Westport, Conn.: Greenwood Press, 1981).

34. Everett C. Ladd, Jr., *Where Have All the Voters Gone?* (New York: Norton, 1978). A detailed, though still partial, answer to this question is given in Walter Dean Burnham, "The Appearance and Disappearance of the American Voter," Chapter Four in the present volume.

35. The evidence on this point, derived from the Census Bureau's biennial reports on registration and voting participation in American elections from 1964 to the present (Series P-20), is overwhelming. A good summary analysis of the uses to which this file can be put—albeit without, I think, a fully adequate highlighting of this point—is Raymond E. Wolfinger and Steven J. Rosenstone, *Who Votes?* (New Haven: Yale University Press, 1980).

36. Arthur T. Hadley, *The Empty Polling Booth* (Englewood Cliffs, N.J.: Prentice-Hall, 1978), especially pp. 67–103. Hadley makes assertions about the socioeconomic sameness of voters and nonvoters that the Census Bureau data simply do not support, but his identification of luck as an important attitudinal discriminant is a most useful contribution to this literature.

37. Anthony Downs, *An Economic Theory of Democracy* (New York: Harper, 1958), especially pp. 260–76. See also Stanley Kelley et al., "Registration and Voting: Putting First Things First," *American Political Science Review* 61 (1967): 359–79.

38. James D. Wright, *Dissent of the Governed* (New York: Academic Press, 1976).

39. Cf. Walter Dean Burnham, "Political Immunization and Political Confessionalism: The United States and Weimar Germany," *Journal of Interdisciplinary History* 3 (Summer 1972): 1–30.

40. See the article by William N. McPhee and Jack Ferguson, "Political Immunization," in William N. McPhee and William A. Glaser, eds., *Public Opinion and Congressional Elections* (New York: Free Press, 1962), pp. 155–79.

41. Richard F. Hamilton, *Class and Politics in the United States* (New York: Wiley, 1972), pp. 399–506.

42. The best empirical discussion of this dominant response to the "facts of life" is Robert E. Lane, *Political Ideology: Why the American Common Man Believes What He Does* (New York: Free Press, 1962), especially pp. 57–81, 413–35.

43. See V. O. Key, Jr., *Southern Politics in State and Nation* (New York: Knopf, 1949), especially pp. 298–311. Compare Maurice Duverger, *Political Parties* (New York: Wiley, 1959), especially pp. 422–27.

44. See for instance, the discussion in Norman H. Nie, Sidney Verba, and John R. Petrocik, *The Changing American Voter,* 2d ed. (Cambridge: Harvard University Press, 1979), pp. 142–44.

45. Position-differential index (PDI) measure taken from Warren E. Miller et al., *American*

National Election Studies Data Sourcebook (Cambridge: Harvard University Press, 1980). For a brief discussion of this measure, see Appendix.

46. Herbert McCloskey et al., "Issue Conflict and Consensus Among Party Leaders and Followers," *American Political Science Review* 54 (1960): 406–27.

47. Everett C. Ladd, Jr., "Liberalism Upside Down: The Inversion of the New Deal Order," *Political Science Quarterly* 91 (Winter 1976–77): 577–600.

48. Joseph A. Califano, Jr., *Governing America* (New York: Simon & Schuster, 1981), pp. 209–11. The whole book casts light on the remarkably sudden collapse of the political-capitalist coalition in 1980–81. Califano's comment about the problem of health-care delivery as organized during the Great Society's term is just one of many such illuminations:

> By 1977 the once diehard opponents of Medicare and Medicaid—the health insurers, hospital administrators, and doctors—were enjoying supping at the public table. In our rush to provide access, the Great Society had let the health industry set the prices, and had acquiesced in its reimbursement systems. Over the intervening decade, the industry had used America's quest for broad access to quality health care to protect and enhance its financial interest and solidify its legislative and regulatory position. The health industry was seated comfortably at a groaning table set by the taxpayers [p. 142].

It should go without saying that this sort of thing—so typically American—will over time give the positive state a bad name. Califano's memoirs also vividly illuminate the truth of the basic proposition that so long as political culture and governmental structure are not fundamentally transformed, you cannot get there from here with left-liberal political intervention, no matter how noble the cause or how genuinely altruistic the policymakers may be. One of the most striking impressions this book makes upon someone reading it in the political atmosphere of late 1981 is that it seems to describe events in a distant and quite irretrievable past—though 1979 was, after all, not so very long ago. From such sensations, no less than from an analysis of more sharply defined empirical evidence, grows the conclusion that the events of 1980–81 form a real turning point in the history of American politics.

49. 304 U.S. 144 (1938), fn. 4 at pp. 152–53. This famous footnote has often been regarded by civil libertarians as a kind of Magna Charta for judicial activism, aimed now at the protection of human rather than corporate rights. It is written in extremely cautious language, as follows (citations omitted):

> There may be narrower scope for operation of the presumption of constitutionality when legislation appears on its face to be within a specific prohibition of the Constitution, such as those of the first ten amendments, which are deemed equally specific when held to be embraced within the Fourteenth.
>
> It is unnecessary to consider now whether legislation which restricts those political processes which can ordinarily be expected to bring about repeal of undesirable legislation, is to be subjected to more exacting judicial scrutiny under the general prohibitions of the Fourteenth Amendment than are most other types of legislation. . . .
>
> Nor need we enquire whether similar considerations enter into the review of statutes directed at particular religious, or national, or racial minorities; whether prejudice against discrete and insular minorities may be a special condition, which tends seriously to curtail the operation of those political processes ordinarily to be relied upon to protect minorities, and which may call for a correspondingly more searching judicial inquiry.

50. This is an essential ingredient in one of the more far-reaching and knowledgeable scholarly critiques of political-capitalist politics in contemporary American: Theodore Lowi, *The End of Liberalism*, 2d ed. (New York: Norton, 1979).

51. A useful analysis of American political subcultures is Daniel J. Elazar, *American Federalism: A View from the States* (New York: Crowell, 1966), pp. 79–140.

52. Burnham, "System of 1896."

53. Cf. Duverger, *Political Parties,* and E. E. Schattschneider, *The Semi-Sovereign People* (New York: Holt, Rinehart & Winston, 1960), especially pp. 114–28.

54. *New York Times,* September 29, 1981, p. A22.

55. Hartz, *Liberal Tradition in America,* pp. 284–309.

Bibliography:
The Writings of
Walter Dean Burnham

In chronological order beginning with the most recent work.

"Printed Sources," in Jerome M. Clubb, William H. Flanigan, and Nancy H. Zingale, eds., *Analyzing Electoral History* (Beverly Hills: Sage, 1981), pp. 39–73.

"The System of 1896: An Analysis," Chapter 5 of Paul Kleppner et al., eds., *The Evolution of American Electoral Systems* (Westport, Conn.: Greenwood Press, 1981), pp. 147–202.

"The 1980 Earthquake: Realignment, Reaction, or What?" in Thomas Ferguson and Joel Rogers, eds., *The Hidden Election* (New York: Pantheon, 1981), pp. 98–140.

"Milestones on the Road to Democracy: Electoral Regimes and Their Relevance to South Africa," in Robert I. Rotberg and John Barratt, eds., *Conflict and Compromise in South Africa* (Lexington, Mass.: Lexington Books, 1980), pp. 77–106.

"American Politics in the 1980s," *Dissent,* Spring 1980, pp. 149–60.

"The Appearance and Disappearance of the American Voter," in Richard Rose, ed., *Electoral Participation: A Comparative Analysis* (Beverly Hills: Sage, 1980), pp. 35–74. This is a revised and somewhat condensed version of an essay by the same name in American Bar Association, *The Disappearance of the American Voter,* (Chicago/Washington: American Bar Association, n.d. [1979]), pp. 125–67.

"Thoughts on the 'Governability Crisis' in the West," *Washington Review of Strategic and International Studies* [now *Washington Quarterly*] 1, no. 3, (1978) 46–57.

American Politics and Public Policy, coedited with Martha Wagner Weinberg (Cambridge: MIT Press, 1978). This volume is in memory of our late colleague Jeffrey L. Pressman. It includes an article by W. D. Burnham, "The 1976 Election: Has the Crisis Been Adjourned?" (pp. 1–25).

"Great Britain: The Death of the Collectivist Consensus?" in Louis Maisel and Joseph

Cooper, eds., *Political Parties: Development and Decay* (Beverly Hills: Sage, 1978), pp. 267–308.

Review essay, "The Politics of Crisis," *Journal of Interdisciplinary History* 8, no. 4 (Spring 1978):747–63.

"Revitalization and Decay: Looking Toward the Third Century of American Electoral Politics" *Journal of Politics* 38 (1976): 146–72.

"Insulation and Responsiveness in Congressional Elections," *Political Science Quarterly* 90 (Fall 1975):411–35.

The American Party Systems, coedited with William N. Chambers, 2d ed., (New York: Oxford University Press, 1975). This edition contains an article by W. D. Burnham, "American Politics in the 1970s: Beyond Party?" (Chapter 11, pp. 308–57). The same article also appears in Louis Maisel and Paul Sacks, eds., *The Future of Political Parties* (Beverly Hills: Sage, 1975), Chapter 8, pp. 238–77.

"The United States: The Politics of Heterogeneity," Chapter 13 of Richard Rose, ed., *Electoral Behavior: A Comparative Handbook* (New York: Free Press, 1974), pp. 653–725.

"Theory and Voting Research: Some Reflections on Converse's 'Change in the American Electorate,'" *American Political Science Review* 68 (1974): 1002–23.

(Editor) *Politics/America: The Cutting Edge of Change* (New York: Van Nostrand, 1973). A selection of articles from *Trans-Action/Society,* with extensive editorial commentary.

"Political Immunization and Political Confessionalism: The United States and Weimar Germany," *Journal of Interdisciplinary History* 3 (Summer 1972):1–30.

"Crisis of American Political Legitimacy," *Society* 10 (November–December 1972).

"A Political Scientist and Voting-Rights Litigation: The Case of the 1966 Texas Registration Statute," *Washington University Law Quarterly,* 1971, pp. 335–58.

Critical Elections and the Mainsprings of American Politics (New York: Norton, 1970).

(with John D. Sprague) "Additive and Multiplicative Models of the Voting Universe: The Case of Pennsylvania, 1960–1968," *American Political Science Review* 64, (1970):471–90.

"The End of American Party Politics," *Trans Action* 7, no. 2 (1969):12–22.

"Election 1968: The Abortive Landslide," *Trans-Action* 6, no. 2 (1968):18–24.

"American Voting Behavior and the 1964 Election," *Midwest Journal of Political Science* 12 (1968):1–40. Also published in Melvin Richter, ed., *Essays in Theory and History: An Approach to the Social Sciences* (Cambridge: Harvard University Press, 1970), pp. 186–220.

"The Changing Shape of the American Political Universe," *American Political Science Review* 59, (1965):7–28.

"The Alabama Senatorial Election of 1962: Return of Inter-Party Competition," *Journal of Politics* 26 (1964):789–829.

Presidential Ballots, 1836–1892 (Baltimore: Johns Hopkins University Press, 1955).

Index

ABC poll (1980 election), 287
Adams, John, 266
Adams, John Quincy, 104
Alford, Robert, 4, 290
Almanac of American Politics, The, 224
American Labor party, 41
American Liberal party, 41
American Political Science Review, 58, 60, 76
American Politics Quarterly, 198
American Socialist party, 99
American State Politics, 9
American Voter, The, 4, 82, 185, 196
Andersen, Kristi, 175, 177, 180
Anderson, John, 285, 286, 302
Australian ballot, 60, 75–78

Baker, Howard, 308
Baker, Lamar, 212
Ballots:
 access devices, 154, 181
 Australian, 60, 75–78
 box stuffing, 122, 128, 131
 fatigue, 74, 75, 75 *tab.*
 secrecy, 130
Barnes, Samuel H., 169
Beard, Charles A., 269
Bell, Daniel, 281
Benson, Lee, 40, 96, 98–99
Blacks:
 disfranchisement, 29, 50, 108
 inclusion in political process, 112, 131, 148, 231
 turnout rates, 263
Bolshevik Revolution 7, 18, 265, 280
Boyd, Richard, 198, 200
Bretton Woods Agreement, 263, 271, 274
Brinton, Crane, 270

Brown, Jr., Edmund Gerald, 233–34
Bryan, William Jennings, 48–50, 98, 109–10
Bush, George, 308
Business Week, 276–77

Califano, Joseph, 300
California:
 elite and voter survey, 297, 297 *tab.,* 298, 298 *tab.,* 314, 314 *tab.*
 1976 election: turnout rates, 238
 San Francisco, 146–47, 147 *tab.,* 179, 179 *tab.,* 180
 tax revolts, 251, 261
Campbell, Angus, 8, 28, 33, 37
"Candidate Evaluations and Turnout," 199
Cantril, Hadley, 254
Capitalism, 19, 278
 crises of, 175, 254, 277
 cultural contradictions, 260, 281–82
 interest groups and, 230
 lack of alternatives to, 293
 laissez-faire, 142, 145, 231, 258, 264
 liberal tradition, 127–29, 196–97, 259, 280
 multinational corporations, 257
 political, 256, 258, 276, 299–300, 310
 (*See also* Industrialization)
Carter, Jimmy, 13, 255, 259, 264, 268, 277, 290, 301, 308
 conservativism of, 251, 285
 discomfort index and, 283, 284 *fig.*
 1976 election, 229, 233–34, 237–39, 282–83
 1980 election, 290–91
 as personality, 235–36
 pro-Carter districts, 304
 public opinion surveys, 286–89
 shift against, 302–3
 style, 243

323